THE EMISSION-LINE UNIVERSE

Emission lines provide a powerful tool for studying the physical properties and chemical compositions of astrophysical objects in the Universe, from the first stars to objects in our Galaxy. The analysis of emission lines allows us to estimate the star formation rate and initial mass function of ionizing stellar populations, and the properties of active galactic nuclei.

This book presents lectures from the eighteenth Winter School of the Canary Islands Astrophysics Institute (IAC). Written by prestigious researchers and experienced observers, it covers the formation of emission lines and the different sources that produce them. It shows how emission lines in different wavelengths, from ultraviolet to near infrared, can provide essential information on understanding the formation and evolution of astrophysical objects. It also includes practical tutorials for data reduction, making this a valuable reference for researchers and graduate students.

Canary Islands Winter School of Astrophysics

Volume XVIII

Editor in Chief
F. Sánchez, *Instituto de Astrofísica de Canarias*

Previous volumes in this series

I. Solar Physics
II. Physical and Observational Cosmology
III. Star Formation in Stellar Systems
IV. Infrared Astronomy
V. The Formation of Galaxies
VI. The Structure of the Sun
VII. Instrumentation for Large Telescopes: a Course for Astronomers
VIII. Stellar Astrophysics for the Local Group: a First Step to the Universe
IX. Astrophysics with Large Databases in the Internet Age
X. Globular Clusters
XI. Galaxies at High Redshift
XII. Astrophysical Spectropolarimetry
XIII. Cosmochemistry: the Melting Pot of Elements
XIV. Dark Matter and Dark Energy in the Universe
XV. Payload and Mission Definition in Space Sciences
XVI. Extrasolar Planets
XVII. 3D Spectroscopy in Astronomy

Participants of the XVIII Canary Islands Winter School.

THE EMISSION-LINE UNIVERSE

XVIII Canary Islands Winter School of Astrophysics

Edited by

JORDI CEPA

Instituto de Astrofísica de Canarias, Tenerife

CAMBRIDGE
UNIVERSITY PRESS

CAMBRIDGE UNIVERSITY PRESS
Cambridge, New York, Melbourne, Madrid, Cape Town, Singapore, São Paulo, Delhi

Cambridge University Press
The Edinburgh Building, Cambridge CB2 8RU, UK

Published in the United States of America by Cambridge University Press, New York

www.cambridge.org
Information on this title: www.cambridge.org/9780521898867

First published 2009

Printed in the United Kingdom at the University Press, Cambridge

A catalogue record for this publication is available from the British Library

Library of Congress Cataloguing in Publication data
Canary Islands Winter School on Astrophysics (18th : 2006 : Instituto de Astrofmsica de Canarias)
The emission-line universe : XVIII Canary Islands Winter School of Astrophysics / edited by Jordi Cepa.
p. cm.
ISBN 978-0-521-89886-7 (hardback)
1. Emission-line galaxies – Congresses. 2. Active galactic nuclei – Congresses. I. Cepa, Jordi.
II. Title.
QB858.3.C365 2006
523.1′12 – dc22 2008026917

ISBN 978-0-521-89886-7 hardback

Contents

Contributors

GRAŻYNA STASIŃSKA, LUTH, Observatoire de Paris-Meudon, France

MAURO GIAVALISCO, Space Telescope Science Institute, USA

PIERO MADAU, Department of Astronomy and Astrophysics, University of California, Santa Cruz, California, USA

DANIEL SCHAERER, Observatoire de Genève, Université de Genève, Sauverny, Switzerland

BRADLEY M. PETERSON, Department of Astronomy, The Ohio State University, Columbus, Ohio, USA

FRANCESCA MATTEUCCI, Dipartamento di Astronomia, Università di Trieste, Italy

STEPHEN S. EIKENBERRY, University of Florida, Florida, USA

SERGIO PASCUAL, Departamento de Astrofísica, Facultad de Ciencias Físicas, Universidad Complutense de Madrid, Spain

BERNABÉ CEDRÉS, Instituto de Astrofísica de Canarias, Tenerife, Spain

MIGUEL SÁNCHEZ-PORTAL, European Space Astronomy Centre, Spain

ANA PÉREZ-GARCÍA, Instituto de Astrofísica de Canarias, Tenerife, Spain

HÉCTOR CASTAÑEDA, Instituto de Astrofísica de Canarias, Tenerife, Spain

ANGEL BONGIOVANNI, Instituto de Astrofísica de Canarias, Tenerife, Spain

Participants

Agüero, Mª	Paz Universidad de Córdoba (Argentina)
Atek, Hakim	Institute d'Astrophysique de Paris (France)
Barro Calvo, Guillermo	Universidad Complutense de Madrid (Spain)
Barway, Sudhanshu	Inter-University Centre for Astronomy and Astrophysics (India)
Casebeer, Darrin A.	University of Oklahoma (USA)
Colavitti, Edoardo	Osservatorio Astronomico di Trieste (Italy)
Cottis, Christopher	University of Leicester (UK)
Delgado Inglada, Gloria	INAOE (Mexico)
Díaz Santos, Tanio	CSIC (Spain)
Fernández Lorenzo, Miriam	Instituto de Astrofísica de Canarias (Spain)
González Pérez, Violeta	IEEC/CSIC (Spain)
Hayes, Matthew	AlbaNova University Centre (Sweden)
Hernández Fernández, Jonathan David	Instituto de Astrofísica de Andalucía (Spain)
Hurley, Rossa	Chalmers University of Technology (Sweden)
Ilić, Dragana	University of Belgrade (Serbia)
Izquierdo Gómez, Jaime	Universidad Complutense de Madrid (Spain)
Köhler, Ralf	Max Planck Institute for Extraterrestrial Physics (Germany)
Kovačević, Jelena	University of Belgrade (Serbia)
Koziel, Dorota	Jagiellonian University (Poland)
Kusterer, Daniel-Jens	Universität Tübingen (Germany)
Lara López, Maritza	Instituto de Astrofísica de Canarias (Spain)
Lazarova, Mariana	University of California, Riverside (USA)
Lee, Janice	University of Arizona (USA)
Marcon Uchida, Monica Midori	Universidade de São Paulo (Brazil)
Mazzalay, Ximena	Córdoba Astronomical Observatory (Argentina)
Montero Dorta, Antonio David	Instituto de Astrofísica de Andalucía (Spain)
Muñoz Marín, Victor Manuel	Instituto de Astrofísica de Andalucía (Spain)
Nava Bencheikh, Aida H.	University of Oklahoma, Norman (USA)
Ovalasen, Jan-Erik	University of Oslo (Norway)
Padilla Torres, Carmen Pilar	Instituto de Astrofísica de Canarias (Spain)
Planelles Mira, Susana	Universidad de Valencia (Spain)
Pović, Mirjana	Instituto de Astrofísica de Canarias (Spain)
Prescott, Moire	University of Arizona (USA)
Raiter, Anna	Nicolaus Copernicus University (Poland)
Ramos Almeida, Cristina	Instituto de Astrofísica de Canarias (Spain)
Ruiz Fernández, Nieves	Instituto de Astrofísica de Andalucía (Spain)
Sampson, Leda	University of Cambridge (UK)
Sánchez, Juan Andrés	Universidad Central de Venezuela (Venezuela)
Simón Díaz, Sergio	Observatoire de Paris-Meudon (LUTH) (France)
Somero, Auni	University of Helsinki (Finland)
Stoklasová, Ivana	Charles University (Czech Republic)

Toloba, Elisa	Universidad Complutense de Madrid (Spain)
Vale Asari, Natalia	Universidade Federal de Santa Catarina (Brazil)
Vavilkin, Tatiana	Stony Brook University (USA)
Villar, Victor	Universidad Complutense de Madrid (Spain)
Wehres, Nadine	Leiden Observatory/Universiteit Groningen (The Netherlands)
Yan, Huirong	Canadian Institute of Theoretical Astrophysics (Canada)

Preface

Emission lines are powerful means to detect faint objects and to study their composition and physical properties. Detecting and studying objects ranging from galactic sources to the most distant galaxies is made possible by using these lines. The aim of the XVIII Winter School is to give a thorough introduction to this emission-line Universe from both theoretical and observational points of view. For this reason, the Winter School contents include not only classical lectures, but also tutorials on data reduction and analysis. This structure enables young researchers to participate actively in current and future research projects, while serving also as a reference book for experienced researchers.

The subject of this School was motivated by the upcoming advent of a new generation of wide-field instruments for large telescopes, specifically optimized for observing emission-line objects in two dimensions. These instruments will boost the study of these kinds of objects by providing large amounts of data, whose digestion will require a theoretical basis as well as specific data-reduction techniques. These powerful facilities will enable the study of very faint emission lines of nearby objects, or conspicuous lines of very distant targets. The former will provide finer details on the chemical composition and characteristics of the gas, while the latter will furnish insight on structure formation and its evolution via scanning of large proper volumes of Universe.

Most cosmological surveys have been based on the continuum emission of the objects of the Universe via broadband imaging and their spectroscopic follow-up. Although only a fraction of the targets will shine in emission, this "emission-line" Universe, which has thus far remained relatively unexplored, will provide information about both the bright and the faint ends of the luminosity function, nicely providing essential pieces of scientific information and complementing the results obtained from classical continuum-based surveys.

The XVIII Winter School includes an introduction to the insight that UV, optical and near-infrared lines can provide on emission-line objects considering physical emission mechanisms, line diagnostics and codes, and then focuses on various types of emission-line galactic objects, identification techniques and applications. A review of characteristics, advantages and disadvantages of emission-line surveys at various wavelengths, both wide and deep, will serve as a starting point to study active galactic nuclei, QSOs, primaeval galaxies, extragalactic star formation and the evolution of the metal content of galaxies. All these topics are tackled considering their cosmic evolution and astrophysical implications. Finally, this Winter School includes hands-on tutorials presenting practical examples of data reduction and analysis of a variety of emission-line objects using diverse observational techniques.

Jordi Cepa
Instituto de Astrofísica de Canarias

Acknowledgements

I want to express my warmest gratitude to the lecturers for their efforts in preparing their lectures and the manuscripts, making them pedagogical and challenging. I also wish to thank our efficient secretaries Nieves Villoslada and Lourdes González for their care in the preparation and organization of the School. Their experience and support have been essential for its success. Also, I am indebted to Ramón Castro for the design and production of the poster, to the technicians of the Servicios Informáticos Comunes for the installation and maintenance of the hardware and communications network, to Carmen del Puerto for her lively Winter School electronic bulletin, to Terry Mahoney for sorting out the subtleties of manuscript production and to Jesús Burgos for his efforts at trying to raise funds from the Ministry of Education (however, I do not thank the Ministry since they gave us nothing at all). I acknowledge Anna Fagan for preparing the final version of this book for Cambridge University Press.

I also wish to thank the Cabildo of the island of Tenerife and the personnel of the Salón de Congresos del Parque Taoro (I am afraid I cannot recall all their names) for their generous support and help.

Finally I wish to thank all participants for the excellent atmosphere of collaboration, enjoyment and learning that they created throughout the School. I will always remember the gift of excellent local wines from the students: not *ora et labora* but *bibere et discere*.

Açúcar!

1. What can emission lines tell us?

GRAŻYNA STASIŃSKA

1.1. Introduction

Emission lines are observed almost everywhere in the Universe, from the Earth's atmosphere (see Wyse & Gilmore 1992 for a summary) to the most-distant objects known (quasars and galaxies), on all scales and at all wavelengths, from the radio domain (e.g. Lobanov 2005) to gamma rays (e.g. Diehl *et al.* 2006). They provide very efficient tools to explore the Universe, measure the chemical composition of celestial bodies and determine the physical conditions prevailing in the regions where they are emitted.

The subject is extremely vast. Here, we will restrict ourselves in wavelength, being mostly concerned with the optical domain, with some excursions to the infrared and ultraviolet domains and, occasionally, to the X-ray region.

We will mainly deal with the mechanisms of line production and with the interpretation of line intensities in various astrophysical contexts. We will discuss neither quasars and Seyfert galaxies, since those are the subject of Chapter 5, nor Lyman-α galaxies, which are extensively covered in Chapter 4 of this book. However, we will discuss diagnostic diagrams used to distinguish active galaxies from other emission-line galaxies and will mention some topics linked with H Lyα. Most of our examples will be taken from recent literature on planetary nebulae, H II regions and emission-line galaxies. Emission-line stars are briefly described in Chapter 7 and a more detailed presentation is given in the book *The Astrophysics of Emission Line Stars* by Kogure & Leung (2007).

The vast subject of molecular emission lines has been left aside. The proceedings of the symposium *Astrochemistry: Recent Successes and Current Challenges* (Lis *et al.* 2006) give a fair introduction to this rapidly expanding field.

In the present text, we will not go into the question of Doppler shifts or line profiles, which tell us about radial velocities and thus about dynamics. This is of course a very important use of emission lines, which would deserve a book of its own. For example, for such objects as planetary nebulae and supernova remnants, emission-line profiles allow one to measure expansion velocities and thus investigate their dynamics. Determining the distribution of radial velocities of planetary nebulae in galactic haloes is a way to probe their kinematics and infer the dark-matter content of galaxies (Romanowsky 2006). Redshift surveys to map the three-dimensional distribution of galaxies in the Universe strongly rely on the use of emission lines (e.g. Lilly *et al.* 2007), which is the most-reliable way to measure redshifts.

We will, however, mention the great opportunity offered by integral-field spectroscopy at high spectral resolution, which provides line intensities and profiles at every location in a given field of view. With appropriate techniques, this allows one to recover the three-dimensional (3D) geometry of a nebula.

The purpose here is not to review all the literature on ionized nebulae, but rather to give clues for understanding the information given by emission lines, to provide some tools for interpreting one's own data, and to argue for the importance of physical arguments and common sense at each step of the interpretation process. Therefore, we will review methods rather than objects and papers. This complements in some sense the text entitled "Abundance determinations in H II regions and planetary nebulae" (Stasińska 2004), to

The Emission-Line Universe, ed. J. Cepa. Published by Cambridge University Press.

which the reader is referred. In order to save space, the topics that have been treated extensively there will not be repeated, unless we wish to present a different approach or add important new material.

In the following, we will assume that the reader is familiar with the first three sections of Stasińska (2004). We also recommend reading Ferland's outstanding (2003) review "Quantitative spectroscopy of photoionized clouds". Those wishing for a more-complete description of the main physical processes occurring in ionized nebulae should consult the textbooks *Physical Processes in the Interstellar Medium* by Spitzer (1978), *Physics of Thermal Gaseous Nebulae* by Aller (1984), *Astrophysics of the Diffuse Universe* by Dopita & Sutherland (2003) and *Astrophysics of Gaseous Nebulae and Active Galactic Nuclei* by Osterbrock & Ferland (2006). For a recent update on X-ray astrophysics, a field that is developing rapidly, one may consult the AIP Conference Proceedings on *X-ray Diagnostics of Astrophysical Plasmas: Theory, Experiment, and Observation* (Smith 2005).

1.2. Generalities

1.2.1 *Line-production mechanisms*

Emission lines arise in diffuse matter. They are produced whenever an excited atom (or ion) returns to lower-lying levels by emitting discrete photons. There are three main mechanisms that produce atoms (ions) in excited levels: recombination, collisional excitation and photoexcitation.

1.2.1.1 Recombination

Roughly two thirds of the recombinations of an ion occur onto excited states from which de-excitation proceeds by cascades down to the ground state. The resulting emission lines are called recombination lines and are labelled with the name of the *recombined* ion, although their intensities are proportional to the abundance of the *recombining* species. The most-famous (and most commonly detected) ones are H I lines (from the Balmer, Paschen, etc. series), which arise from recombination of H^+ ions; He I lines ($\lambda 5876, \ldots$), which arise from recombination of He^+; and He II lines ($\lambda 4686, \ldots$), which arise from recombination of He^{++} ions. Recombination lines from heavier elements are detected as well (e.g. C II $\lambda 4267$, O II $\lambda 4651, \ldots$), but they are weaker than recombination lines of hydrogen by several orders of magnitude, due to the much lower abundances of those elements.

The energy e_{ijl} emitted per unit time in a line l due to the recombination of the ion j of an element X^i can be written as

$$e_{ijl} = n_e n(X_i^j) e_{ijl}^0 T_e^{-\alpha}, \tag{1.1}$$

where e_{ijl}^0 is a constant and the exponent α is of the order of 1. Thus, recombination line intensities increase with decreasing temperature, as might be expected.

1.2.1.2 Collisional excitation

Collisions with thermal electrons lead to excitation onto levels that are low enough to be attained. Because the lowest-lying level of hydrogen is at 10.2 eV, collisional excitation of hydrogen lines is effective only at electron temperatures T_e larger than $\sim 2 \times 10^4$ K. On the other hand, heavy elements such as nitrogen, oxygen and neon have low-lying levels that correspond to fine-structure splitting of the ground level. Those can be excited at any temperature that can be encountered in a nebula, giving rise to infrared lines.

At "typical" nebular temperatures of 8000 – 12 000 K, levels with excitation energies of a few eV can also be excited, giving rise to optical lines. Slightly higher temperatures are needed to excite levels corresponding to ultraviolet lines.

In the simple two-level approximation, when each excitation is followed by a radiative de-excitation, the energy e_{ijl} emitted per unit time in a line l due to collisional excitation of an ion j of an element X^i can be written as

$$\begin{aligned} e_{ijl} &= n_\mathrm{e} n(\mathrm{X}_i^j) q_{ijl} h\nu_l \\ &= 8.63 \times 10^{-6} n_\mathrm{e} n(\mathrm{X}_i^j) \Omega_{ijl} / (\omega_{ijl} T_\mathrm{e}^{-0.5} \mathrm{e}^{(\chi_{ijl}/kT_\mathrm{e})} h\nu_l), \end{aligned} \quad (1.2)$$

where Ω_{ijl} is the collision strength, ω_{ijl} is the statistical weight of the upper level and χ_{ijl} is the excitation energy.

Collisionally excited lines (CELs) are traditionally separated into forbidden, semi-forbidden and permitted lines, according to the type of electronic transition involved. Observable forbidden lines have transition probabilities of the order of $10^{-2}\,\mathrm{s}^{-1}$ (or less for infrared lines), semi-forbidden lines have probabilities of the order of $10^2\,\mathrm{s}^{-1}$, and permitted lines have probabilities of the order of $10^8\,\mathrm{s}^{-1}$. This means that the critical density (i.e. the density at which the collisional and radiative de-excitation rates are equal) of forbidden lines is much smaller than those of intercombination lines and of permitted lines. Table 2 of Rubin (1989) lists critical densities for optical and infrared lines. Table 1 of Hamman *et al.* (2001) gives critical densities for ultraviolet lines.

Resonance lines are special cases of permitted lines: they are the longest-wavelength lines arising from ground levels. Examples of resonance lines are H Lyα and C IV λ1550.

1.2.1.3 Fluorescent excitation

Permitted lines can also be produced by photoexcitation due to stellar light or to nebular recombination lines. The Bowen lines (Bowen 1934) are a particular case of fluorescence, where O III is excited by the He II Lyα line and returns to the ground level by cascades giving rise to O III λ3133, 3444 as well as to the line O III λ304, which in turn excites a term of N III.

The interpretation of fluorescence lines is complex, and such lines are not often used for diagnostics of the nebulae or their ionizing radiation. On the other hand, it is important to know which lines can be affected by fluorescence, in order to avoid improper diagnostics assuming pure recombination (see Escalante & Morisset 2005). Order-of-magnitude estimates can be made with the simple approach of Grandi (1976).

Quantitative analysis of fluorescence lines requires heavy modelling. It can be used to probe the He II radiation field in nebulae (Kastner & Bhatia 1990). The X-ray-fluorescence iron line has been used to probe accretion flows very close to massive black holes (Fabian *et al.* 2000).

1.2.1.4 Some hints

Each line can be produced by several processes, but usually there is one that dominates. There are, however, cases in which secondary processes may not be ignored. For example, the contribution of collisional excitation to H Lyα is far from negligible at temperatures of the order of 2×10^4 K. The contribution of recombination to the intensities of [O II] $\lambda\lambda$7320, 7330 is quite important at low temperatures (below, say, 5000 K) and becomes dominant at the lowest temperatures expected in H II regions (see Stasińska 2005).

In the appendix, we give tables of forbidden, semi-forbidden and resonance lines for ions of C, N, O, Ne, S, Cl and Ar that can be observed in H II regions and planetary nebulae.

These are extracted from the atomic-line list maintained by Peter van Hoof, which is available at http://www.pa.uky.edu/~peter/atomic. This database contains identification of about one million allowed, intercombination and forbidden atomic transitions with wavelengths in the range from 0.5 Å to 1000 µm.

W. C. Martin and W. L. Wiese produced a very useful atomic physics "compendium of basic ideas, data, notations and formulae" that is available at: http://www.physics.nist.gov/Pubs/AtSpec/index.html.

1.2.1.5 Atomic data

In the interpretation of emission lines, atomic data play a crucial role. Enormous progress has been made in atomic physics during recent years, but not all relevant data are available yet or known with sufficient accuracy. The review by Kallman & Palmeri (2007) is the most-recent critical compilation of atomic data for emission-line analysis and photoionization modelling of X-ray plasmas. A recent assessment of atomic data for planetary nebulae is given by Bautista (2006).

Many atomic databases are available on the Internet.

- A compilation of databases for atomic and plasma physics: http://plasma-gate.weizmann.ac.il/DBfAPP.html.
- Reference data: http://physics.nist.gov/PhysRefData/~physical.
- Ultraviolet and X-ray radiation at: http://www.arcetri.astro.it/science/chianti/chianti.html.
- Atomic data for astrophysics (but only up to 2000): http://www.pa.uky.edu/~verner/atom.html.
- Atomic data from the Opacity Project: http://cdsweb.u-strasbg.fr/topbase/topbase.html.
- Atomic data from the IRON project: http://cdsweb.u-strasbg.fr/tipbase/home.html.

1.2.2 *The transfer of radiation*

This section is not to provide a detailed description of radiative-transfer techniques, but simply to mention the problems and the reliability of the methods that are used.

1.2.2.1 The transfer of Lyman-continuum photons emitted by the ionizing source

The photons emitted by a source of radiation experience geometrical dilution as they leave the source. They may be absorbed on their way by gas particles, predominantly by hydrogen and helium. The first photons to be absorbed are those which have energies slightly above the ionization threshold, due to the strong dependence of the photoionization cross section on frequency (roughly proportional to ν^{-3}). This gives rise to a "hardening" of the radiation field as one approaches the outer edge of an ionized nebula. Photons may also be absorbed (and scattered) by dust grains.

1.2.2.2 The transfer of the ionizing photons produced by the nebula

Recombination produces photons that can in turn ionize the nebular gas. These photons are emitted in all directions, so their transfer is not simple. Authors have developed several kinds of approximation to deal with this.

The "on-the-spot approximation" (or OTS) assumes that all the photons recombining to the ground state are reabsorbed immediately and at the locus of emission. This is approximately true far from the ionizing source, where the population of neutral species is sufficiently large to ensure immediate reabsorption. However, in the zones of high ionization (or high "excitation" as is often improperly said), this is not true. Computationally, the OTS assumption allows one simply to discount all the recombinations to the excited levels. This creates a spurious temperature structure in the nebula, with the temperature being overestimated in the high-excitation zone, due to the fact that the stellar ionizing radiation field is "harder" than the combined stellar plus recombination radiation field. The effect is not negligible, about 1000–2000 K in nebulae of solar chemical composition. Because of this, for some kinds of problems it might be preferable to use a simple one-dimensional (1D) photoionization code with reasonable treatment of the diffuse radiation rather than a 3D code using the OTS approximation.

The OTS approximation, however, has its utility in dynamical simulations incorporating radiation transfer, because it significantly decreases the computational time.

Note that the OTS approximation is valid on a global scale. In the integrated volume of a Strömgren sphere, the total number of ionizing photons of the source is exactly balanced by the total number of recombinations to excited levels. This is a useful property for analytic estimations, since it implies that the total Hα luminosity of a nebula that absorbs all the ionizing photons (and is devoid of dust) is simply proportional to $Q_{\rm H}$, the total number of ionizing hydrogen photons.[†]

In increasing order of accuracy (and complexity), then comes the "outward-only approximation", which was first proposed in 1967 by Tarter in his thesis, very early in the era of photoionization codes (see Tarter *et al.* (1969) for a brief description). Here, the ionizing radiation produced in the nebula is computed at every step, but artificially concentrated in the outward-directed hemisphere, where it is distributed isotropically. This gives a relatively accurate description of the nebular ionizing radiation field, since photons that are emitted inwards tend to travel without being absorbed until they reach the symmetrical point relative to the central source. The great advantage of this approximation is that it allows the computation of a model without having to iterate over the entire volume of the nebula. The code PHOTO, used by Stasińska, and the code NEBU, used by Péquignot and by Morisset, are based on this approximation. The code CLOUDY, by Ferland, uses the outward-only approximation in a radial-only mode. It appears, from comparisons of benchmark models (e.g. Péquignot *et al.* 2001), that the global results of models constructed with codes that treat the transfer of diffuse radiation completely (e.g. the code NEBULA by Rubin) are quite similar. Note that the full outward-only approximation allows one to compute the ionizing radiation field in the shadows from optically thick clumps by artificially suppressing the stellar radiation field blocked by the clump.

Codes treating the transfer of ionizing continuum photons exactly, iterating over the entire nebula, interestingly appeared also at the beginning of the era of photoionization codes (Harrington 1968, Rubin 1968). At that time, computers were slow and had little memory, and only spherical or plane-parallel geometries could be treated by such codes.

With the present computational capacities, one can do much better and treat the transfer problem accurately for any geometry, by using Monte Carlo methods. The first such code is MOCASSIN, by Ercolano (see Ercolano *et al.* 2003). One advantage of Monte Carlo methods is that they allow one to treat the transfer accurately also in extremely dusty nebulae (Ercolano *et al.* 2005), while the outward-only approximation breaks down in such cases.

[†] This property is formally true only for a pure-hydrogen nebula, but it so happens that absorption by helium and subsequent recombination produces line photons that ionize hydrogen and compensate rather well for the photons absorbed by helium.

The transfer of resonance-line radiation produced in the nebula is the most difficult to treat accurately, at least in classical approaches to the transfer. This is because it is generally treated in the "escape-probability" approximation. The effect of line transfer is crucial in optically thick X-ray plasmas such as the central regions of active galactic nuclei (AGNs). The code TITAN by Dumont treats the transfer of line radiation in an "exact" manner, using the "accelerated lambda iteration" method (Dumont *et al.* 2003).

1.2.2.3 The non-ionizing lines emitted by the nebula

In general, non-ionizing photons in dust-free nebulae escape as soon as they have been emitted (except perhaps in AGNs, where column densities are higher). Resonance lines constitute an exception: they may be trapped a long time in the nebula, due to multiple scattering by atoms that, under nebular conditions, are predominantly in their ground levels.

In dusty objects, line photons suffer absorption and scattering by dust grains on their path out of the nebula. Resonance lines, whose path length can be multiplied by enormous factors due to atomic scattering, are then preferentially affected by dust extinction. Their observed luminosities then represent only a lower limit to the total energy produced by these lines.

Another case where the diagnostic potential is expected to be reduced due to transfer effects is that of infrared fine-structure lines of abundant ions, such as [O III] $\lambda 88$, 52 μm, which can become optically thick in massive H II regions (Rubin 1978). However, due to a combination of independent reasons, this appears not to be the case even in the extreme situations explored by Abel *et al.* (2003).

1.3. Empirical diagnostics based on emission lines

1.3.1 *Electron temperature and density*

It is well known, and mentioned in all textbooks, that some line ratios (e.g. the ratios of the lines labelled A1 and N2 in Table 1.11 in the appendix) are strongly dependent on the temperature, since they have different excitation energies. If the critical densities for collisional de-excitations are larger than the density in the medium under study, these line ratios depend only on the temperature and are ideal temperature indicators. The most frequently used is the [O III] $\lambda 4363/5007$ ratio.

On the other hand, in collisionally excited lines that arise from levels of similar excitation energies, their ratios depend only on the density. The commonest density indicator in the optical is the [S II] $\lambda 6716/6731$ ratio. Other ones can easily be found by browsing in Table 1.11. Rubin (1989) gives a convenient list of optical and infrared line-density indicators showing the density range where each of them is useful.

Similar plasma diagnostics are now available in the X-ray region (Porquet & Dubau 2000, Delahaye *et al.* 2006, see also Porter & Ferland 2006).

1.3.2 *Ionic and elemental abundances*

There are basically four methods to derive the chemical composition of ionized nebulae. The first one, generally thought to be the "royal way", is through tailored photoionization modelling. The second is by comparison of given objects with a grid of models. These two methods will be discussed in the next section. In this section, which deals with purely empirical methods, we will discuss the other two: direct methods, which obtain an abundance using information directly from the spectra, and statistical methods, which use relations obtained from families of objects.

1.3.2.1 Direct methods

In these methods, one first derives *ionic* abundance ratios directly from observed line ratios of the relevant ions:

$$\frac{I_{ijl}}{I_{i'j'l'}} = \frac{\int n(\mathrm{X}_i^j) n_{\mathrm{e}} \epsilon_{ijl}(T_{\mathrm{e}}, n_{\mathrm{e}}) \mathrm{d}V}{\int n(\mathrm{X}_{i'}^{j'}) n_{\mathrm{e}} \epsilon_{i'j'l'}(T_{\mathrm{e}}, n_{\mathrm{e}}) \mathrm{d}V}, \tag{1.3}$$

where the Is are the intensities and the ϵs are given by $e_{ijl} = \epsilon_{ijl} n(\mathrm{X}_i^j) n_{\mathrm{e}}$.

Therefore

$$\frac{\int n(\mathrm{X}_i^j) n_{\mathrm{e}}\, \mathrm{d}V}{\int n(\mathrm{X}_{i'}^{j'}) n_{\mathrm{e}}\, \mathrm{d}V} = \frac{I_{ijl}/I_{i'j'l'}}{\epsilon_{ijl}(T_l, n_l)/\epsilon_{i'j'l'}(T_{l'}, n_{l'})}, \tag{1.4}$$

where T_l and n_l are, respectively, the electron temperature and density representative of the emission of the line l.

Assuming that the chemical composition is uniform in the nebula, one obtains the element abundance ratios:

$$\frac{n(\mathrm{X}_i)}{n(\mathrm{X}_{i'})} = \frac{I_{ijl}/I_{i'j'l'}}{\epsilon_{ijl}(T_l, n_l)/\epsilon_{i'j'l'}(T_{l'}, n_{l'})} \mathrm{ICF}, \tag{1.5}$$

where ICF is the ionization correction factor,

$$\mathrm{ICF} = \frac{\int n(\mathrm{X}_i^j)/n(\mathrm{X}_i) n_{\mathrm{e}}\, \mathrm{d}V}{\int n(\mathrm{X}_{i'}^{j'})/n(\mathrm{X}_{i'}) n_{\mathrm{e}}\, \mathrm{d}V}. \tag{1.6}$$

In a case where several ions of the same element are observed, one can use a "global" ICF adapted to the ions that are observed (e.g. ICF($O^+ + O^{++}$) for planetary nebulae in which oxygen may be found in higher ionization states). Note that in H II regions (except those ionized by hot Wolf–Rayet stars) ICF($O^+ + O^{++}$) = 1.

The application of direct methods requires a correct evaluation of the T_ls and n_ls as well as a good estimate of the ionization correction factor.

For some ions, the T_ls can be obtained from emission-line ratios such as [O III] λ4363/5007 and [N II] λ5755/6584. For the remaining ions, the T_ls are derived using empirical relations with T(4363/5007) or T(5755/6584) obtained from grids of photoionization models. The most-popular empirical relations are those listed by Garnett (1992). A newer set of relations, based on a grid of models that reproduces the properties of H II galaxies, is given by Izotov *et al.* (2006). It must be noted, however, that observations show larger dispersion about those relations than predicted by photoionization models. It is not clear whether this is due to underestimated observational error bars, or to additional processes not taken into account by photoionization models. At high metallicities,† the relevance of any empirical relation among the various T_ls is even more questionable, due to the existence of large temperature gradients in the nebulae, which are strongly dependent on the physical conditions.

† Throughout, the word "metallicity" is used with the meaning of "oxygen abundance". This is common practice in nebular studies. Although oxygen is not a metal according to the definition given by chemistry, in nebular astronomy the word metal is often used to refer to any element with relative atomic mass ≥12. The use of the O/H abundance ratio to represent the "metallicity" – as was first done by Peimbert (1978) – can be justified by the facts that oxygen represents about half of the total mass of the "metals" and that it is the major actor – after hydrogen and helium – in the emission spectra of nebulae. Note that, for stellar astronomers, the word "metallicity" is related to the iron abundance, rather than to the oxygen abundance, so the two uses of the word "metallicity" are not strictly compatible, since the O/Fe ratio changes during the course of chemical evolution.

The ionization correction factors that are used are based either on ionization potential considerations or on formulae obtained from grids of photoionization models. For H II galaxies, a set of ICFs is given by Izotov *et al.* (2006). For planetary nebulae, a popular set of ionization correction factors is that from Kingsburgh & Barlow (1994), which is based on a handful of unpublished photoionization models. Stasińska (2007, in preparation) gives a set of ICFs for planetary nebulae based on a full grid of photoionization models. It must be noted, however, that theoretical ICFs depend on the model stellar atmospheres that are used in the photoionization models. Despite the tremendous progress in the modelling of stellar atmospheres in recent years, it is not yet clear whether predicted spectral energy distributions (SEDs) in the Lyman continuum are correct.

Finally, note that the line-of-sight ionization structure, in the case of observations that sample only a small fraction of the entire nebula, is different from the integrated ionization structure. This is especially important to keep in mind when dealing with trace ionization stages.

A case of failure of T_e-based abundances: metal-rich giant H II regions. Until recently, it was not possible to measure the electron temperature in metal-rich H II regions. The usual temperature diagnostics involve weak auroral lines, which easily fall below the detection threshold at low temperatures. With very large telescopes, such temperature-sensitive line ratios as [O III] λ4363/5007, [N II] λ5755/6584 and [S III] λ6312/9532 can now be measured even at high metallicities (e.g. Kennicutt *et al.* 2003, Bresolin *et al.* 2005, Bresolin 2007). However, due to the large temperature gradients expected to occur in high-metallicity nebulae, which are a consequence of the extremely efficient cooling in the O^{++} zone due to the infrared [O III] lines, [O III] λ4363/5007 does not represent the temperatures of the O^{++} zone. As a consequence, the derived abundances can be strongly biased, as shown by Stasińska (2005). The magnitude of the bias depends on the physical properties of the H II region and on which observational temperature indicators are available.

A further problem in the estimation of T_e at high metallicity is the contribution of recombination to the intensities of collisionally excited lines, which becomes important at low values of T_e. For example, the contribution of recombination from O^{++} to the intensity of [O II] can be very important. It can be corrected for by using the formula given in Liu *et al.* (2000), provided that the temperature characteristic of the emission of the recombination line is known. If the temperature is measured using ratios of CELs only, that is not the case.

1.3.2.2 Statistical methods

In many cases, the weak [O III] λ4363 or [N II] λ5755 lines are not available because either the temperature is too low or the spectra are of low signal-to-noise ratio, or else the data consist of narrow-band images in the strongest lines only. Then, one may use the so-called "strong-line methods" to derive abundances. Such methods are only statistical, in the sense that they allow one to derive the metallicity of an H II region only on the assumption that this H II region shares the same properties as those of the H II regions used to calibrate the method. In practice, such methods work rather well for giant H II regions, since it appears that giant H II regions form a narrow sequence (see e.g. McCall *et al.* 1985), in which the hardness of the ionizing radiation field and the ionization parameter are closely linked to the metallicity. Indeed, an increased metallicity enhances the metal line blocking of the emergent stellar flux in the extreme ultraviolet and softens the ionizing spectrum. In addition, the pressure exerted on the nebular gas increases with the strength of the stellar winds, which are related to metallicity, and this in turn decreases the ionization parameter (Dopita *et al.* 2006).

Unlike direct methods for abundance determinations, statistical methods have to be calibrated. The reliability of these methods depends not only on the choice of an adequate indicator, but also on the quality of the calibration. This calibration can be done using grids of ab-initio photoionization models (McGaugh 1991), using a few tailored photoionization models (Pagel *et al.* 1979), abundances derived from direct methods (Pilyugin & Thuan 2005), or objects other than H II regions thought to have the same chemical composition (Pilyugin 2003).

The oldest and still most-popular statistical method is the one based on oxygen lines. Pagel *et al.* (1979) introduced the ([O II] λ3727 + [O III] λ4959,5007)/Hβ ratio (later referred to as R_{23} or O_{23}) to estimate O/H. This method has been calibrated many times, with results that may differ by about 0.5 dex. McGaugh (1994) and later Pilyugin (2000, 2001) refined the method to account for the ionization parameter.

Many other metallicity indicators have been proposed: [O III] λ5007/[N II] λ6584 (O_3N_2) by Alloin *et al.* (1979); [N II] λ6584/Hβ (N_2), by Storchi-Bergmann *et al.* (1994); ([S III] λ9069 + [S II] λ6716, 6731)/Hα (S_{23}) by Vílchez & Esteban (1996); [N II] λ6584/[O II] λ3727 (N_2O_2) by Dopita *et al.* (2000); [Ar III] λ7135/[O III] λ5007 (Ar_3O_3) and [S III] λ9069/[O III] λ3869 (S_3O_3) by Stasińska (2006); and [Ne III] λ9069/[O II] λ3727 (Ne_3O_2) by Nagao *et al.* (2006). The metallicity indicators proposed until 2000 have been compared by Pérez-Montero & Díaz (2005). However, all those methods will have to be recalibrated when the emission-line properties of the most-metal-rich H II regions are well understood, which is not the case at present.

A few comments are in order. First, any method based on the ratio of an optical CEL and a recombination line (e.g. O_{23} or S_{23}) is bound to be double-valued, as illustrated e.g. by Figure 7 of Stasińska (2002). This is because, at low metallicities, such ratios increase with increasing metallicity, whereas at high metallicities they decrease with increasing metallicity due to the greater cooling by infrared lines, which lowers the temperature below the excitation threshold of optical CELs. In such circumstances, external arguments must be relied upon to find out whether the object under study is on the "high-abundance" or "low-abundance" branch. The commonest argument is based on the [N II] λ6584 line. The reason why this argument works is that the N/O ratio is observed to increase as O/H increases, at least at high metallicities. Besides, high-metallicity H II regions tend to have lower ionization parameters, favouring low-excitation lines such as [N II] λ6584. The biggest problem is at intermediate metallicities, where the maxima of O_{23} and S_{23} occur and the metallicity is very ill-determined. By using both O_{23} and S_{23} indices at the same time, it would perhaps be possible to reduce the uncertainty.

Methods that use the [N II] λ6584 lines have another potential difficulty. The chemical evolution of galaxies changes the N/O ratio in a complicated and non-universal way. Therefore, a calibration is not necessarily relevant for the group of objects under study.

Perhaps the most satisfactory methods, on the theoretical side, are the ones using the Ar_3O_3 or S_3O_3 indicators, since these indicators are monotonic and work for well-understood reasons, which are directly linked to metallicity.

Conversely, the Ne_3O_2 index, which is seen to decrease as metallicity decreases, behaves in such a way *only* because metal-richer giant H II regions happen to be excited by a softer radiation field and have a lower ionization parameter. This is a very indirect metallicity indicator!

It is important to be aware that, in principle, strong-line methods can be safely used only when applied to the same category of objects as was used for the calibration. The meaning of the results in the case of integrated spectra of galaxies, for example, is far from

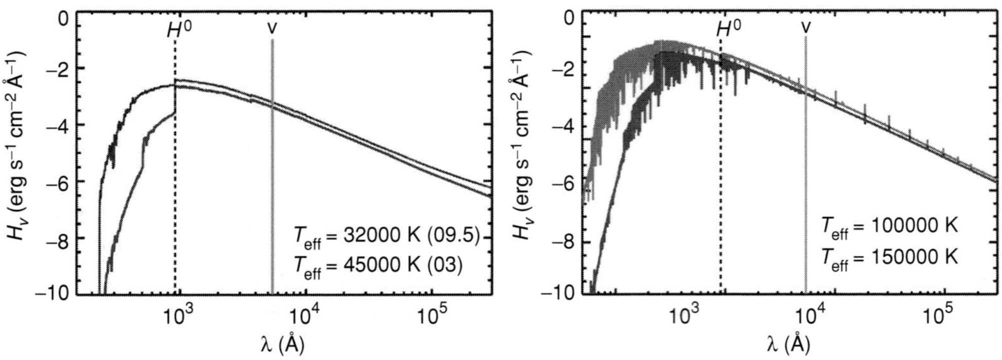

FIGURE 1.1. Spectral energy distributions (SEDs) for effective temperatures corresponding to massive stars (left) and central stars of planetary nebulae (right). The dotted line indicates the position of the ionization potential of hydrogen; the full line indicates the wavelength of the V filter.

obvious in an absolute sense. Such spectra contain the light from H II regions differing in chemical composition and extinction as well as the light from the diffuse ionized interstellar medium. In addition, inclination effects may be important. A few studies have addressed these issues from an observational point of view (Zaritsky *et al.* 1994, Kobulnicky *et al.* 1999, Moustakas & Kennicutt 2006), but clearly the subject is not closed.

A further step in strong-line abundance determinations has been made by using ratios of line equivalent widths (EWs) instead of intensities (Kobulnicky *et al.* 2003). The advantage of using equivalent widths is that they are almost insensitive to interstellar reddening, which allows one to apply the method even when reddening corrections are not available, especially at redshifts larger than 1.6. The reason why equivalent widths work well for integrated spectra of galaxies is that there is empirically a very close correlation between line intensities and equivalent widths, meaning that, statistically, stellar and nebular properties as well as the reddening are closely interrelated.

1.3.3 *Estimation of the effective temperature of the ionizing stars*

$T_\star$ from the Zanstra method. This method, proposed by Zanstra (1931), makes use of the fact that the number of stellar quanta in the Lyman continuum, normalized with respect to the stellar luminosity at a given wavelength, is an increasing function of the effective temperature. This is illustrated in Figure 1.1 (based on modern stellar model atmospheres). In practice, it is the luminosity of the Hβ line which is the counter of Lyman-continuum photons. This assumes that all the Lyman-continuum photons are absorbed by hydrogen. This assumption breaks down in the case of density-bounded nebulae or of nebulae containing dust mixed with the ionized gas. In real nebulae, some Lyman-continuum photons are also absorbed by He^0 and He^+. However, recombination of these ions produces photons that are able to ionize hydrogen, so the basic assumption of the Zanstra method is generally remarkably well fulfilled. Of course, the value of the effective temperature, $T_\star$, obtained by the Zanstra method will depend on the model atmosphere used in the derivation.

For very hot stars, such as the central stars of planetary nebulae, one can also define a He^+ Zanstra temperature, using the He II λ4686 flux as a measure of the number of photons with energies above 54.4 eV.

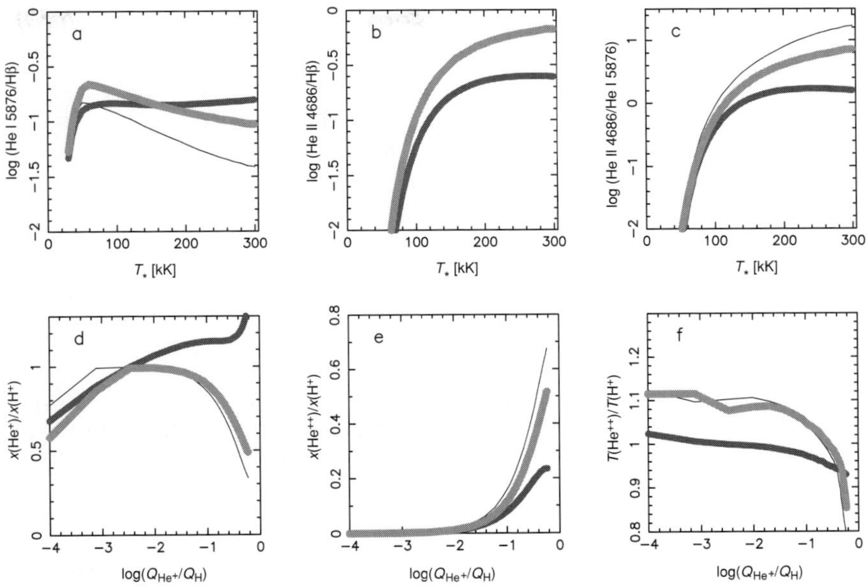

FIGURE 1.2. Sequences of photoionization models for spherical, uniform nebulae ionized by stars of various effective temperatures. Thin curve: ionization parameter $U = 10^{-2}$, He/H = 0.1. Thicker curve: $U = 10^{-3}$, He/H = 0.1. Thickest curve: $U = 10^{-2}$, He/H = 0.15. For simplicity, the models were run with a black-body central star.

A few properties, illustrated by the models shown in Figure 1.2, are worth noting.

- He I λ5876/Hβ measures $T_\star$ only in a small range of temperatures ($T\star < 40$ kK for our models with a black-body central star as seen in Figure 1.2a, this limit being of course different for realistic stellar atmospheres).
- Owing to the competition between H^0 and He^+ to absorb photons with energies above 54.4 eV, He II λ4686/Hβ saturates at $T_\star > 150$ kK and depends on U at $T_\star > 100$ kK (Figure 1.2b).
- He II λ4686/Hβ is independent of He/H (Figure 1.2b), while He II λ4686/He I λ5876 depends on He/H (Figure 1.2c). Indeed, He II λ4686 is a counter of photons, as long as the He^{++} Strömgren sphere is smaller than the H^+ Strömgren sphere.
- The temperatures in the H^0 and He^+ zones are not equal (Figure 1.2f).

T_* from the observed ionization structure. Since the ionization structure of a nebula depends on T_*, one can think of using the line ratios of two successive ions to infer T_* (e.g. Kunze *et al.* 1996). However, in such an approach, the effect of the ionization parameter must be considered as well.

To alleviate this problem, Vílchez & Pagel (1988) introduced a "radiation softness parameter", which they defined as

$$\eta = \frac{O^+/O^{++}}{S^+/S^{++}} \tag{1.7}$$

and expected to be independent of U to a first approximation. This, however, was not sufficient for an accurate determination of $T_\star$ (or even for ranking the effective temperatures of different objects).

Morisset (2004) constructed a grid of photoionization models with various values of U (at various metallicities) to allow a proper estimate of $T_\star$. The grid was constructed using the WM-BASIC model atmospheres of Pauldrach *et al.* (2001).

$T_\star$ from energy-balance methods. This method was first proposed by Stoy (1933). It makes use of the fact that the heating rate of a nebula is a function of the effective temperature. Since, in thermal equilibrium, the heating rate is compensated for by the cooling rate, a measure of the cooling rate allows one to estimate $T_\star$. Most of the cooling is done through collisionally excited lines. Therefore, an estimation of $T_\star$ can be obtained from the formula

$$\sum L_{\rm CEL}/L({\rm H}\beta) = f(T_\star). \tag{1.8}$$

Pottasch & Preite-Martinez (1983) proposed a calibration of this method for planetary nebulae.

A similar argument led Stasińska (1980) to propose a diagram to estimate the average effective temperature of the ionizing stars of giant H II regions. This diagram plots the value of [O III] λ4363/5007 as a function of the metallicity O/H, for photoionization models corresponding to various values of $T_\star$.

1.3.4 *Determining the star-formation rate*

The star-formation rate is an important quantity to measure in galaxies. It can be obtained in many ways: from the ultraviolet continuum, far-infrared continuum, radio continuum, recombination lines and forbidden lines. The latter two are sensitive to the most recent star formation (less than a few Myr ago, which is the lifetime of the most-massive stars). Note that, while the luminosity in a line measures the absolute value of the recent star-formation rate, the equivalent width of the line measures the ratio of the present to past star-formation rate. Any technique must be calibrated using simulations of stellar populations in which the basic parameters are the stellar initial mass function, the star-formation history and the metallicity. These simulations are based on libraries of stellar evolutionary tracks and of stellar atmospheres. Kennicutt (1998) and Schaerer (2000) give exhaustive reviews on the question. Here, we simply mention a few issues regarding the estimation of star-formation rates using emission lines.

Determining the star-formation rate using Hβ or Hα. As with any determination of a physical parameter using Hβ, the basic assumption is that all the Lyman-continuum photons are absorbed by the gas. Also, the effect of extinction arising from intervening dust must be properly accounted for (see Kewley *et al.* 2002). As shown by Schaerer (2000), the derived star-formation rate strongly depends on the choice of stellar initial mass function and upper stellar mass limit.

Determining the star-formation rate using [O II] λ3727. It may seem strange to use such a line to measure the star-formation rate, instead of simply using Hα. In addition to the provisos on Hα, the [O II] λ3727 obviously must depend on the metallicity and the ionization parameter. The main reason for attempting to use the [O II] λ3727 line in spite of this is that it can be observed in the optical range at larger redshifts than can Hα and Hβ. However, it is an observational fact that the [O II] λ3727/Hβ ratio strongly varies among emission-line galaxies, even discounting objects containing an active nucleus (see Figure 1.3). Therefore, the use of [O II] λ3727 as a star-formation-rate indicator is extremely risky.

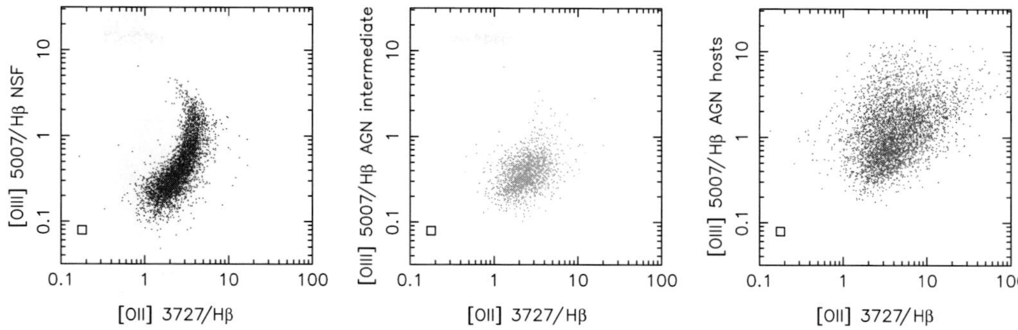

FIGURE 1.3. Diagrams using data on galaxies from the Sloan Digital Sky Survey (taken from Stasińska 2006) to show how strongly the [O II] $\lambda3727/\mathrm{H}\beta$ ratio varies among normal star-forming galaxies (left), galaxies containing an active nucleus (right) and hybrid galaxies (middle).

1.3.5 *How can one distinguish normal galaxies from AGN hosts?*

After the discovery of spiral galaxies with a very bright nucleus emitting strong and broad (several thousands of km s^{-1}) emission lines (Seyfert 1941), it became clear that these galactic nuclei† were the locus of violent, non-stellar activity (Burbidge *et al.* 1963, Osterbrock & Parker 1965), perhaps of the same nature as found in quasars. Heckman (1980) performed a spectroscopic survey of the nuclei of a complete sample of 90 galaxies, and found that low-ionization nuclear emission-line regions (LINERs) were quite common and seemed to be the scaled-down version of Seyfert nuclei. Baldwin *et al.* (1981) were the first to propose spectroscopic diagnostics based on emission-line ratios to distinguish normal star-forming galaxies from AGNs. The most famous is the [O III] $\lambda5007/\mathrm{H}\beta$ versus [N II] $\lambda6584/\mathrm{H}\alpha$ diagram, often referred to as the BPT diagram (for Baldwin, Phillips & Terlevich). The physics underlying such a diagram is that photons from AGNs are harder than those from the massive stars that power H II regions. Therefore, they induce more heating, implying that optical collisionally excited lines will be brighter with respect to recombination lines than in the case of ionization by massive stars only. Veilleux & Osterbrock (1987) proposed additional diagrams: [O III] $\lambda5007/\mathrm{H}\beta$ versus [S II] $\lambda6725/\mathrm{H}\alpha$ and [O III] $\lambda5007/\mathrm{H}\beta$ versus [O I] $\lambda6300/\mathrm{H}\alpha$. As had previously been found by McCall *et al.* (1985), giant H II regions form a very narrow sequence in these diagrams.

It was a great surprise, after the first thousands of galaxy spectra from the Sloan Digital Sky Survey (SDSS) (York *et al.* 2000) had been released, to find that a proper subtraction of the stellar continuum in galaxies (Kauffmann *et al.* 2003) allowed one to see a second sequence in the BPT diagram, in the direction opposite to that of the star-forming sequence. Thus, emission-line galaxies in the BPT diagrams are distributed in two wings, which look like the wings of a flying seagull (see Figure 1.4).

Just a few years before, Kewley *et al.* (2001) had constructed a grid of photoionization models in order to determine a theoretical upper limit to the ionization by massive stars in the BPT diagram. This upper limit, later referred to as the "Kewley line", proved well to the right of the star-forming wing from the SDSS. Kauffmann *et al.* (2003) shifted this line to the left to define an empirical limit between normal star-forming galaxies and AGN hosts (the "Kauffmann" line). Stasińska *et al.* (2006) found this limit to be still too "generous", and proposed a more-restrictive one, based on a grid of photoionization

† The term Seyfert galaxy was used for the first time by de Vaucouleurs (1960).

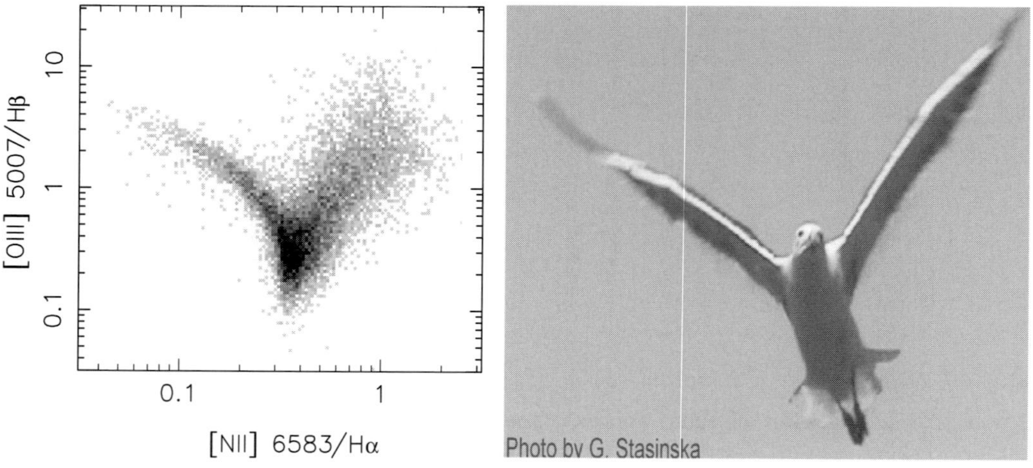

FIGURE 1.4. Galaxies from the Sloan Digital Sky Survey in the BPT diagram are distributed in two wings: the wings of a flying seagull.

models aimed at reproducing the upper envelope of the left wing of the seagull. Its equation is

$$y = (-30.787 + 1.1358x + 0.27297x^2)\tanh(5.7409x) - 31.093, \tag{1.9}$$

where $y = \log([\text{O III}]\ \lambda5007/\text{H}\beta)$ and $x = \log([\text{N II}]\ \lambda6584/\text{H}\alpha)$.

Note that, among the three diagnostic diagrams proposed by Veilleux and Osterbrock (1989), [O III] λ5007/Hβ versus [N II] λ6584/Hα is the most efficient, due to the greater N/O in galaxies of high metallicity. Then, heating by an AGN boosts the [N II] line and creates a clear separation of the two wings. One then understands why the [O III] λ5007/Hβ versus [O II] λ3727/Hβ diagram, which is used as a surrogate of the BPT diagram at redshifts of ∼0.2 (Lamareille *et al.* 2004), is much less efficient at separating AGN hosts from normal star-forming galaxies.

Stasińska *et al.* (2006) also noted that a classification based on [N II] λ6584/Hα only is also feasible, and has the advantage of being applicable to a much larger number of galaxies, since [O III] λ5007 and Hβ are not needed.

Finally, these authors proposed another classification diagram, plotting $D_n(4000)^\dagger$ as a function of the EW of [O II] λ3727 (the DEW diagram). The rationale for this diagram is that, if, in a galaxy with a large $D_n(4000)$, emission lines can be seen, they must be due to some other cause than photoionization by massive stars. The hope is to be able to use this diagram to distinguish AGNs at redshifts higher than 0.2. However, the fact that higher redshifts correspond to younger ages of the stellar populations is an issue that requires further investigation.

To end this section, let us remark that the term "LINER" is nowadays often employed to designate the galaxies that are found to the lower right of the BPT diagram. This is an unfortunate deviation from the original meaning, since it is likely that, in these galaxies, the emission lines come not only from the nucleus, but also from a much larger zone of the galaxy. The original "LINERS" discovered by Heckman would not appear in the BPT diagram due to the low luminosity of the active nucleus, unless there were significant line emission from the rest of the galaxy.

† $D_n(4000)$ is the discontinuity observed at 4000 Å in the spectra of galaxies; it increases with stellar metallicity and age.

1.4. Photoionization modelling

We now turn to the "royal way" of analysing emission-line spectra: photoionization modelling. We will show that photoionization modelling is an art that requires not only photoionization codes, but also a certain dose of common sense.

1.4.1 *Photoionization codes*

Most of the codes listed below have been intercompared at the Lexington conference "Spectroscopic Challenges of Photoionized Plasmas" in 2000. The results of the code comparisons can be found in Péquignot *et al.* (2001).

1.4.1.1 One-dimensional photoionization codes

- CLOUDY by Gary Ferland and associates computes models for ionized nebulae and photodissociation regions (PDRs). It is regularly updated, well documented and widely used. It is available at http://www.nublado.org.
- MAPPINGS by Michael Dopita plus Kewley, Evans, Groves, Sutherland, Binette, Allen and Leitherer computes models for photoionized nebulae and for planar shocks. It can be found at http://www.ifa.hawaii.edu/˜kewley/Mappings.
- XSTAR by Tim Kallman computes models for photoionized regions with special attention to the treatment of X-rays. It can be found at http://heasarc.nasa.gov/lheasoft/xstar/xstar.html.

1.4.1.2 Three-dimensional photoionization codes

- CLOUDY-3D by Christophe Morisset is a pseudo-3D code based on CLOUDY. It allows quick modelling of 3D nebulae and visualization (including computation and visualization of line profiles). It can be found at http://132.248.1.102/Cloudy_3D.
- MOCASSIN by Barbara Ercolano is a full 3D Monte Carlo photoionization code that also treats dust transfer in an accurate manner. It is available from be@star.ucl.ac.uk (see also http://hea-www.harvard.edu/˜bercolano).

1.4.1.3 Other codes

Many other, independent photoionization codes are mentioned in the literature (some of them benchmarked in Péquignot *et al.* 2001), but have not been made available for public access. This is especially the case of hydrodynamical codes that include the physics of photoionization and the computation of line emission. In such codes, of course, the physics of radiation is treated in a simplified manner, since a simultaneous treatment of the macrophysics and the microphysics requires tremendous computing power.

1.4.2 *Why do we construct photoionization models?*

There can be many different reasons for building photoionization models. For example, one might want to

- check the sensitivity of observable properties to input parameters
- compute a grid of models for easy interpretation of a certain class of objects
- calculate ionization-correction factors
- derive the chemical composition of a given nebula
- estimate characteristics of the ionizing source
- probe stellar-atmosphere model predictions in the far ultraviolet.

1.4.3 *How should one proceed?*

Each of the problems above requires a specific approach. It is always worth spending some time on finding the best way to achieve one's goal. For example, if one wants to derive the chemical composition of a nebula by means of emission-line fitting, it is not sufficient to find one solution. One must explore the entire range of possible solutions, given the observational constraints. This is not always easy.

If the aim is to interpret one object, or a given class of objects, the first step is to collect all the observational constraints needed for this purpose. This includes monochromatic images as well as line intensities in various wavelength ranges and with different apertures. Also it is important to characterize the ionizing sources as well as possible from the observations (visual magnitude, spectral type in the case of a single star, age of the ionizing stellar population in the case of a large collection of coeval stars).

Then, one must define a strategy. How will one explore a parameter space? How will one deal with error bars? How will one test the validity of a model?

Finally, one must evaluate the result of the investigation. Was the goal achieved? If not, what does this imply? In fact, this aspect is often overlooked; nevertheless, it is potentially very instructive and is an incentive for progress in the field.

In the following we present some commented examples.

1.4.4 *Abundance derivation by tailored model fitting*

The general procedure would be something like the following sequences.

(1) Define the input parameters: the characteristics of the ionizing radiation field (luminosity, spectral energy distribution); the density distribution of the nebular gas; the chemical composition of the nebular gas; the distance.

(2) Use an appropriate photoionization code and compute a model.

(3) Compare the outputs with the observations (corrected for extinction by intervening dust): the total observed Hα flux; the Hα surface-brightness distribution and the angular size of the Hα-emitting zone; the line intensities etc. ...

(4) Go back to (1) and iterate until the observations are satisfactorily reproduced. Here "satisfactorily" means that *all* the observational data are reproduced within acceptable limits, those limits taking into account both the observational errors and the approximations of the model. These "limits" should be set by a critical analysis of the situation, *before* running the models. It may be that one finds no solution. This is by no means a defeat. It is actually an important result too, which tells us something about the physics of the object. But, in order to be useful, such a result must be convincing, in other words the reason why no solution can be found must be clearly explained.

For a good-quality model fitting, it is important to do the following.

(1) Use as many observational constraints as possible (not only line-intensity ratios).

(2) Keep in mind that the importance of the constraint has nothing to do with the strengths of the lines. For example He II λ4686/Hβ, which is of the order of a few per cent at most in H II regions, indicates the presence of photons with energy above 54.4 eV, which are not expected in main-sequence, massive stars (unless they are part of an X-ray binary system); [O III] λ4363/5007 and [N II] λ5755/6584 indicate whether the energy budget is well reproduced. It is too often ignored that photoionization models compute not only the ionization structure of the nebulae, but also compute their temperature, which is essential in predicting the strengths of the emission lines.

(3) Recall that some constraints are not independent. For example, if [O III] λ5007/Hβ is fitted by the model, then [O III] λ4959/Hβ should be fitted as well because the [O III] λ5007/4959 ratio is fixed by atomic physics, since both lines originate from the same level.† In no case can the fact that both lines are fitted at the same time be taken as a success of the model. On the other hand, if they are not, this may indicate an observational problem, e.g. that the strong [O III] λ5007 line is saturated. As a matter of fact, many observers use precisely this ratio as a check of the accuracy of their observations.

(4) Choose a good estimator for the "goodness of fit", e.g. avoid using a χ^2-minimization technique without being convinced that this is the most-appropriate test in the case under study. All the observables should be fitted (within limits defined a priori).

(5) Try to visualize the model–result comparison as much as possible. Examples among many possibilities can be found in Stasińska & Schaerer (1999) or in Stasińska *et al.* (2004).

Outcomes from model-fitting. A priori, the most satisfactory situation is when *all* the observations are fitted within the error bars. This may imply (but does not necessarily show) that the model abundances are the real abundances. If the constraints are insufficient, the model abundances may actually be very different from the true ones.

Quite often, some of the observations cannot be fitted. This means either that the observations are not as good as was thought, or that the model does not represent the object well. Some assumptions in the modelling may be incorrect. For example the nebula has a geometry different from the assumed one, the stellar ionizing radiation field is not well described, or an important heating mechanism is missing from the model. In such a situation, the chemical composition is generally not known with the desired accuracy, a fact too often overlooked.

An example of photoionization modelling without a satisfactory solution: the most-metal-poor galaxy I ZW 18 (Stasińska & Schaerer 1999).

The observational constraints were provided by narrow-band imaging in nebular lines and the stellar continuum, together with optical spectra giving nebular line intensities. The ionizing-radiation field was given by a stellar-population synthesis model aimed at reproducing the observed features from hot Wolf–Rayet stars. The conclusion from the modelling exercise was that, even taking into account strong deviations from the adopted spectral energy distribution of the ionizing radiation and the effect of postulated additional X-rays, the photoionization models yield an [O III] λ4363/5007 ratio that is too low by about 30%.‡ This significant discrepancy cannot be solved by invoking expected inaccuracies in the atomic data. The missing energy is of the same order of magnitude as the one provided by the stellar photons.

Interestingly, this paper was rejected by a first referee on the ground that "the temperature of the nebula, as indicated by the ratio [O III] λ4363/5007, is in all trial

† The lack of variation of the [O III] λ5007/4959 ratio among quasars at various redshifts shows that the fine-structure constant, α, does not depend on cosmic time (Bahcall *et al.* 2003). A far-reaching application of emission-line astrophysics!

‡ Note that the observed geometrical constraint provided by the images was important in reaching this conclusion, since the authors also showed that, by assuming a simpler geometry with no central hole – but in complete disagreement with the observations – one could obtain a higher electron temperature, due to a reduced H Lyα cooling.

models out of the observed range by 30%. The authors seem to have tried everything possible [...] but still ... no secure results". A second referee accepted the paper without changes, emphasizing that "The results are of highest interest for future studies of ionized regions, mainly by the identification of a serious problem in the interpretation of the [O III] λ4363/5007 ratio, which cannot be consistently reproduced by the models"!

Another example of photoionization modelling without a satisfactory solution: the giant H II region NGC 588 in the galaxy M33 (Jamet *et al.* 2005).

In this case, the observational constraints were provided by a detailed characterization of the ionizing stars (Jamet *et al.* 2004), narrow-band imaging, long-slit optical spectra and far-infrared line fluxes from the ISO satellite. In the analysis, aperture effects were properly taken into account. The temperature derived from the [O III] λ4363/5007 ratio is larger than the one derived from the [O III] λ88μ/5007 ratio by 3000 K. It was found that no photoionization model could explain the observed temperature diagnostics. Unless the measured [O III] λ88μ flux is in error by a factor of two, this implies that the oxygen abundance in this object is uncertain by about a factor of two.

Note that both this study and the former one used a 1D photoionization code, while the study of the nebular surface-brightness distribution indicates that the nebula is not spherical. The nebular geometry was taken into account in the discussion, and the conclusions reached in both cases are robust.

An example of photoionization modelling with insufficient constraints: the chemical composition of the Galactic bulge planetary nebula M2–5.

This is an object for which no direct temperature diagnostic was available. Using photoionization modelling, Ratag (in a 1992 thesis) and Ratag *et al.* (1997) claimed an oxygen abundance of one fourth Solar. However, as shown in Stasińska (2002), models fulfilling *all* the observational constraints can be constructed with oxygen abundances as different as $O/H = 1.2 \times 10^{-3}$ and 2.4×10^{-4}! This is linked to the "double-value problem" discussed in Section 1.2.2.2. Because of only crude knowledge of the physics of nebulae and insufficient exploration of the parameter space, Ratag had drawn a possibly erroneous conclusion.

1.4.5 *Abundance derivation using grids of models*

One is often interested in determining the metallicities of a large sample of objects, the extreme case being emission-line galaxies from the SDSS, which number tens of thousands. A tailored model fitting is out of reach in such cases. Apart from statistical methods, discussed in Section 1.2.2.2, one may consider building a vast, finely meshed grid of photoionization models and use them to derive the metallicities of observed emission-line galaxies either by interpolation or by Bayesian methods. Such an approach has been used by Charlot & Longhetti (2000) and by Brinchmann *et al.* (2004). This method is powerful and appealing, since it can be completely automated. However, it is not the ultimate answer to the problem of abundance determination and it must be used with some circumspection. As shown by Yin *et al.* (2007), when applied to a sample of galaxies in which the [O III] λ4363 line is observed, allowing direct abundance determinations, this method returns significantly higher metallicities than does the direct method. Yin *et al.* argue that the reason lies in the abundances chosen in the model grid. Since the procedure uses simultaneously all the strong lines to derive O/H, any offset between the real N/O and the N/O adopted in the model grid induces an offset in the derived O/H.

1.4.6 *Testing model atmospheres of massive stars using H II regions*

In an ongoing study of the Galactic H II region M43 and its ionizing star (Simón-Díaz *et al.*, in preparation), one of the aims is to use the nebula to probe the spectral energy distribution modelled for its ionizing star. The nebula has a rather simple structure for an H II region, being apparently round. The characteristics of the ionizing star (effective temperature and gravity) were obtained by fitting the stellar H and He optical lines using the stellar-atmosphere code FASTWIND (Puls *et al.* 2005). The stellar luminosity was then obtained, knowing the distance. The predicted stellar spectral energy distribution was used as an input to the photoionization code CLOUDY (great care was taken in the treatment of the absorption edges, which are not the same in CLOUDY and FASTWIND). The nebula was assumed to be spherically symmetrical and the density distribution was chosen to reproduce the observed Hα surface brightness distribution. The starting nebular abundances were obtained using the classical T_e-based method. The observational constraints are the de-reddened line intensities at various positions in the nebula.

It was found that the distribution of emission-line ratios across the nebula cannot be reproduced by the model. While this could imply that the tested model atmosphere does not predict the energy distribution correctly, another hypothesis must be investigated first: that the geometry assumed for the nebula is wrong. In spite of its roundish appearance, it is possible that M43 is not a sphere, but a blister seen face on. The photoionization model constructed would then correspond to its "spherical impostor", using the expression coined by Morisset *et al.* (2005). As shown by Morisset & Stasińska (2006), observation of the emission-line profiles would allow one to distinguish between the two geometries. In the meantime, photoionization computations using the pseudo-3D code CLOUDY_3D of Morisset (2006) are being carried out, to provide models for the blister geometry.

1.4.7 *A photoionization study of an aspherical nebula using a three-dimensional code: the planetary nebula NGC 7009*

The use of 3D photoionization modelling is generally difficult, due to the number of parameters involved in describing the geometrical structure of a nebula and to the long computational time required to run a model. A 1D code is often sufficient to identify a physical problem that needs a solution, as seen in Section 1.3.4. On the other hand, a 3D code is essential when tackling a problem in which the effect of geometry is important. Such is the case, for example, for the low-ionization knots observed in planetary nebulae, which have been said to show nitrogen enhancement by factors of 2–5 (e.g. Balick *et al.* 1994). Authors of subsequent studies argued that this conclusion might be wrong, due to the use of an improper ionization-correction scheme (it was assumed that N/O = N^+/O^+). Gonçalves *et al.* (2006) examined this problem in the case of NGC 7009, a planetary nebula with two conspicuous symmetrical knots. Their aim was to explore the possibility that the enhanced [N II] emission observed in the knots could be due to ionization effects, by building a 3D photoionization model of homogeneous chemical composition reproducing the observed geometry and spectroscopic peculiarities. They simulated observations in various regions of the nebula, and found that the N^+/O^+ ratio varies strongly with position and can explain (at least partly) the apparent nitrogen enhancement in the knots.

1.4.8 *The interpretation of data from integral field spectroscopy*

Integral field spectroscopy is becoming a major tool to study nebulae and galaxies. A wealth of spatially resolved data will routinely become available. The question now is that of which tools should be used to interpret such data. Obviously, starting by building 3D photoionization models is not a good way to proceed. As argued above, 3D modelling is difficult and time-consuming, and a proper evaluation of the benefits of this kind of approach is recommended before undertaking such an endeavour. On the other hand, one may be tempted to use results from grids of published 1D photoionization models or observational diagnostic diagrams relevant for the integrated nebular light. This is risky. For example, one might attribute the increase of [N II]/Hα or [S II]/Hα in certain zones to non-stellar ionization while it might merely be a line-of-sight effect in a region close to an ionization front.

1.5. Questions pending

There are a few important aspects that have not been mentioned above, for the sake of a more-linear presentation. Most of them have been treated extensively in Stasińska (2004), and will be mentioned here only briefly, with some updates if necessary.

1.5.1 *Correction for reddening, underlying stellar absorption and aperture effects*

Before being analysed in terms of abundances or star-formation rates or being compared with the results of photoionization models, the intensities of observed lines must be corrected for various effects. The presence of dust between the zone of emission and the observer attenuates the collected radiation and modifies its colour. It appears that the dust-extinction curve, which was once considered universal, is actually a one-parameter function characterized by the value of the total-to-selective extinction $R_V = A_V/E(B-V)$ (Fitzpatrick 1999, 2004). The canonical value of R_V, which is generally used for extinction corrections, is 3.1 or 3.2. However, the measurement of R_V using 258 Galactic O stars yields a distribution around this value with a dispersion of ±0.5. Values of R_V as small as 1.6 or as large as 5 are found (Patriarchi *et al.* 2003).

The extinction curve of a single star caused by interstellar dust is different from the obscuration curve of an extended nebula (Calzetti 2001) because in the latter case the observed radiation includes some scattered light. From a sample of starburst galaxies observed from the far ultraviolet to the near infrared, an obscuration curve for galaxies was derived (Calzetti 1997, Calzetti *et al.* 2000).

The amount of reddening is usually determined by fitting the observed Balmer decrement to the theoretical one at the appropriate temperature. For emission-line galaxies, in giant H II regions (and also in unresolved planetary nebulae), the observed Balmer lines are affected by the underlying absorption from the stellar component. This absorption can be determined empirically together with the extinction by fitting several Balmer lines to the theoretical Balmer decrement (Izotov *et al.* 1994). It can also be modelled, using spectral-synthesis techniques to reproduce the stellar absorption features other than Balmer lines and reading out from the model spectrum that best fits the data the stellar absorption at the Balmer lines' wavelengths (Cid Fernandes *et al.* 2005). This procedure applied to a sample of 20 000 galaxies from the SDSS (see Stasińska *et al.* 2006 for a description of the sample) shows a spectacular correlation between absorption equivalent widths at the Balmer-line wavelengths and the discontinuity at 4000 Å, $D_n(4000)$, as shown for the first time here (Figure 1.5). Since $D_n(4000)$ is relatively easy to measure, this provides an empirical way to estimate the underlying stellar absorption at Hα and Hβ. The data shown in Figure 1.5 can be modelled by the expressions

$$\mathrm{EW}_{\mathrm{abs}}(\mathrm{H}\alpha) = 9\exp(-1.2x) + 1.3 \tag{1.10}$$

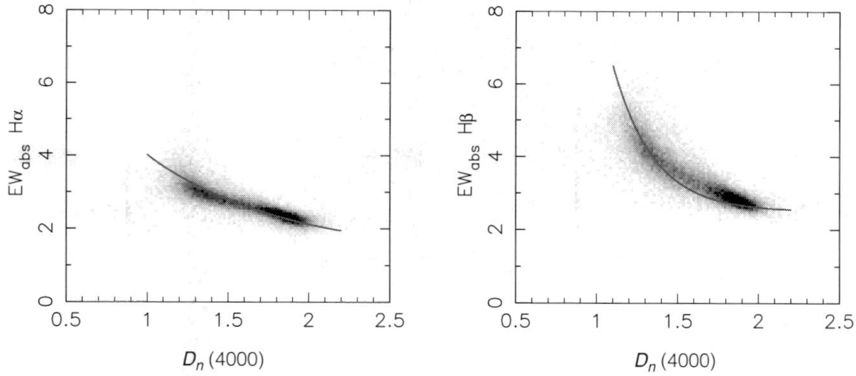

FIGURE 1.5. The stellar absorption equivalent widths at Hα (left) and Hβ (right) as a function of the discontinuity at 4000 Å. They have been determined in a sample of galaxies extracted from the SDSS by stellar-population analysis using the method of Cid-Fernandes *et al.* (2005). The curves correspond to the fits given by Equations (1.10) and (1.11).

and

$$\mathrm{EW_{abs}}(\mathrm{H}\beta) = 400\exp(-4.x + 0.5) + 2.5, \qquad (1.11)$$

where $x = D_n(4000)$.

Another issue arises when the objects under study are more extended than the observing beam. This fact must be taken into account in the analysis. For example, spectra obtained with several instruments must be carefully combined. This is especially important – and difficult – when merging ultraviolet or infrared data with optical ones, since in that case flux-calibration problems may arise. In the case of observations where a large number of lines is available, it is theoretically possible to combine all the data by comparing each observation with the results of a tailored photoionization model seen through an appropriate aperture. Most of the time, however, aperture corrections are performed in an approximate way, neglecting the effect of the ionization structure of the nebulae on the line fluxes recorded through small apertures.

A further problem, when dealing with spectra of entire galaxies, is the physical meaning of the derived abundance, especially in the presence of abundance gradients. This issue, which is also relevant to the simple "strong-line-indicators" approach, has not yet been fully addressed.

1.5.2 *Escape of ionizing radiation*

In most nebular studies it is assumed that the nebulae are ionization-bounded. This affects such issues as the estimation of stellar temperatures via the Zanstra method, the estimation of star-formation rates in galaxies and the outcome of tailored photoionization modelling.

There is, however, growing evidence that many nebulae are density-bounded in at least some directions. This has been known for a long time for planetary nebulae. It is now recognized to be the case also for many giant H II regions (Beckman *et al.* 2002, Stasińska & Izotov 2003, Castellanos *et al.* 2003).

1.5.3 *The importance of the stellar energy distribution*

The outcome of photoionization models depends critically on the ionizing stellar energy distribution (SED) adopted, especially as regards the nebular ionization structure. There are two main issues in this respect. One is the description of the ionizing spectral energy distribution of an individual star. The other is how one deals with the case of ionization by a group of stars.

Model atmospheres for hot stars. The atmospheres of stars are generally extended and experience important deviations from local thermodynamic equilibrium (LTE). Hundreds of thousands of metallic lines produce line-blanketing and line-blocking effects, and are at the origin of radiation driven winds. Modern stellar-atmosphere codes are able to handle these aspects, although each one works with some degree of approximation at least in one aspect.

The most broadly used in photoionization modelling of H II regions are TLUSTY (Hubeny & Lanz 1995), WM-BASIC (Pauldrach *et al.* 2001) and CMFGEN (Hillier & Miller 1998). What matters for the modelling of ionized nebulae is the SED. There are sizable differences among the predictions from these codes in the ionizing continuum. It is not clear, at present, which model predictions are best suited for photoionization modelling. Authors of a few studies (e.g. Morisset *et al.* 2004a, 2004b) have used observed H II regions to test the predictions of the model atmospheres. However, this is a difficult task, and the answer is not yet clear. Simón-Díaz *et al.* (2007) are attacking this problem anew, with both a theoretical and an observational approach.

The central stars of planetary nebulae can reach much higher effective temperatures than massive stars. Temperatures of over 2×10^5 K have been found. When they are close to becoming white dwarfs, these stars have tiny atmospheres for which the plane-parallel approximation is reasonable. A grid of plane-parallel line-blanketed non-LTE (NLTE) atmospheres has been computed by Rauch (2003) using the code PRO2 (Werner & Dreizler 1993). At earlier stages of the planetary nebulae, the effect of winds is expected to be significant. They are important for planetary nebulae central stars of [WR] type, which share the same atmospheric properties as massive Wolf–Rayet stars. The same kind of approach as for H II regions has been used for planetary nebulae to test the ionizing SEDs predicted by model atmospheres, e.g. by Stasińska *et al.* (2004), who used the Potsdam code for expanding atmospheres in NLTE described in Hamann & Koesterke (1998).

Some grids of model atmospheres computed with the above codes are available, for example at the CLOUDY website.

Ionization by a group of stars. In the case of giant H II regions, where the ionization is provided by many stars, one generally relies on SEDs predicted by stellar-population synthesis codes. Starburst 99 (Leitherer 1999, Smith *et al.* 2002) and PEGASE (Fioc & Rocca-Volmerange 1997, 1999) are public-access codes that are able to do this. The integrated SED of an entire stellar population is obviously quite different from the SED of the component stars, but the quality of the SED for photoionization studies is related to the reliability of the predictions in the Lyman continuum of the stellar-atmosphere codes used to compute the stellar libraries.

There is, however, another important issue, which was pointed out by Cerviño *et al.* (2000). It is the fact that, in real H II regions, statistical sampling strongly affects the ionizing SED, and SEDs predicted by traditional stellar-population synthesis models may be far from real. Practical considerations as regards the outcome from photoionization models can be found in Cerviño *et al.* (2003) and a detailed discussion on uncertainties in stellar synthesis models and "survival strategies" is given by Cerviño & Luridiana (2005).

One of the unsolved problems is the question of the often-observed He II emission in H II galaxies. Traditional models (Stasińska & Izotov 2003) have difficulties in accounting for it properly. While there are several possible explanations (X-ray binaries, X-rays from hot-star atmospheres, etc.), sampling effects may also be an issue that needs to be considered.

Finally, in the case of a group of ionizing stars, one might wonder whether the effect on the H II region strongly depends on whether the stars are concentrated or distributed in the nebula. This question is being addressed by Ercolano *et al.* (2007).

1.5.4 *Dust*

The question of the effect of dust in ionized nebulae has been reviewed amply in Stasińska (2004). Among publicly available photoionization codes, CLOUDY, MAPPINGS and MOCASSIN include a treatment of dust grains coupled with the gas particles.

When it comes to entire galaxies, a proper treatment of dust should take into account the large-scale geometrical distribution of dust inside the galaxies. Charlot & Longhetti (2001) made a first step by including a different prescription for the absorption of photons from H II regions and from older stars, as proposed by Charlot & Fall (2000). Panuzzo *et al.* (2003) implemented results from photoionization simulations with CLOUDY in models of dusty galaxies computed with the code GRASIL2 (Silva *et al.* 1998), which include a sound treatment of all the aspects of dust reprocessing, providing a more realistic description of the integrated spectrum of a galaxy.

1.5.5 *Temperature fluctuations and the ORL/CEL discrepancy*

Ever since the seminal work by Peimbert (1967), there has existed the suspicion that elemental abundances derived from optical collisional emission lines (CELs) might be plagued by the presence of temperature fluctuations in nebulae. If such temperature fluctuations exist, then the highest-temperature zones will favour the lines of higher excitation, and as a result the temperature derived from [O III] λ4363/5007 will over-estimate the temperature of the O^{++} zone, leading to an underestimate of elemental abundances with respect to hydrogen.

Forty years later, nebular specialists (see e.g. the recent reviews by Esteban 2002, Liu 2002, Ferland 2003, Peimbert & Peimbert 2003) are still debating this issue, which is of fundamental importance for the reliability of abundance determinations of nebulae. We are talking here of an effect of typically 0.2 dex in metal abundances relative to hydrogen. One can summarize the debate in terms of four questions.

(1) Is there evidence for the existence of temperature fluctuations at a level that significantly biases the abundances from CELs?
(2) If so, is there a reliable procedure to correct for this bias?
(3) What are the possible causes of such temperature fluctuations, if they exist?
(4) Is the problem the same in H II regions and in planetary nebulae?

At present, there are no generally accepted answers to these questions. In the following, we simply collect some elements for the discussion, by supplementing what can be found in the above reviews and in Stasińska (2004).

(1) Direct evidence for temperature fluctuations large enough to be of sizable consequence for abundance determinations is scarce. It is provided by Hubble Space Telescope temperature mapping of the Orion nebula (O'Dell *et al.* 2003) or spectroscopy (Rubin *et al.* 2003). Indirect evidence comes from the comparison of temperatures measured by various means (such as [O III] λ4363/5007 and the Balmer or Paschen jump). There are many examples, both in H II regions and in planetary nebulae, indicating a typical r.m.s. temperature fluctuation, t^2, of 0.04 (see references in the review quoted above). However, a recent determination of Balmer-jump temperatures in H II galaxies (Guseva *et al.* 2007) gives t^2 close to zero, on

average. Note that these measurements concern metal-poor H II regions. Another piece of evidence comes from the discrepancy of abundances derived from ORLs and CELs for the same ion. In this case, the implied t^2 varies considerably, from less than 0.02 to almost 0.1, with the values for H II regions being small and those for planetary nebulae spanning a wide range.

(2) Peimbert (1967) and Peimbert & Costero (1969) devised a scheme, based on Taylor-series expansion of the analytic expressions giving the emissivities, to correct for the abundance bias, with the help of several simplifying assumptions. It can, however, be shown, on the basis of a two-zone toy model (Stasińska 2002), that a proper description of temperature fluctuations and any correction scheme for abundance determinations require at least three parameters, not just two as in Peimbert & Costero (1969). Besides, due to the restricted number of observational constraints, the application of the Peimbert & Costero scheme implies some assumptions that are incorrect, such as the equality of $t^2(O^{++})$ and $t^2(H^+)$.

(3) Some possible causes for the presence of temperature fluctuations have been listed by Torres-Peimbert & Peimbert (2003) and Peimbert & Peimbert (2006). These include shocks, magnetic reconnection, dust heating, etc. However, the energy requirements to obtain t^2 values of about 0.04 are very large, typically a factor of two greater than the energy gains from gas photoionization. Note that it is not so difficult to obtain a t^2 of 0.02–0.04 for the entire nebular volume, since the main cooling lines change across the nebulae. What is difficult is to obtain such a large t^2 in the O^{++} or O^+ zones only, which is what would be implied by the ORL/CEL discrepancy. Using a "hot-spot representation" of temperature fluctuations, Binette *et al.* (2001) propose a sound quantification of energy requirements for the observed t^2 and find that "the combined mechanical luminosity of stellar winds, champagne flows and photoevaporation flows from proplyds is insufficient to account for the temperature fluctuations of typical H II regions." Note that, at high metallicities, the natural temperature gradients in H II regions easily lead to a formal value of $t^2(O^{++})$ of 0.04 or even more, because of the effect of strong [O III] λ88, 52-μm cooling. The difficulty is to obtain such a large $t^2(O^{++})$ at metallicities smaller than $12 + \log(O/H) = 8.7$.

(4) There is now a large database of high-quality observations of metal recombination lines; see references in Liu (2006) for planetary nebulae and in García-Rojas & Esteban (2006) for H II regions. Analysing the behaviour of the abundance-discrepancy factors in both types of objects led García-Rojas & Esteban (2006) to conclude that the source for the ORL/CEL abundance discrepancy in H II regions must be different from that in planetary nebulae. Since the paper by Liu *et al.* (2000), it has been argued that, in planetary nebulae, the ORL/CEL abundance discrepancy is probably due to the presence of hydrogen-poor inclusions in the body of the nebula (see e.g. Liu 2002, 2003, 2006, Tsamis *et al.* 2004). These metal-rich clumps would be too cool to excite the CELs. Quite a small mass of matter in the clumps is sufficient to reproduce the observed ORL/CEL discrepancies. The nature of these clumps could be photoevaporating planetesimals. Tsamis *et al.* (2003) and Tsamis & Péquignot (2005) suggest that the ORL/CEL discrepancy in H II regions might also be due to hydrogen-poor inclusions (obviously of different astrophysical origins). This point of view is not universally accepted; see García-Rojas & Esteban (2006) for a different opinion. Yet, it seems that the galaxy chemical-enrichment scenario proposed by Tenorio-Tagle (1996), in which the oxygen-rich products from supernova explosions end up in metal-rich droplets falling back onto the galaxies after an excursion in the intergalactic medium, would be able to explain

quantitatively the ORL/CEL abundance discrepancies observed in H II regions (Stasińska *et al.* 2007).

The debate on temperature fluctuations and ORL/CEL discrepancies is far from over. Yet its outcome will have important implications for the use of emission lines in the derivation of elemental abundances in galaxies. Peimbert *et al.* (2006) are already proposing to use recombination-line abundances to calibrate the strong-line methods for abundance derivations. They argue that recombination-line abundances are not affected by any temperature bias, which is true in the case of a homogeneous chemical composition of the nebulae. If abundance inhomogeneities are present, one must investigate the astrophysical significance of the abundances measured. Stasińska *et al.* (2007) have examined this issue in the case of H II regions.

1.5.6 *Shocks and related issues*

Photoionization is not the only process leading to the formation of emission lines. Cooling flows (Cox & Smith 1976, Fabian & Nulsen 1977) and interstellar shocks can also produce them. Shocks are ubiquitous in galaxies, being caused by jets (associated with brown dwarfs, protostars or massive black holes), winds (stellar winds, winds from protostars and AGNs) or supersonic turbulence. Dopita & Sutherland (1995, 1996) provide a grid of models of line emission produced by pure shocks. However, the efficiency of ionization by stellar photons is such that, whenever hot stars are present, they are likely to dominate the ionization budget of the surrounding gas. The same can probably be said for radiation arising from accretion onto a massive black hole. However, it is true that, in order to explain observed emission-line ratios, the possible contribution of shock heating to photoionization must be examined.

The effects of shocks that leave a signature on emission-line spectra in photoionized nebulae can be summarized as follows.

(1) Shocks generate compression and thus high gas densities.
(2) This compression locally reduces the ionization parameter, and thus enhances the low-ionization emission lines such as [O II], [N II], [S II] and [O I].
(3) Shock heating produces very high temperatures (of the order of millions of Kelvins), which lead to collisional ionization and the production of such ions as O^{3+} and He^{++}.
(4) This high-temperature gas emits X-rays, which contribute to the ionization in front of the shock and behind it.
(5) In the region between the X-ray-emitting zone and the low-excitation zone, the temperature of the gas passes through intermediate values of several tens of thousands of Kelvin, boosting the emission of ultraviolet CELs and of auroral lines. The [O III] λ4363/5007 ratio can be significantly increased.

However, it must be realized that effects (2), (3) and possibly (5) are spectroscopically indistinguishable from the effects of photoionization by an external X-ray source. Insofar as the enhancement of the low-ionization lines is concerned, it can simply be due to a small ionization parameter without the necessity of invoking shock heating; see e.g. the explanation proposed by Stasińska & Izotov (2003) for the behaviour of the [O I] lines in H II galaxies.

Finally, dynamical heating and cooling as well as non-equilibrium photoionization may be an issue in certain cases, when the dynamical times are shorter than the radiative time. Only a few works have started exploring these avenues, which require significant computing power and therefore cannot use the same approach to real nebulae as the ones discussed in the present text. An enlightening introduction to the role of dynamics in photoionized nebulae is given by Henney (2006).

Appendix: Lists of useful lines and how to deal with them

In this appendix, we list all the forbidden, semi-forbidden and resonance lines from C, N, O, Ne, S, Cl and Ar ions which have ionization potentials smaller than 126 eV. These lines as well as all the information on them were extracted from the atomic line list compiled by Peter van Hoof and accessible at http://www.pa.uky.edu/˜peter/atomic. Permitted lines are far too numerous to be listed here. They are available in the atomic line list. However, it is more convenient, for a first approach, to use the identifications made by Zhang *et al.* (2005) on their deep, high-resolution spectrum of the very-high-excitation planetary nebula NGC 7027. This spectrum covers the entire range between 3300 and 9130 Å.

Table 1.1 gives the electronic configurations and ionization potentials of the C, N, O, Ne, S, Cl and Ar ions. Those ions corresponding to the same isoelectronic sequence have been listed in the same column. Such ions have similar families of lines and provide the same plasma diagnostics. Note that H II regions contain only traces (if any) of ions with ionization potentials larger than 54 eV. On the other hand, planetary nebulae, which are generally ionized by hotter stars, can have an important fraction of Ne V and ions of similar ionization potentials.

Tables 1.2–1.9 list, for the ions of C, N, O, Ne, S, Cl and Ar, respectively, the vacuum wavelength of each line, the identification, the type of transition (TT), the electronic configurations for the lower and upper level, the spectroscopic terms for the lower and upper levels (J–J), the transition probability and the energies of the lower and upper levels.

Note that it is more common to use the wavelength in air than the wavelength in vacuum. The IAU standard for conversion from air to vacuum wavelengths is given in Morton (1991). For vacuum wavelengths λ_v in ångström units, convert to air wavelength λ_a via $\lambda_a = \lambda_v/(1.0 + 2.735182 \times 10^{-4} + 131.4182/\lambda_v^2 + 2.76249 \times 10^8/\lambda_v^4)$.

TABLE 1.1. Electronic configurations and ionization potentials (in eV) of ions of C, N, O, Ne, S, Cl and Ar with ionization potentials up to 126 eV

	2s	2s^2	2p	2p^2	2p^3	2p^4	2p^5	2p^6
C	C IV 64.49	C III 47.89	C II 24.38	C I 11.26				
N	N V 97.89	N IV 77.47	N III 47.45	N II 29.60	N I 14.53			
O		O V 113.9	O IV 77.41	O III 54.93	O II 35.12	O I 13.61		
Ne				Ne V 126.21	Ne IV 97.11	Ne III 63.45	Ne II 40.46	Ne I 21.56
	3s	3s^2	3p	3p^2	3p^3	3p^4	3p^5	3p^6
S	S VI 88.05	S V 72.68	S IV 47.30	S III 34.83	S II 23.33	S I 10.36		
Cl	Cl VII 114.19	Cl VI 93.03	Cl V 67.70	Cl IV 53.46	Cl III 39.61	Cl II 23.81	Cl I 12.97	
Ar		Ar VII 124.4	Ar VI 91.01	Ar V 75.04	Ar IV 59.81	Ar III 40.74	Ar II 27.63	Ar I 15.76

The identification in the tables is represented as Fe I for allowed transitions and Fe I] for intercombination transitions and [Fe I] for forbidden transitions.

The symbols for the transition types are E1, allowed transitions; M1, magnetic-dipole forbidden transitions; and E2, electric-quadrupole forbidden transitions.

The transition probabilities have been taken from the atomic line list of Peter van Hoof, or from the Chianti Atomic Database at http://www.arcetri.astro.it/science/chianti/chianti.html if followed by an asterisk.

TABLE 1.2. Lines of C ions

λ_{vac} (Å)	Ion	TT	Terms	J–J	A_{ki} (s^{-1})	Levels (cm^{-1})
4622.864	[C I]	M1	^3P–^1S	1–0		16.40–21648.01
8729.52	[C I]	E2	^1D–^1S	2–0		10192.63–21648.01
9811.01	[C I]	E2	^3P–^1D	0–2		0.00–10192.63
9826.82	[C I]	M1	^3P–^1D	1–2		16.40–10192.63
9852.96	[C I]	M1	^3P–^1D	2–2		43.40–10192.63
2304000	[C I]	E2	^3P–^3P	0–2		0.00–43.40
3704000	[C I]	M1	^3P–^3P	1–2	2.65(−7)	16.40–43.40
6100000	[C I]	M1	^3P–^3P	0–1	7.93(−8)	0.00–16.40
1334.5323	C II	E1	$^2P^o$–^2D	$\frac{1}{2}$–$\frac{3}{2}$	2.38(+8)	0.00–74932.62
1335.6627	C II	E1	$^2P^o$–^2D	$\frac{3}{2}$–$\frac{3}{2}$	4.74(+7)	63.42–74932.62
1335.7077	C II	E1	$^2P^o$–^2D	$\frac{3}{2}$–$\frac{5}{2}$	2.84(+8)	63.42–74930.10
2322.69	[C II]	M2	$^2P^o$–^4P	$\frac{1}{2}$–$\frac{5}{2}$		0.00–43053.60
2324.21	C II]	E1	$^2P^o$–^4P	$\frac{1}{2}$–$\frac{3}{2}$		0.00–43025.30
2325.40	C II]	E1	$^2P^o$–^4P	$\frac{1}{2}$–$\frac{1}{2}$		0.00–43003.30
2326.11	C II]	E1	$^2P^o$–^4P	$\frac{3}{2}$–$\frac{5}{2}$		63.42–43053.60
2327.64	C II]	E1	$^2P^o$–^4P	$\frac{3}{2}$–$\frac{3}{2}$		63.42–43025.30
2328.84	C II]	E1	$^2P^o$–^4P	$\frac{3}{2}$–$\frac{1}{2}$		63.42–43003.30
1576800	[C II]	M1	$^2P^o$–$^2P^o$	$\frac{1}{2}$–$\frac{3}{2}$	2.290(−6)*	0.00–63.42
4550000	[C II]	M1	^4P–^4P	$\frac{1}{2}$–$\frac{3}{2}$	2.39(−7)	43003.30–43025.30
1906.683	[C III]	M2	^1S–$^3P^o$	0–2		0.00–52447.11
1908.734	C III]	E1	^1S–$^3P^o$	0–1		0.00–52390.75
1249200	[C III]	E2	$^3P^o$–$^3P^o$	0–2		52367.06–52447.11
1774300	[C III]	M1	$^3P^o$–$^3P^o$	1–2	2.450(−6)*	52390.75–52447.11
4221000	[C III]	M1	$^3P^o$–$^3P^o$	0–1	2.39(−7)	52367.06–52390.75
1548.203	C IV	E1	^2S–$^2P^o$	$\frac{1}{2}$–$\frac{3}{2}$	2.65(+8)	0.00–64591.00
1550.777	C IV	E1	^2S–$^2P^o$	$\frac{1}{2}$–$\frac{1}{2}$	2.64(+8)	0.00–64483.80

TABLE 1.3. Lines of N ions

λ_{vac} (Å)	Ion	TT	Terms	J–J	A_{ki} (s^{-1})	Levels (cm^{-1})
3467.4898	[N I]	M1	$^4S^o$–$^2P^o$	$\frac{3}{2}$–$\frac{3}{2}$	6.210(−3)*	0.00–28839.31
3467.5362	[N I]	M1	$^4S^o$–$^2P^o$	$\frac{3}{2}$–$\frac{1}{2}$	2.520(−3)*	0.00–28838.92
5199.3490	[N I]	E2	$^4S^o$–$^2D^o$	$\frac{3}{2}$–$\frac{3}{2}$	2.260(−5)*	0.00–19233.18
5201.7055	[N I]	E2	$^4S^o$–$^2D^o$	$\frac{3}{2}$–$\frac{5}{2}$	5.765(−6)*	0.00–19224.46
10400.587	[N I]	M1	$^2D^o$–$^2P^o$	$\frac{5}{2}$–$\frac{3}{2}$	5.388(−2)*	19224.46–28839.31
10401.004	[N I]	E2	$^2D^o$–$^2P^o$	$\frac{5}{2}$–$\frac{1}{2}$	3.030(−2)*	19224.46–28838.92
10410.021	[N I]	M1	$^2D^o$–$^2P^o$	$\frac{3}{2}$–$\frac{3}{2}$	2.435(−2)*	19233.18–28839.31
10410.439	[N I]	M1	$^2D^o$–$^2P^o$	$\frac{3}{2}$–$\frac{1}{2}$	4.629(−2)*	19233.18–28838.92
2137.457	[N II]	M2	^3P–$^5S^o$	0–2		0.00–46784.56
2139.683	N II]	E1	^3P–$^5S^o$	1–2		48.67–46784.56
2143.450	N II]	E1	^3P–$^5S^o$	2–2		130.80–46784.56
3063.728	[N II]	M1	^3P–^1S	1–0	4.588(−6)*	48.67–32688.64
3071.457	[N II]	E2	^3P–^1S	2–0		130.80–32688.64
5756.240	[N II]	E2	^1D–^1S	2–0	9.234(−1)*	15316.19–32688.64
6529.04	[N II]	E2	^3P–^1D	0–2	1.928(−6)*	0.00–15316.19
6549.85	[N II]	M1	^3P–^1D	1–2	9.819(−4)*	48.67–15316.19
6585.28	[N II]	M1	^3P–^1D	2–2	3.015(−3)*	130.80–15316.19
764500	[N II]	E2	^3P–^3P	0–2		0.00–130.80
1217600	[N II]	M1	^3P–^3P	1–2	7.460(−6)*	48.67–130.80
2055000	[N II]	M1	^3P–^3P	0–1	2.080(−6)*	0.00–48.67
1744.351	[N III]	M2	$^2P^o$–^4P	$\frac{1}{2}$–$\frac{5}{2}$		0.00–57327.90
1746.823	N III]	E1	$^2P^o$–^4P	$\frac{1}{2}$–$\frac{3}{2}$		0.00–57246.80
1748.646	N III]	E1	$^2P^o$–^4P	$\frac{1}{2}$–$\frac{1}{2}$		0.00–57187.10
1749.674	N III]	E1	$^2P^o$–^4P	$\frac{3}{2}$–$\frac{5}{2}$		174.40–57327.90
1752.160	N III]	E1	$^2P^o$–^4P	$\frac{3}{2}$–$\frac{3}{2}$		174.40–57246.80
1753.995	N III]	E1	$^2P^o$–^4P	$\frac{3}{2}$–$\frac{1}{2}$		174.40–57187.10
573400	[N III]	M1	$^2P^o$–$^2P^o$	$\frac{1}{2}$–$\frac{3}{2}$	4.770(−5)*	0.00–174.40
710000	[N III]	E2	^4P–^4P	$\frac{1}{2}$–$\frac{5}{2}$		57187.10–57327.90
1233000	[N III]	M1	^4P–^4P	$\frac{3}{2}$–$\frac{5}{2}$	8.63(−6)	57246.80–57327.90
1675000	[N III]	M1	^4P–^4P	$\frac{1}{2}$–$\frac{3}{2}$	4.78(−6)	57187.10–57246.80
1483.321	[N IV]	M2	^1S–$^3P^o$	0–2		0.00–67416.30
1486.496	N IV]	E1	^1S–$^3P^o$	0–1		0.00–67272.30
482900	[N IV]	E2	$^3P^o$–$^3P^o$	0–2		67209.20–67416.30
694000	[N IV]	M1	$^3P^o$–$^3P^o$	1–2	4.027(−5)*	67272.30–67416.30
1585000	[N IV]	M1	$^3P^o$–$^3P^o$	0–1	4.52(−6)	67209.20–67272.30

TABLE 1.4. Lines of O ions

λ_{vac} (Å)	Ion	TT	Terms	J–J	A_{ki} (s^{-1})	Levels (cm^{-1})
2973.1538	[O I]	M1	^{3}P–^{1}S	1–0		158.26–33792.58
5578.8874	[O I]	E2	^{1}D–^{1}S	2–0		15867.86–33792.58
6302.046	[O I]	M1	^{3}P–^{1}D	2–2		0.00–15867.86
6365.536	[O I]	M1	^{3}P–^{1}D	1–2		158.26–15867.86
6393.500	[O I]	E2	^{3}P–^{1}D	0–2		226.98–15867.86
440573	[O I]	E2	^{3}P–^{3}P	2–0		0.00–226.98
631850	[O I]	M1	^{3}P–^{3}P	2–1	8.91(−5)	0.00–158.26
1455350	[O I]	M1	^{3}P–^{3}P	1–0	1.75(−5)	158.26–226.98
2470.966	[O II]	M1	^{4}S°–^{2}P°	$\frac{3}{2}$–$\frac{1}{2}$	2.380(−2)*	0.00–40470.00
2471.088	[O II]	M1	^{4}S°–^{2}P°	$\frac{3}{2}$–$\frac{3}{2}$	5.800(−2)*	0.00–40468.01
3727.092	[O II]	E2	^{4}S°–^{2}D°	$\frac{3}{2}$–$\frac{3}{2}$	1.810(−4)*	0.00–26830.57
3729.875	[O II]	E2	^{4}S°–^{2}D°	$\frac{3}{2}$–$\frac{5}{2}$	3.588(−5)*	0.00–26810.55
7320.94	[O II]	E2	^{2}D°–^{2}P°	$\frac{5}{2}$–$\frac{1}{2}$	5.630(−2)*	26810.55–40470.00
7322.01	[O II]	M1	^{2}D°–^{2}P°	$\frac{5}{2}$–$\frac{3}{2}$	1.074(−1)*	26810.55–40468.01
7331.68	[O II]	M1	^{2}D°–^{2}P°	$\frac{3}{2}$–$\frac{1}{2}$	9.410(−2)*	26830.57–40470.00
7332.75	[O II]	M1	^{2}D°–^{2}P°	$\frac{3}{2}$–$\frac{3}{2}$	5.800(−2)*	26830.57–40468.01
4995000	[O II]	M1	^{2}D°–^{2}D°	$\frac{5}{2}$–$\frac{3}{2}$	1.320(−7)*	26810.55–26830.57
1657.6933	[O III]	M2	^{3}P–^{5}S°	0–2		0.00–60324.79
1660.8092	O III]	E1	^{3}P–^{5}S°	1–2		113.18–60324.79
1666.1497	O III]	E1	^{3}P–^{5}S°	2–2		306.17–60324.79
2321.664	[O III]	M1	^{3}P–^{1}S	1–0		113.18–43185.74
2332.113	[O III]	E2	^{3}P–^{1}S	2–0		306.17–43185.74
4364.436	[O III]	E2	^{1}D–^{1}S	2–0	4.760(−1)*	20273.27–43185.74
4932.603	[O III]	E2	^{3}P–^{1}D	0–2	7.250(−6)*	0.00–20273.27
4960.295	[O III]	M1	^{3}P–^{1}D	1–2	6.791(−3)*	113.18–20273.27
5008.240	[O III]	M1	^{3}P–^{1}D	2–2	2.046(−2)*	306.17–20273.27
326612	[O III]	E2	^{3}P–^{3}P	0–2		0.00–306.17
518145	[O III]	M1	^{3}P–^{3}P	1–2	1.010(−4)*	113.18–306.17
883560	[O III]	M1	^{3}P–^{3}P	0–1	2.700(−5)*	0.00–113.18
1393.621	[O IV]	M2	^{2}P°–^{4}P	$\frac{1}{2}$–$\frac{5}{2}$		0.00–71755.50
1397.232	O IV]	E1	^{2}P°–^{4}P	$\frac{1}{2}$–$\frac{3}{2}$		0.00–71570.10
1399.780	O IV]	E1	^{2}P°–^{4}P	$\frac{1}{2}$–$\frac{1}{2}$		0.00–71439.80
1401.164	O IV]	E1	^{2}P°–^{4}P	$\frac{3}{2}$–$\frac{5}{2}$		386.25–71755.50
1404.813	O IV]	E1	^{2}P°–^{4}P	$\frac{3}{2}$–$\frac{3}{2}$		386.25–71570.10
1407.389	O IV]	E1	^{2}P°–^{4}P	$\frac{3}{2}$–$\frac{1}{2}$		386.25–71439.80
258903	[O IV]	M1	^{2}P°–^{2}P°	$\frac{1}{2}$–$\frac{3}{2}$	5.310(−4)*	0.00–386.25
316800	[O IV]	E2	^{4}P–^{4}P	$\frac{1}{2}$–$\frac{5}{2}$		71439.80–71755.50
539400	[O IV]	M1	^{4}P–^{4}P	$\frac{3}{2}$–$\frac{5}{2}$	1.03(−4)	71570.10–71755.50
767000	[O IV]	M1	^{4}P–^{4}P	$\frac{1}{2}$–$\frac{3}{2}$	4.97(−5)	71439.80–71570.10
1213.809	[O V]	M2	^{1}S–^{3}P°	0–2		0.00–82385.30
225800	[O V]	E2	^{3}P°–^{3}P°	0–2		81942.50–82385.30
326100	[O V]	M1	^{3}P°–^{3}P°	1–2	3.913(−4)*	82078.60–82385.30
735000	[O V]	M1	^{3}P°–^{3}P°	0–1	4.53(−5)	81942.50–82078.60

TABLE 1.5. Lines of Ne ions

λ_{vac} (Å)	Ion	TT	Terms	J–J	A_{ki} (s^{-1})	Levels (cm^{-1})
128135.48	[Ne II]	M1	$^2P^\circ$–$^2P^\circ$	$\frac{3}{2}$–$\frac{1}{2}$	8.590(−3)*	0.00–780.42
1793.802	[Ne III]	E2	^3P–^1S	2–0		0.00–55747.50
1814.730	[Ne III]	M1	^3P–^1S	1–0		642.88–55747.50
3343.50	[Ne III]	E2	^1D–^1S	2–0	2.550	25838.70–55747.50
3870.16	[Ne III]	M1	^3P–^1D	2–2	1.653(−1)*	0.00–25838.70
3968.91	[Ne III]	M1	^3P–^1D	1–2	5.107(−2)*	642.88–25838.70
4013.14	[Ne III]	E2	^3P–^1D	0–2	1.555(−5)*	920.55–25838.70
108630.5	[Ne III]	E2	^3P–^3P	2–0	2.524(−8)*	0.00–920.55
155550.5	[Ne III]	M1	^3P–^3P	2–1	5.971(−3)*	0.00–642.88
360135	[Ne III]	M1	^3P–^3P	1–0	1.154(−3)*	642.88–920.55
4713.2	[Ne IV]	M1	$^2D^\circ$–$^2P^\circ$	$\frac{5}{2}$–$\frac{3}{2}$	4.857(−1)*	41210–62427
4715.4	[Ne IV]	E2	$^2D^\circ$–$^2P^\circ$	$\frac{5}{2}$–$\frac{1}{2}$	1.142(−1)*	41210–62417
4718.8	[Ne IV]	M1	$^2D^\circ$–$^2P^\circ$	$\frac{3}{2}$–$\frac{3}{2}$	5.881(−1)*	41235–62427
4721.0	[Ne IV]	M1	$^2D^\circ$–$^2P^\circ$	$\frac{3}{2}$–$\frac{1}{2}$	4.854(−1)*	41235–62417
1575.1	[Ne V]	M1	^3P–^1S	1–0		411.23–63900.00
1592.6	[Ne V]	E2	^3P–^1S	2–0		1109.47–63900.00
2976	[Ne V]	E2	^1D–^1S	2–0	2.834	30294.00–63900.00
3301.0	[Ne V]	E2	^3P–^1D	0–2	5.460(−5)*	0.00–30294.00
3346.4	[Ne V]	M1	^3P–^1D	1–2	1.222(−1)*	411.23–30294.00
3426.5	[Ne V]	M1	^3P–^1D	2–2	3.504(−1)*	1109.47–30294.00
90133.3	[Ne V]	E2	^3P–^3P	0–2		0.00–1109.47
143216.8	[Ne V]	M1	^3P–^3P	1–2	4.585(−3)*	411.23–1109.47
243175	[Ne V]	M1	^3P–^3P	0–1	1.666(−3)*	0.00–411.23

TABLE 1.6. Lines of S ions

λ_{vac} (Å)	Ion	TT	Terms	J–J	A_{ki} (s^{-1})	Levels (cm^{-1})
4590.5464	[S I]	M1	^3P–^1S	1–0		396.06–22179.95
7727.172	[S I]	E2	^1D–^1S	2–0		9238.61–22179.95
10824.140	[S I]	M1	^3P–^1D	2–2		0.00–9238.61
11308.950	[S I]	M1	^3P–^1D	1–2		396.06–9238.61
11540.722	[S I]	E2	^3P–^1D	0–2		573.64–9238.61
174325.4	[S I]	E2	^3P–^3P	2–0		0.00–573.64
252490	[S I]	M1	^3P–^3P	2–1	1.40(−3)	0.00–396.06
563111	[S I]	M1	^3P–^3P	1–0	3.02(−4)	396.06–573.64
1250.5845	S II	E1	$^4S^\circ$–^4P	$\frac{3}{2}$–$\frac{1}{2}$	4.43(+7)	0.00–79962.61
1253.8111	S II	E1	$^4S^\circ$–^4P	$\frac{3}{2}$–$\frac{3}{2}$	4.40(+7)	0.00–79756.83

TABLE 1.6. (Cont.)

λ_{vac} (Å)	Ion	TT	Terms	J–J	A_{ki} (s^{-1})	Levels (cm^{-1})
1259.5190	S II	E1	$^4S^o$–^4P	$\frac{3}{2}$–$\frac{5}{2}$	4.34(+7)	0.00–79395.39
4069.749	[S II]	M1	$^4S^o$–$^2P^o$	$\frac{3}{2}$–$\frac{3}{2}$	2.670(−1)*	0.00–24571.54
4077.500	[S II]	M1	$^4S^o$–$^2P^o$	$\frac{3}{2}$–$\frac{1}{2}$	1.076(−1)*	0.00–24524.83
6718.29	[S II]	E2	$^4S^o$–$^2D^o$	$\frac{3}{2}$–$\frac{5}{2}$	3.338(−4)*	0.00–14884.73
6732.67	[S II]	E2	$^4S^o$–$^2D^o$	$\frac{3}{2}$–$\frac{3}{2}$	1.231(−3)*	0.00–14852.94
10289.55	[S II]	M1	$^2D^o$–$^2P^o$	$\frac{3}{2}$–$\frac{3}{2}$	1.644(−1)*	14852.94–24571.54
10323.32	[S II]	M1	$^2D^o$–$^2P^o$	$\frac{5}{2}$–$\frac{3}{2}$	1.938(−1)*	14884.73–24571.54
10339.24	[S II]	M1	$^2D^o$–$^2P^o$	$\frac{3}{2}$–$\frac{1}{2}$	1.812(−1)*	14852.94–24524.83
10373.34	[S II]	E2	$^2D^o$–$^2P^o$	$\frac{5}{2}$–$\frac{1}{2}$	7.506(−2)*	14884.73–24524.83
2141000	[S II]	M1	$^2P^o$–$^2P^o$	$\frac{1}{2}$–$\frac{3}{2}$	9.16(−7)	24524.83–24571.54
3146000	[S II]	M1	$^2D^o$–$^2D^o$	$\frac{3}{2}$–$\frac{5}{2}$	3.452(−7)*	14852.94–14884.73
1704.3928	[S III]	M2	^3P–$^5S^o$	0–2		0.00–58671.92
1713.1137	[S III]	E1	^3P–$^5S^o$	1–2		298.68–58671.92
1728.9415	[S III]	E1	^3P–$^5S^o$	2–2		833.06–58671.92
3722.69	[S III]	M1	^3P–^1S	1–0	1.016	298.68–27161.00
3798.25	[S III]	E2	^3P–^1S	2–0		833.06–27161.00
6313.8	[S III]	E2	^1D–^1S	2–0	3.045	11322.70–27161.00
8831.8	[S III]	E2	^3P–^1D	0–2	2.570(−4)*	0.00–11322.70
9071.1	[S III]	M1	^3P–^1D	1–2	3.107(−2)*	298.68–11322.70
9533.2	[S III]	M1	^3P–^1D	2–2	1.856(−1)*	833.06–11322.70
120038.8	[S III]	E2	^3P–^3P	0–2	5.008(−8)*	0.00–833.06
187130.3	[S III]	M1	^3P–^3P	1–2	1.877(−3)*	298.68–833.06
334810	[S III]	M1	^3P–^3P	0–1	8.429(−4)*	0.00–298.68
1387.459	[S IV]	M2	$^2P^o$–^4P	$\frac{1}{2}$–$\frac{5}{2}$		0.00–72074.20
1398.050	S IV]	E1	$^2P^o$–^4P	$\frac{1}{2}$–$\frac{3}{2}$		0.00–71528.20
1404.800	S IV]	E1	$^2P^o$–^4P	$\frac{1}{2}$–$\frac{1}{2}$		0.00–71184.50
1406.019	S IV]	E1	$^2P^o$–^4P	$\frac{3}{2}$–$\frac{5}{2}$		951.43–72074.20
1416.897	S IV]	E1	$^2P^o$–^4P	$\frac{3}{2}$–$\frac{3}{2}$		951.43–71528.20
1423.831	S IV]	E1	$^2P^o$–^4P	$\frac{3}{2}$–$\frac{1}{2}$		951.43–71184.50
105104.9	[S IV]	M1	$^2P^o$–$^2P^o$	$\frac{1}{2}$–$\frac{3}{2}$	7.750(−3)*	0.00–951.43
112400	[S IV]	E2	^4P–^4P	$\frac{1}{2}$–$\frac{5}{2}$		71184.50–72074.20
183200	[S IV]	M1	^4P–^4P	$\frac{3}{2}$–$\frac{5}{2}$	2.63(−3)	71528.20–72074.20
291000	[S IV]	M1	^4P–^4P	$\frac{1}{2}$–$\frac{3}{2}$	9.13(−4)	71184.50–71528.20
1188.281	[S V]	M2	^1S–$^3P^o$	0–2		0.00–84155.20
88400	[S V]	E2	$^3P^o$–$^3P^o$	0–2		83024.00–84155.20
131290	[S V]	M1	$^3P^o$–$^3P^o$	1–2	5.96(−3)	83393.50–84155.20
270600	[S V]	M1	$^3P^o$–$^3P^o$	0–1	9.07(−4)	83024.00–83393.50

TABLE 1.7. Lines of Cl ions

λ_{vac} (Å)	Ion	TT	Terms	J–J	A_{ki} (s^{-1})	Levels (cm^{-1})
113333.52	[Cl I]	M1	$^2P^\circ$–$^2P^\circ$	$\frac{3}{2}$–$\frac{1}{2}$	1.24(−2)	0.00–882.35
3587.055	[Cl II]	E2	^3P–^1S	2–0		0.00–27878.02
3678.902	[Cl II]	M1	^3P–^1S	1–0	1.297	696.00–27878.02
6163.54	[Cl II]	E2	^1D–^1S	2–0	2.416	11653.58–27878.02
8581.05	[Cl II]	M1	^3P–^1D	2–2	1.038(−1)*	0.00–11653.58
9126.10	[Cl II]	M1	^3P–^1D	1–2	2.872(−2)*	696.00–11653.58
9383.41	[Cl II]	E2	^3P–^1D	0–2	1.134(−5)*	996.47–11653.58
100354	[Cl II]	E2	^3P–^3P	2–0	6.053(−7)*	0.00–996.47
143678	[Cl II]	M1	^3P–^3P	2–1	7.566(−3)*	0.00–696.00
332810	[Cl II]	M1	^3P–^3P	1–0	1.462(−3)*	696.00–996.47
3343.81	[Cl III]	M1	$^4S^\circ$–$^2P^\circ$	$\frac{3}{2}$–$\frac{3}{2}$		0.00–29906.00
3354.17	[Cl III]	M1	$^4S^\circ$–$^2P^\circ$	$\frac{3}{2}$–$\frac{1}{2}$		0.00–29813.60
5519.25	[Cl III]	E2	$^4S^\circ$–$^2D^\circ$	$\frac{3}{2}$–$\frac{5}{2}$		0.00–18118.40
5539.43	[Cl III]	E2	$^4S^\circ$–$^2D^\circ$	$\frac{3}{2}$–$\frac{3}{2}$		0.00–18052.40
8436.3	[Cl III]	M1	$^2D^\circ$–$^2P^\circ$	$\frac{3}{2}$–$\frac{3}{2}$		18052.40–29906.00
8483.5	[Cl III]	M1	$^2D^\circ$–$^2P^\circ$	$\frac{5}{2}$–$\frac{3}{2}$		18118.40–29906.00
8502.5	[Cl III]	M1	$^2D^\circ$–$^2P^\circ$	$\frac{3}{2}$–$\frac{1}{2}$		18052.40–29813.60
8550.5	[Cl III]	E2	$^2D^\circ$–$^2P^\circ$	$\frac{5}{2}$–$\frac{1}{2}$		18118.40–29813.60
1082000	[Cl III]	M1	$^2P^\circ$–$^2P^\circ$	$\frac{1}{2}$–$\frac{3}{2}$	7.09(−6)	29813.60–29906.00
1515000	[Cl III]	M1	$^2D^\circ$–$^2D^\circ$	$\frac{3}{2}$–$\frac{5}{2}$	3.10(−6)	18052.40–18118.40
3119.51	[Cl IV]	M1	^3P–^1S	1–0		492.50–32548.80
3204.52	[Cl IV]	E2	^3P–^1S	2–0		1342.90–32548.80
5324.47	[Cl IV]	E2	^1D–^1S	2–0		13767.60–32548.80
7263.4	[Cl IV]	E2	^3P–^1D	0–2		0.00–13767.60
7532.9	[Cl IV]	M1	^3P–^1D	1–2		492.50–13767.60
8048.5	[Cl IV]	M1	^3P–^1D	2–2		1342.90–13767.60
74470	[Cl IV]	E2	^3P–^3P	0–2		0.00–1342.90
117590	[Cl IV]	M1	^3P–^3P	1–2	8.29(−3)	492.50–1342.90
203000	[Cl IV]	M1	^3P–^3P	0–1	2.15(−3)	0.00–492.50
67090	[Cl V]	M1	$^2P^\circ$–$^2P^\circ$	$\frac{1}{2}$–$\frac{3}{2}$	2.98(−2)	0.00–1490.50
72400	[Cl V]	E2	^4P–^4P	$\frac{1}{2}$–$\frac{5}{2}$		(86000 + v)–(87381 + v)
118600	[Cl V]	M1	^4P–^4P	$\frac{3}{2}$–$\frac{5}{2}$	9.70(−3)	(86538 + v)–(87381 + v)
186000	[Cl V]	M1	^4P–^4P	$\frac{1}{2}$–$\frac{3}{2}$	3.50(−3)	(86000 + v)–(86538 + v)
58200	[Cl VI]	E2	$^3P^\circ$–$^3P^\circ$	0–2		(97405 + x)–(99123 + x)
85800	[Cl VI]	M1	$^3P^\circ$–$^3P^\circ$	1–2	2.13(−2)	(97958 + x)–(99123 + x)
180800	[Cl VI]	M1	$^3P^\circ$–$^3P^\circ$	0–1	3.04(−3)	(97405 + x)–(97958 + x)

TABLE 1.8. Lines of Ar ions

λ_{vac} (Å)	Ion	TT	Terms	J–J	A_{ki} (s^{-1})	Levels (cm^{-1})
69852.74	[Ar II]	M1	$^2P^o$–$^2P^o$	$\frac{3}{2}$–$\frac{1}{2}$	5.28(−2)	0.00–1431.58
3006.10	[Ar III]	E2	^3P–^1S	2–0		0.00–33265.70
3110.08	[Ar III]	M1	^3P–^1S	1–0		1112.18–33265.70
5193.27	[Ar III]	E2	^1D–^1S	2–0		14010.00–33265.70
7137.8	[Ar III]	M1	^3P–^1D	2–2		0.00–14010.00
7753.2	[Ar III]	M1	^3P–^1D	1–2		1112.18–14010.00
8038.8	[Ar III]	E2	^3P–^1D	0–2		1570.28–14010.00
63682.9	[Ar III]	E2	^3P–^3P	2–0		0.00–1570.28
89913.8	[Ar III]	M1	^3P–^3P	2–1	3.09(−2)	0.00–1112.18
218291	[Ar III]	M1	^3P–^3P	1–0	5.19(−3)	1112.18–1570.28
2854.48	[Ar IV]	M1	$^4S^o$–$^2P^o$	$\frac{3}{2}$–$\frac{3}{2}$	2.134	0.00–35032.60
2868.99	[Ar IV]	M1	$^4S^o$–$^2P^o$	$\frac{3}{2}$–$\frac{1}{2}$	8.702(−1)*	0.00–34855.50
4712.69	[Ar IV]	E2	$^4S^o$–$^2D^o$	$\frac{3}{2}$–$\frac{5}{2}$	1.906(−3)*	0.00–21219.30
4741.49	[Ar IV]	E2	$^4S^o$–$^2D^o$	$\frac{3}{2}$–$\frac{3}{2}$	2.277(−2)*	0.00–21090.40
7172.5	[Ar IV]	M1	$^2D^o$–$^2P^o$	$\frac{3}{2}$–$\frac{3}{2}$	8.915(−1)*	21090.40–35032.60
7239.4	[Ar IV]	M1	$^2D^o$–$^2P^o$	$\frac{5}{2}$–$\frac{3}{2}$	6.559(−1)*	21219.30–35032.60
7264.7	[Ar IV]	M1	$^2D^o$–$^2P^o$	$\frac{3}{2}$–$\frac{1}{2}$	6.660(−1)*	21090.40–34855.50
7333.4	[Ar IV]	E2	$^2D^o$–$^2P^o$	$\frac{5}{2}$–$\frac{1}{2}$	1.179(−1)*	21219.30–34855.50
564700	[Ar IV]	M1	$^2P^o$–$^2P^o$	$\frac{1}{2}$–$\frac{3}{2}$	4.937(−5)*	34855.50–35032.60
776000	[Ar IV]	M1	$^2D^o$–$^2D^o$	$\frac{3}{2}$–$\frac{5}{2}$	2.287(−5)*	21090.40–21219.30
2691.63	[Ar V]	M1	^3P–^1S	1–0		763.23–37915.50
2786.55	[Ar V]	E2	^3P–^1S	2–0		2028.80–37915.50
4626.22	[Ar V]	E2	^1D–^1S	2–0		16299.60–37915.50
6135.12	[Ar V]	E2	^3P–^1D	0–2		0.00–16299.60
6436.51	[Ar V]	M1	^3P–^1D	1–2		763.23–16299.60
7007.3	[Ar V]	M1	^3P–^1D	2–2		2028.80–16299.60
49290.2	[Ar V]	E2	^3P–^3P	0–2		0.00–2028.80
79015.8	[Ar V]	M1	^3P–^3P	1–2	2.73(−2)	763.23–2028.80
131021.9	[Ar V]	M1	^3P–^3P	0–1	7.99(−3)	0.00–763.23
978.58	[Ar VI]	M2	$^2P^o$–^4P	$\frac{1}{2}$–$\frac{5}{2}$		0.00–102189.20
45295	[Ar VI]	M1	$^2P^o$–$^2P^o$	$\frac{1}{2}$–$\frac{3}{2}$	9.68(−2)	0.00–2207.74
49237	[Ar VI]	E2	^4P–^4P	$\frac{1}{2}$–$\frac{5}{2}$		100158.20–102189.20
81010	[Ar VI]	M1	^4P–^4P	$\frac{3}{2}$–$\frac{5}{2}$	3.04(−2)	100954.80–102189.20
125530	[Ar VI]	M1	^4P–^4P	$\frac{1}{2}$–$\frac{3}{2}$	1.14(−2)	100158.20–100954.80
865.16	[Ar VII]	M2	^1S–$^3P^o$	0–2		0.00–115585.00
40310	[Ar VII]	E2	$^3P^o$–$^3P^o$	0–2		113104.00–115585.00
59500	[Ar VII]	M1	$^3P^o$–$^3P^o$	1–2	6.41(−2)	113904.00–115585.00
125000	[Ar VII]	M1	$^3P^o$–$^3P^o$	0–1	9.21(−3)	113104.00–113904.00

TABLE 1.9. Lines from $2p^5$ and $3p^5$ ions; the first column gives the type of line, as defined in Figure 1.6

	[Ne II]	[Cl I]	[Ar II]
F_1	128135.48	113333.52	69852.74

TABLE 1.10. Lines from $2p^4$ and $3p^4$ ions

	[O I]	[Ne III]	[S I]	[Cl II]	[Ar III]
T_1	2973.1538	1814.730	4590.5464	3678.902	3110.08
A_1	5578.8874	3343.50	7727.172	6163.54	5193.27
N_1	6302.046	3870.16	10824.140	8581.05	7137.8
N_2	6365.536	3968.91	11308.950	9126.10	7753.2
N_3	6393.500	4013.14	11540.722	9383.41	8038.8
F_1	440573	108630.5	174325.4	100354	63682.9
F_2	631850	155550.5	252490	143678	89913.8
F_3	1455350	360135	563111	332810	218291

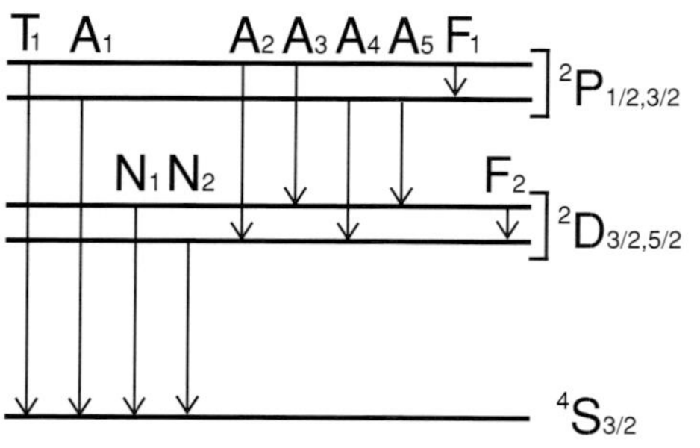

FIGURE 1.6. Simplified energy-level diagrams with transition types indicated. A stands for "auroral", N for "nebular", T for "transauroral" and F for "fine structure". Left: ions p^4 (Table 1.10) and p^2 (Table 1.12). Right: ions p^3 (Table 1.11).

In Tables 1.2 to 1.9, it is easy to see which ions are available for each element in a given wavelength range. For example, there is no line of N III in the optical. One can also see which lines arise from the same upper level, and therefore have intensity ratios that depend only on atomic physics, not on the conditions inside the nebula. For example, for [O III], this is the case for the 4969 and 5007-Å lines, whose intensity ratio is given by the ratio of transition probabilities divided by the ratio of wavelengths.

Tables 1.10–1.14 group the lines according to the electronic configuration of their ions. In each case, the first column specifies the type of line, as defined in Figure 1.6. Figure 1.6 shows simplified energy-level diagrams, which allow one to understand which line pairs are temperature or density indicators. Of course, in practice, one uses only the lines with the largest transition probabilities. For example, [O III] $\lambda4363/4931$ could be a temperature indicator like [O III] $\lambda4663/5007$, but the intensity of the $\lambda4931$ line is less than one thousandth of that of $\lambda5007$ (as inferred from Table 1.4, comparing the transition probabilities).

TABLE 1.11. Lines from $2p^3$ and $3p^3$ ions

	[N I]	[O II]	[Ne IV]	[S II]	[Cl III]	[Ar IV]
T_1	3467.4898	2470.966	4713.2	4069.749	3343.81	2854.48
T_2	3467.5362	2471.088	4715.4	4077.500	3354.17	2868.99
N_1	5199.3490	3727.092	4718.8	6718.29	5519.25	4712.69
N_2	5201.7055	3729.875	4721.0	6732.67	5539.43	4741.49
A_1	10400.587	7320.94		10289.55	8436.3	7172.5
A_2	10401.004	7322.01		10323.32	8483.5	7239.4
A_3	10410.021	7331.68		10339.24	8502.5	7264.7
A_4	10410.439	7332.75		10373.34	8550.5	7333.4
F_1		4995000		2141000	1082000	564700
F_2				3146000	1515000	776000

TABLE 1.12. Lines from $2p^2$ and $3p^2$ ions

	[C I]	[N II]	[O III]	[Ne V]	[S III]	[Cl IV]	[Ar V]
T_1	4622.864	3063.728	2321.664	1575.1	3722.69	3119.51	2691.63
A_1	8729.52	5756.240	4364.436	2976	6313.8	5324.47	4626.22
N_1	9811.01	6529.04	4932.603	3301.0	8831.8	7263.4	6135.12
N_2	9826.82	6549.85	4960.295	3346.4	9071.1	7532.9	6436.51
N_3	9852.96	6585.28	5008.240	3426.5	9533.2	8048.5	7007.3
F_1	2304000	764500	326612	90133.3	120038.8	74470	49290.2
F_2	3704000	1217600	518145	143216.8	187130.3	117590	79015.8
F_3	6100000	2055000	883560	243175	334810	203000	131021.9

TABLE 1.13. Lines from 2p and 3p ions

	[C II]	[N III]	[O IV]	[S IV]	[Cl V]
F_1	1576800	573400	258903	105104.9	67090
F_2					72400
F_3					118600
F_4					186000

TABLE 1.14. Lines from $2s^2$ and $3s^2$ ions

	[C III]	[N IV]	[O V]
F_1	1249200	482900	225800
F_2	1774300	694000	326100
F_3	4221000	1585000	735000

Acknowledgments

I thank Jordi Cepa for having invited me to share my experience on emission lines with the participants at the XVIII Canary Island Winter School "The Emission-Line Universe", and I thank the participants for their interest. Much of the material presented here has benefited from discussions during recent years with my colleagues, collaborators and friends: Fabio Bresolin, Miguel Cerviño, Roberto Cid-Fernandes, Barbara Ercolano, César Esteban, Gary Ferland, Jorge García-Rojas, Rosa González Delgado, Sławomir Górny, William Henney, Artemio Herrero, Yuri Izotov, Luc Jamet, Valentina Luridiana, Abilio Mateus, Christophe Morisset, Antonio Peimbert, Manuel Peimbert, Miriam Peña, Enrique Pérez, Michael Richer, Mónica Rodríguez, Daniel Schaerer, Sergio Simón-Díaz, Laerte Sodré, Ryszard Szczerba, Guillermo Tenorio-Tagle, Silvia Torres-Peimbert, Yiannis Tsamis and José Vílchez. Special thanks are due to Jorge García-Rojas, Ángel López-Sánchez, Sergio Simón-Díaz, and Natalia Vale Asari for their detailed proof-reading and comments on this text. I apologize to the many researchers in the field who did not find their names in the list of references. It is unfortunately impossible in such a short text to do justice to all the work that has contributed to our understanding of the subject. While preparing this review, I have made extensive use of the Nasa ADS database. I have also used the atomic line list maintained by Peter van Hoof, from which the tables in the appendix were extracted, with the great help of Sergio Simón-Díaz. In the body of this contribution, I have indicated several websites that I think might be of interest to the reader.

REFERENCES

Abel, N. *et al.* 2003, PASP, 115, 188

Aller, L. H. 1984, *Physics of Thermal Gaseous Nebulae* (Dordrecht: Kluwer)

Alloin, D., Collin-Souffrin, S., Joly, M., Vigroux, L. 1979, A&A 78, 200

Bahcall, J. N., Steinhardt, C. L., Schlegel, D. 2004, ApJ, 600, 520

Baldwin, J., Phillips, M. M., Terlevich, R. 1981, PASP, 93, 5

Bautista, M. A. 2006, IAUS, 234, 203

Beckman, J. E., Zurita, A., Rozas, M., Cardwell, A., Relaño, M. 2002, RMxAC, 12, 213

Binette, L., Luridiana, V., Henney, W. J. 2001, RMxAC, 10, 19

Bowen, I. S. 1934, PASP, 46, 146

Bresolin, F., Schaerer, D., González Delgado, R. M., Stasińska, G. 2005, A&A, 441, 981

Bresolin, F. 2007, ApJ, 656, 186

Brinchmann, J., Charlot, S., White, S. D. M., Tremonti, C., Kauffmann, G., Heckman, T., Brinkmann, J. 2004, MNRAS, 351, 1151

Burbidge, G. R., Burbidge, E. M., Sandage, A. R. 1963, RvMP, 35, 947

Calzetti, D. 1997, AJ, 113, 162

Calzetti, D., Armus, L., Bohlin, R. C., Kinney, A. L., Koornneef, J., Storchi-Bergmann, T. 2000, ApJ, 533, 682

Calzetti, D. 2001, PASP, 113, 1449

Castellanos, M., Díaz, Á. I., Tenorio-Tagle, G. 2002, ApJ, 565, L79

Cerviño, M., Luridiana, V., Castander, F. J. 2000, A&A, 360, L5

Cerviño, M., Luridiana, V., Pérez, E., Vílchez, J. M., Valls-Gabaud, D. 2003, A&A, 407, 177

Cerviño, M., Luridiana, V. 2005, astro, arXiv:astro-ph/0510411

Charlot, S., Fall, S. M. 2000, ApJ, 539, 718

Charlot, S., Longhetti, M. 2001, MNRAS, 323, 887

Chen, G. X., Pradhan, A. K. 2000, ApJ, 536, 420

Cid Fernandes, R., Mateus, A., Sodré, L., Stasińska, G., Gomes, J. M. 2005, MNRAS, 358, 363

Cox, D. P., Smith, B. W. 1976, ApJ, 203, 361

Delahaye, F., Pradhan, A. K., Zeippen, C. J. 2006, J. Phys. B 39, 3465

de Vaucouleurs, G. 1960, AJ, 65, 51

Diehl, R., Prantzos, N., von Ballmoos, P. 2006, Ncl. Phys. A 777, 70

Dopita, M. A., Kewley, L. J., Heisler, C. A., Sutherland, R. S. 2000, ApJ, 542, 224

Dopita, M. A., Sutherland, R. S. 1995, ApJ, 455, 468

Dopita, M. A., Sutherland, R. S. 1996, ApJS, 102, 161

Dopita, M. A. *et al.* 2006, ApJ, 647, 244

Dopita, M. A., Sutherland, R. S. 2003, *Astrophysics of the Diffuse Universe* (Berlin: Springer)

Dumont, A.-M., Collin, S., Paletou, F., Coupé, S., Godet, O., Pelat, D. 2003, A&A, 407, 13

Ercolano, B. *et al.* 2007, MNRAS, 379, 945

Ercolano, B., Barlow, M. J., Storey, P. J., Liu, X.-W. 2003, MNRAS, 340, 1136

Ercolano, B., Barlow, M. J., Storey, P. J. 2005, MNRAS, 362, 1038

Escalante, V., Morisset, C. 2005, MNRAS, 361, 813

Esteban, C. 2002, RMxAC, 12, 56

Fabian, A. C., Nulsen, P. E. J. 1977, MNRAS, 180, 479

Fabian, A. C., Iwasawa, K., Reynolds, C. S., Young, A. J. 2000, PASP, 112, 1145

Ferland, G. J. 2003, ARA&A, 41, 517

Fioc, M., Rocca-Volmerange, B. 1997, A&A, 326, 950

Fioc, M., Rocca-Volmerange, B. 1999, astro, arXiv:astro-ph/9912179

Fitzpatrick, E. L. 1999, PASP, 111, 63

Fitzpatrick, E. L. 2004, ASPC, 309, 33

García-Rojas, J., Esteban, C. 2006, astro, arXiv:astro-ph/0610903

Garnett, D. R. 1992, AJ, 103, 1330

Gonçalves, D. R., Ercolano, B., Carnero, A., Mampaso, A., Corradi, R. L. M. 2006, MNRAS, 365, 1039

Guseva, N. G., Izotov, Y. I., Papaderos, P., Fricke, K. J. 2007, astro, arXiv:astro-ph/0701032

Hamann, F., Korista, K. T., Ferland, G. J., Warner, C., Baldwin, J. 2002, ApJ, 564, 592

Harrington, J. P. 1968, ApJ, 152, 943

Heckman, T. M. 1980, A&A, 87, 152

Henney, W. J. 2006, astro, arXiv:astro-ph/0602626

Hillier, D. J., Miller, D. L. 1998, ApJ, 496, 407

Hubeny, I., Lanz, T. 1995, ApJ, 439, 875

Izotov, Y. I., Stasińska, G., Meynet, G., Guseva, N. G., Thuan, T. X. 2006, A&A, 448, 955

Izotov, Y. I., Thuan, T. X., Lipovetsky, V. A. 1994, ApJ, 435, 647

Jamet, L., Pérez, E., Cerviño, M., Stasińska, G., González Delgado, R. M., Vílchez, J. M. 2004, A&A, 426, 399

Jamet, L., Stasińska, G., Pérez, E., González Delgado, R. M., Vílchez, J. M. 2005, A&A, 444, 723

Kallman, T. R., Palmeri, P. 2006, ArXiv Astrophysics e-prints, arXiv:astro-ph/0610423

Kallman, T. R., Palmeri, P. 2007, RvMP, 79, 79

Kastner, S. O., Bhatia, A. K. 1990, ApJ, 362, 745

Kauffmann, G. *et al.* 2003, MNRAS, 346, 1055

Kennicutt, R. C., Jr. 1998, ARA&A, 36, 189

Kennicutt, R. C., Jr., Bresolin, F., Garnett, D. R. 2003, ApJ, 591, 801

Kewley, L. J., Dopita, M. A., Sutherland, R. S., Heisler, C. A., Trevena, J. 2001, ApJ, 556, 121

Kewley, L. J., Geller, M. J., Jansen, R. A., Dopita, M. A. 2002, AJ, 124, 3135

Kingsburgh, R. L., Barlow, M. J. 1994, MNRAS, 271, 257

Kobulnicky, H. A., Kennicutt, R. C., Jr., Pizagno, J. L. 1999, ApJ, 514, 544

Kobulnicky, H. A., Phillips, A. C. 2003, ApJ, 599, 1031

Kogure, T., Leung, K.-Ch. 2007, *The Astrophysics of Emission Line Stars* (Dordrecht: Kluwer)

Kunze, D. *et al.* 1996, A&A, 315, L101

Lamareille, F., Mouhcine, M., Contini, T., Lewis, I., Maddox, S. 2004, MNRAS, 350, 396

Leitherer, C. *et al.* 1999, ApJS, 123, 3

Lilly, S. J. *et al.* 2007, astro, arXiv:astro-ph/0612291

Lis, D. C., Blake, G. A., Herbst, E. 2005, *Astrochemistry: Recent Successes and Current Challenges* (Cambridge: Cambridge University Press)

Liu, X.-W., Storey, P. J., Barlow, M. J., Danziger, I. J., Cohen, M., Bryce, M. 2000, MNRAS, 312, 585

Liu, X.-W. 2002, RMxAC, 12, 70

Liu, X.-W. 2003, IAUS, 209, 339

Liu, X.-W. 2006, IAUS, 234, 219

Lobanov, A. P. 2005, MSAIS, 7, 12

McCall, M. L., Rybski, P. M., Shields, G. A. 1985, ApJS, 57, 1

McGaugh, S. S. 1991, ApJ, 380, 140

McGaugh, S. S. 1994, ApJ, 426, 135

Morisset, C., Schaerer, D., Martín-Hernández, N. L., Peeters, E., Damour, F., Baluteau, J.-P., Cox, P., Roelfsema, P. 2002, A&A, 386, 558

Morisset, C., Schaerer, D., Bouret, J.-C., Martins, F. 2004, A&A, 415, 577

Morisset, C. 2004, ApJ, 601, 858

Morisset, C., Stasińska, G., Peña, M. 2005, RMxAC, 23, 115

Morisset, C. 2006, IAUS, 234, 467

Morisset, C., Stasińska G. 2006, RMxAA, 42, 153

Morton, D. C. 1991, ApJS, 77, 119

Moustakas, J., Kennicutt, R. C., Jr., 2006, ApJ, 651, 155

Nagao, T., Maiolino, R., Marconi, A. 2006, A&A, 459, 85

O'Dell, C. R., Peimbert, M., Peimbert, A. 2003, AJ, 125, 2590

Osterbrock, D. E., Ferland, G. J. 2006, *Astrophysics of Gaseous Nebulae and Active Galactic Nuclei*, 2nd. edn. (Sausalito, CA: University Science Books)

Osterbrock, D. E., Parker, R. A. R. 1965, ApJ, 141, 892

Pagel, B. E. J., Edmunds, M. G., Blackwell, D. E., Chun, M. S., Smith, G. 1979, MNRAS, 189, 95

Panuzzo, P., Bressan, A., Granato, G. L., Silva, L., Danese, L. 2003, A&A, 409, 99

Patriarchi, P., Morbidelli, L., Perinotto, M. 2003, A&A, 410, 905

Pauldrach, A. W. A., Hoffmann, T. L., Lennon, M. 2001, A&A, 375, 161

Peimbert, M. 1967, ApJ, 150, 825

Peimbert, M., Costero, R. 1969, BOTT, 5, 3

Peimbert, M. 1975, ARA&A, 13, 113

Peimbert, M., Peimbert, A. 2003, RMxAC, 16, 113

Peimbert, M., Peimbert, A. 2006, IAUS, 234, 227

Peimbert, M., Peimbert, A., Esteban, C., García-Rojas, J., Bresolin, F., Carigi, L., Ruiz, M. T., Lopez-Sanchez, A. R. 2006, astro, arXiv:astro-ph/0608440

Péquignot, D. *et al.* 2001, ASPC, 247, 533

Pérez-Montero, E., Díaz, A. I., 2005, MNRAS, 361, 1063

Pilyugin, L. S. 2000, A&A, 362, 325

Pilyugin, L. S. 2001, A&A, 369, 594

Pilyugin, L. S. 2003, A&A, 399, 1003

Pilyugin, L. S., Thuan, T. X. 2005, ApJ, 631, 231

Porquet, D., Dubau, J. 2000, A&AS, 143, 495

Porter, R., Ferland, G. 2006, AAS, 209, #34.01

Pottasch, S. R., Preite-Martinez, A. 1983, A&A, 126, 31

Puls, J., Urbaneja, M. A., Venero, R., Repolust, T., Springmann, U., Jokuthy, A., Mokiem, M. R. 2005, A&A, 435, 669

Puls, J., Markova, N., Scuderi, S., Stanghellini, C., Taranova, O. G., Burnley, A. W., Howarth, I. D. 2006, A&A, 454, 625

Ratag, M. A., Pottasch, S. R., Dennefeld, M., Menzies, J. 1997, A&AS, 126, 297

Rauch, T. 2003, IAUS, 209, 191

Romanowsky, A. J. 2006, *Elliptical Galaxy Halo Masses from Internal Kinematics* (Paris: EAS Publications), pp. 119–126

Rubin, R. H. 1968, ApJ, 153, 761

Rubin, R. H. 1989, ApJS, 69, 897

Rubin, R. H., Martin, P. G., Dufour, R. J., Ferland, G. J., Blagrave, K. P. M., Liu, X.-W., Nguyen, J. F., Baldwin, J. A. 2003, MNRAS, 340, 362

Schaerer, D. 2000, in *Building Galaxies; from the Primordial Universe to the Present*, ed. F. Hammer, T. X. Thuan, V. Cayatte, B. Guiderdoni, & J. T. Thanh Van (Singapore: Word Scientific), p. 389

Seyfert, C. K. 1941, PASP, 53, 231

Silva, L., Granato, G. L., Bressan, A., Danese, L. 1998, ApJ, 509, 103

Simón-Díaz, S., Stasińska, G., García-Rojas, J., Morisset, C., López-Sánchez, A. R., Esteban, C. 2007, astro, arXiv:astro-ph/0702363

Smith, L. J., Norris, R. P. F., Crowther, P. A. 2002, MNRAS, 337, 1309

Smith, R. K. (ed.), 2005, *X-Ray Diagnostics of Astrophysical Plasmas: Theory, Experiment, and Observation* (New York: AIP)

Spitzer, L. 1978, *Physical Processes in the Interstellar Medium* (New York: Wiley-Interscience)

Stasińska, G., 1980, A&A, 84, 320

Stasińska, G., 2002, RMxAC, 12, 62

Stasińska, G., 2004, in *Cosmochemistry. The Melting Pot of the Elements*, ed. C. Esteban, R. J. García López, A. Herrero & F. Sánchez (Cambridge: Cambridge University Press), pp. 115–170

Stasińska, G., Gräfener, G. Peña, M., Hamann, W.-R., Koesterke, L., Szczerba, R. 2004, A&A, 413, 329

Stasińska, G., Izotov, Y. 2003, A&A, 397, 71

Stasińska, G., 2005, A&A, 434, 507

Stasińska, G., Cid Fernandes, R., Mateus, A., Sodré, L., Asari, N. V. 2006, MNRAS, 371, 972

Stasińska, G., Schaerer, D. 1999, A&A, 351, 72

Stasińska, G., Tenorio-Tagle, G., Rodriguez, M., Henney, W. 2007, A&A, 471, 193

Storchi-Bergmann, T., Calzetti, D., Kinney, A. L. 1994, ApJ, 429, 572

Stoy, R. H., 1933, MNRAS, 93, 588

Tarter, C. B., Salpeter, E. E. 1969, ApJ, 156, 953

Torres-Peimbert, S., Peimbert, M. 2003, IAUS, 209, 363

Tsamis, Y. G., Barlow, M. J., Liu, X.-W., Danziger, I. J., Storey, P. J. 2003, MNRAS, 338, 687

Tsamis, Y. G., Barlow, M. J., Liu, X.-W., Storey, P. J., Danziger, I. J. 2004, MNRAS, 353, 953

Tsamis, Y. G., Péquignot, D. 2005, MNRAS, 364, 687

Veilleux, S., Osterbrock, D. E. 1987, ApJS, 63, 295

Vílchez, J. M., Esteban, C. 1996, MNRAS, 280, 720

Vílchez, J. M., Pagel, B. E. J. 1988, MNRAS, 231, 257

Werner, K., Dreizler, S. 1993, AcA, 43, 321

Wyse, R. F. G., Gilmore, G. 1992, MNRAS, 257, 1

Yin, S. Y., Liang, Y. C., Hammer, F., Brinchmann, J., Zhang, B., Deng, L. C., Flores, H. 2007, A&A, 462, 535

York, D. G., *et al.* 2000, AJ, 120, 1579

Zaritsky, D., Kennicutt, R. C., Jr., Huchra, J. P. 1994, ApJ, 420, 87

Zhang, Y., Liu, X.-W., Luo, S.-G., Péquignot, D., Barlow, M. J. 2005, A&A, 442, 249

2. The observer's perspective: Emission-line surveys

MAURO GIAVALISCO

2.1. Introduction

The goal of emission-line surveys is to identify a preselected class of astrophysical sources by means of emission lines in their spectral energy distribution. The techniques vary, depending on the sources, the spectral region and even the exact wavelength at which the observations are made, the desired sensitivity of the survey, and the volume of space that it aims to cover. But the basic methodology remains the same: exploiting the fact that emission lines in the spectrum of astronomical sources radiate significantly more luminosity over a relatively small wavelength interval than the continuum emission does over the same, or even larger, intervals in nearby portions of the spectrum. The excess luminosity, quantitatively expressed as the equivalent width of the line, enhances the signal-to-noise ratio of measures of the line flux for those sources whose emission lines satisfy the selection criteria of the survey, allowing the observer to cull them from otherwise similar sources.

Operationally, emission-line surveys exploit the presence of emission lines to substantially improve the detection rate of a particular class of astrophysical sources for which other methods of investigation would be significantly more inefficient or even outright impossible. The typical targets of these surveys are either sources that are too faint to detect by means of their continuum emission or sources that are rare and/or inconspicuous, and hence very difficult to recognize from all the other sources of similar apparent luminosity and/or morphology that crowd images made using the continuum emission. When observed in the light of some emission line, the targeted sources prominently stand out from the general counts. Examples of sources include quasars and other types of active galactic nuclei (AGNs), galactic or extragalactic planetary nebulae, and star-forming galaxies, especially at high redshifts.

Emission-line surveys are also used when it is necessary to select a class of (line-emitting) sources either with the same radial velocity or located at the same redshift, i.e. in a preassigned volume of space. In these cases, the scientific goal of the surveys is either to obtain information about the spatial distribution of the sources or to use their spatial location to infer some other property.

2.2. Notations, definitions, and other useful concepts

Before we address the methodology of the narrow-band surveys and the typical instrumental configurations and observational strategies, it is useful to review a few basic concepts commonly encountered in the design, planning, and execution of emission-line surveys, including source properties, instrumental characteristics, and the typical observational setup.

2.2.1 *Redshift*

The redshift z is a parameter that quantifies the cosmic expansion as a function of time in a way that is convenient when working with spectral features. It is the ratio of the

The Emission-Line Universe, ed. J. Cepa. Published by Cambridge University Press.

cosmic scale factor $a(t)$ at the cosmic time t to the scale factor $a(t_0)$ at the current time t_0:

$$1 + z = \frac{a(t_0)}{a(t)}.$$

In the following we will use, when needed, a world model with flat curvature, filled with cold dark matter (CDM) and dark energy in the proportion of $\Omega_m = 0.3$ and $\Omega_\Lambda = 0.7$, and characterized by Hubble constant $H_0 = 70$ km s^{-1} Mpc^{-1} (1 Mpc $= 3.086 \times 10^{24}$ cm).

As a result of the cosmic expansion, an emission line with rest-frame wavelength λ_e from a source at redshift z will be observed by an observer at redshift $z = 0$ at wavelength

$$\lambda_o = (1 + z)\lambda_e.$$

Thus, for example, the Lyα emission line, whose rest-frame wavelength $\lambda_{Ly\alpha} = 1216$ Å is in the far-ultraviolet (UV) portion of the spectrum, if received from a source at redshift $z = 3$ (e.g. a star-forming galaxy or an AGN), is observed at $\lambda_o = 4864$ Å, namely in the B band of the optical window. In this spectral region the night sky is very dark and modern charged-coupled-device (CCD) detectors are very sensitive, making the observations of cosmologically distant ($z \gg 0$) galaxies relatively straightforward for 4-meter- and 8-meter-class telescopes. Another strong and very useful emission line, because it is much less sensitive to dust obscuration than Lyα, is Hα, whose rest-frame is $\lambda_e = 6563$ Å. Unfortunately, even for a moderate value for the redshift, the line is observed in the near-infrared (IR) region of the spectrum, where the sky is very bright, about 10^5 times brighter than in the B band in typical broad filters; for spectroscopic observations and narrow-band imaging the background depends on the presence of strong OH emission lines in the target band. For example, the line is observed in the Z band (0.9 μm) at $z \sim 0.37$, in the J band (1.2 μm) at $z \sim 0.9$, in the H band (1.6 μm) at $z \sim 1.44$, and in the K band (2.2 μm) at $z \sim 2.35$. Furthermore, members of the current generation of near-IR detectors are smaller and with worse noise characteristics than those working at optical wavelengths, which affects the sensitivity of narrow-band imaging and spectroscopy.

2.2.2 *Narrow-band filters and dispersers*

The spectral elements commonly used in emission-line surveys are imaging filters with relatively narrow transmittance curves, commonly referred to as narrow-band filters. These are typically interference filters, consisting of thin reflective film layers deposited on the surface of an optical glass filter, that block by destructive interference all wavelengths of the transmitted light except those in a narrow interval centered around the central wavelength of the filter. Widths of the transmittance curve, typically measured as the "full-width at half maximum" (FWHM) of the line, range from a few times 10^2 Å (medium-band filters), via ≈50–100 Å (narrow-band filters) to a few ångstiröm units (very-narrow-band filters). Transmittance at the peak can be very high, up to ≈90%, in medium- and narrow-band elements, but it is often significantly less, ∼10% to ∼50%, in very-narrow-band filters.

Other common dispersing spectral elements used in emission-line surveys are the objective prism, the diffraction grating, and the grism.

The objective prism, or simply prism, is typically used in slitless spectroscopic surveys to obtain very-low-dispersion spectra of targets. The transmittance of the prism is a function of the wavelength, and can reach very high values in some portion of the transmittance band, up to ~80% or more. The dispersion of the prism is very low, typically of the order of $\lambda/\Delta\lambda \sim 10$–20.

The diffraction grating, or grating, is a dispersing spectral element based on the interference pattern of the reflected light created by a large number of grooves on its surface. Its transmittance can reach high peak values and the dispersion $\lambda/\Delta\lambda$ can reach values of several thousands. To avoid superposition of spectra of different orders from the same source and of spectra of the same order from different sources, the grating is typically used with a slit.

The grism is the combination of a prism with a grating. Typically, its dispersion is intermediate between those of the previous two cases. It also is used with or without a slit (especially at lower resolution).

2.2.3 *The signal-to-noise ratio*

The most important metric to assess the sensitivity of observations of any type is the ratio of the signal that is being measured to the noise, i.e. the uncertainty, or error, that affects the measure, $S_N = S/N$. This ratio quantifies how many times the strength of the measured signal S exceeds the uncertainty N that affects the measure. When using photon-counting devices, such as the detectors commonly adopted in optical and near-infrared observations, the uncertainty in the measured number of photons N is simply $\Delta S_\gamma = \sqrt{S_\gamma}$, because photons obey the Poisson statistics. In this case, a good approximation of the signal-to-noise ratio can be expressed as

$$S_N = \frac{S_\gamma}{\sqrt{S_\gamma + B_\gamma + N_\gamma^2}},$$

where S_γ is the number of photons from the source in the desired spectral band (either the imaging filter band or the spectral-resolution element), B_γ is the number of photons from any type of background emission, such as the sky, telescope thermal emission, or unwanted emission from the source itself (e.g. the continuum emission if one is interested only in the line emission), and N_γ is the equivalent number of photons from additional sources of noise coming from the detector.

If the detector noise dominates the noise budget

$$S_N \propto S_\gamma \propto T_{\rm exp}$$

($T_{\rm exp}$ is the exposure time). In this case the sensitivity of the observations increases linearly with the observing time, and an increased investment in observing time pays dividends in the same proportion. This is often the case with narrow-band imaging, especially with very narrow filters and/or in spectral regions where the sky is particularly quiet and/or when the observations are made from space.

If the detector noise can be neglected,

$$S_N \propto \sqrt{T_{\rm exp}}$$

since both S_γ and B_γ are linearly proportional to $T_{\rm exp}$. In this case the gain in sensitivity with exposure time is much slower. For example, doubling the exposure time improves S_N only by a meager factor of $\sqrt{2} \sim 1.4$ or ~0.38 on the magnitude scale.

It is clear from these simple considerations that it is very important to be strategic when designing the observations for an emission-line survey: one needs to clearly understand the nature of the limiting noise and use the minimum amount of resources to achieve the desired sensitivity, given the problem at hand.

2.2.4 *Source parameters*

A source with luminosity L (in units of erg s^{-1} in the cgs system, which we use here) placed at redshift z is observed as having flux F (in units of erg s^{-1} cm^{-2}),

$$F = \frac{L}{4\pi D_{\mathrm{L}}^2(z)},$$

where $D_{\mathrm{L}}^2(z)$ is the luminosity distance of the source, which depends on the parameters of the world model adopted. In the standard CDM world with zero curvature (standard CDM, adopted here), the primary parameters that determine $D_{\mathrm{L}}^2(z)$ are the Hubble constant H_0 and the matter and dark-energy density parameters Ω_{m} and Ω_{Λ}.

It is customary in astronomy to express flux and luminosity density as a function either of wavelength or of frequency. We have

- f_ν: flux density, in units of $\mathrm{erg\,s^{-1}\,cm^{-2}\,Hz^{-1}}$;
- l_ν: luminosity density, in units of $\mathrm{erg\,s^{-1}\,Hz^{-1}}$;
- f_λ: flux density, in units of $\mathrm{erg\,s^{-1}\,cm^{-2}}$ Å^{-1}; and
- l_λ: luminosity density, in units of $\mathrm{erg\,s^{-1}}$ Å^{-1}.

To convert the flux and luminosity density from one unit system to the other, one must remember that the luminosity in the differential element of wavelength or frequency must be independent of the unit system, namely

$$f_\nu \, \mathrm{d}\nu = f_\lambda \, \mathrm{d}\lambda.$$

It can be shown that the flux from a source placed at redshift z and its luminosity density are related by

$$f_\nu = \frac{l_\nu(1+z)}{4\pi D_{\mathrm{L}}^2(z)}$$

if the flux is expressed as a function of frequency ν, and by

$$f_\lambda = \frac{l_\lambda}{4\pi D_{\mathrm{L}}^2(z)(1+z)}$$

if it is expressed as a function of wavelength λ.

Finally, recall that

$$\nu = \frac{\lambda}{c},$$

where $c = 2.9979 \times 10^{10}$ cm s^{-1} is the speed of light, and that 1Å $= 10^{-8}$ cm.

Later, when we use magnitudes, we will adopt either the AB scale, defined as

$$m_{\mathrm{AB}} = -2.5 \log(f_\nu) - 48.6$$

(Oke & Gunn 1977), or the ST scale, defined as

$$m_{\mathrm{ST}} = -2.5 \log(f_\lambda) - 21.1$$

(Walsh 1995).

2.2.5 *Line width*

Emission lines from astrophysical sources have finite width in wavelength space, namely they are not infinitely narrow. The physics of the ionized nebulae and emission-line production will not be discussed here. For our purposes, keep in mind that the observed width of emission lines reflects the kinematics of the gas in the emitting regions. If the integrated light from more than one discrete region is observed, the relative motions of the individual regions also contribute to the observed line width which, in fact, can be used as a tracer of the dynamics of these sources. In this case, the resulting width of the line is a combination of the line width of the individual regions with that produced by the motions of the regions.

The line width observed in a spectrum is a combination of the intrinsic width of the line with the finite resolution of the spectrograph, which, to a good approximation, add in quadrature, namely

$$\Delta\lambda_{\rm O} = \sqrt{\Delta\lambda^2_{\rm Inst} + \Delta\lambda^2_{\rm intr}}.$$

In general, if the emission-line regions can be described as composed of individual regions whose relative motions are of the order of Δv, and if the central wavelength of the emission line is λ, the velocity width of the emission line is given (assuming Gaussian line profiles) by

$$\frac{\Delta\lambda}{\lambda} = \frac{\Delta v}{c},$$

where $c = 2.9979 \times 10^{10}$ cm s^{-1} is the speed of light. For example, if $\Delta v = 200\,{\rm km\,s^{-1}}$, the typical gravitational motions (either rotation or velocity dispersion) in a galaxy with $M \sim M^*$ ($M^* = 10^{12}\,{\rm M_\odot}$ and $\lambda = 5500$ Å (the V band), the contribution to the line width (to be added in quadrature to the finite resolution of the spectrograph) is of the order of

$$\Delta\lambda \approx 4\,\text{Å}.$$

Larger velocity fields, about one magnitude larger, can be observed in quasar and AGN spectra.

2.2.6 *Equivalent width*

Another key parameter to consider in the design of emission-line surveys is the equivalent width of the line, often designated $E_{\rm w}$, which is the width of a rectangular line of equal flux whose flux density is the same as that of the underlying continuum at the central wavelength of the line. According to this definition, if f_λ is the continuum flux density at the line-central wavelength, in units of erg s^{-1} cm^{-2} Å^{-1}, and $F_{\rm l}$ is the flux of the line, in units of erg s^{-1} cm^{-2}, the equivalent width is given by

$$E_{\rm w} = \frac{F_{\rm l}}{f_\lambda},$$

and its units are ångström units. Basically, the equivalent width specifies how large a portion of the continuum emission (of the same flux density as that at the center of the line) needs to be so that its integrated flux equals that of the line, and provides a quantitative way to express the strength of the line relative to the continuum emission.

The concept of equivalent width applies to both emission and absorption lines as well. In the case of an absorption line, its equivalent width is the portion of continuum whose integrated flux would fill the missing flux in the line. Historically, the equivalent width of an absorption line is defined as positive, that of emission lines as negative. Since in this work only emission lines are considered, for the sake of simplicity we will define the equivalent width as positive.

It is important to realize that the sensitivity of emission-line surveys, whether narrow-band imaging or spectroscopy, is limited by both flux and equivalent width. The sensitivity in flux is easy to understand, since even a line with infinite equivalent width, i.e.

a source with only line emission and no continuum, cannot be detected if the line flux is below the sensitivity of the survey.

The sensitivity in equivalent width comes from the fact that, to recognize the presence of an emission line, the difference of flux contributed by the line over that contributed by the continuum (over some finite wavelength interval, either the width of the bandpass or the spectral-resolution element – see later for more details) must be larger than the photometric error at that flux level.

2.3. Methodology of emission-line surveys

In broad terms, two main methodologies are generally adopted for emission-line surveys, namely imaging through filters with a relatively narrow transmittance curve, designed to select the targeted emission line, and spectroscopy, either with a slit or slitless.

2.3.1 *Narrow-band imaging*

Narrow-band imaging surveys are generally carried out either with traditional transmitting filters or with tunable filters, such as Fabry–Pérot interferometers. One of the advantages of this technique is that the acquisition and reduction of imaging data, as well as source identification, are generally relatively straightforward. Imaging also has the further advantage of producing spatial maps of the line emission within the sources, as well as the spatial distribution of the sources themselves. However, only the limited range of wavelengths transmitted through the adopted bandpass can be probed at any one time, resulting in the selection of sources within a very narrow range of radial velocity or redshift.

The images can reach relatively deep flux levels, because the background noise from the sky is greatly suppressed by the narrow bandpass of the spectral element compared with broad-band images of similar exposure time (continuum images). However, the final sensitivity of narrow-band imaging to faint emission lines depends on how well the continuum emission at the wavelength of the line can be measured and subtracted. The measuring of the continuum requires additional imaging in spectral regions coincident with or adjacent to that of the targeted line. This means that either broad-band images (in at least one bandpass) or additional narrow-band images need to be obtained together with the primary narrow-band ones. Figure 2.1 shows the transmission curve of a typical narrow-band filter used in surveys for high-redshift galaxies together with that of the accompanying broad-band filter and the line-emission spectrum of the night sky (from Hu *et al.* 2004). (The inset in Figure 2.4 shows the case of two broad-band filters used to estimate the continuum.)

Figure 2.2 shows the identification of Lyα-emitting galaxies at redshift $z = 5.66$ by narrow-band imaging centered at $\lambda = 8160$ Å by Taniguchi *et al.* (2003). A galaxy with strong Lyα emission is detected with high signal-to-noise ratio in the narrow-band filter and is undetected in the broad-band passbands, except in the I-band filter, where most of the observed flux is also coming from the line. Figure 2.3 shows the two-dimensional spectrum of the galaxy (top), as well as the extracted spectrum (middle), where only the emission line is observed.

Sources with strong emission lines (large equivalent width) are identified, for example, using color–magnitude diagrams in which the (narrow-broad) color is plotted versus the broad-band magnitude to show the line-flux excess over the continuum. Sources with weak emission lines can be detected only if the line contribution to the total flux recorded in the narrow-band image is larger than the uncertainty in the continuum

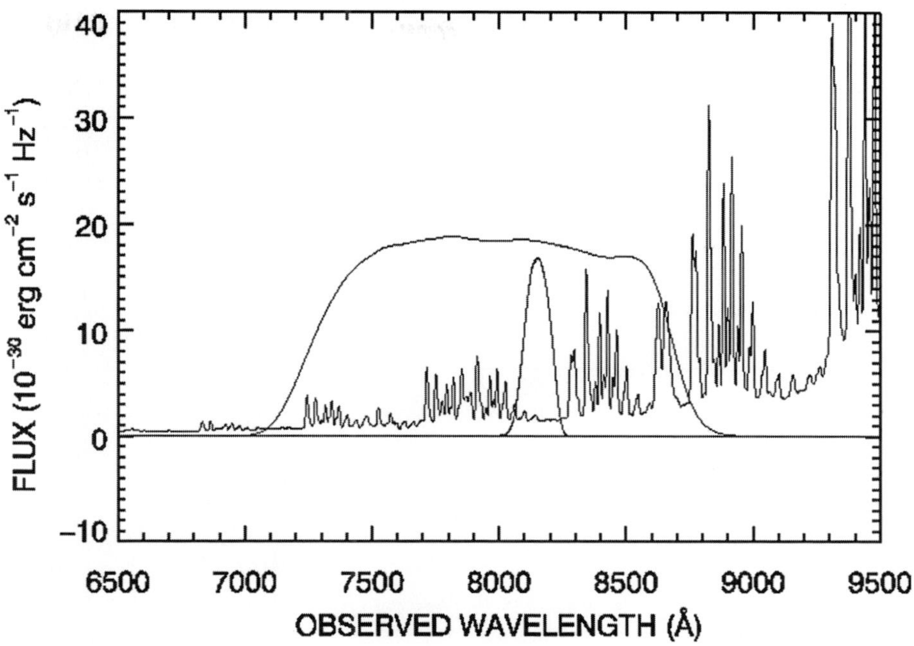

FIGURE 2.1. The transmission curve of the narrow-band filter centered at $\lambda = 8160$ Å used by Hu *et al.* (2004) in their survey for Lyα emitters at redshift $z = 5.7$ together with that of the Cousins I broad-band filter. Also plotted is the spectrum of the night sky obtained with the LRIS spectrograph on the Keck telescope at Mauna Kea. Note how the narrow-band filter has been strategically chosen to limit the sky background by avoiding strong night-sky emission lines. The filter is also very nicely located near the center of the broad bandpass, which minimizes errors arising from the subtraction of the continuum contribution to the narrow-band images.

contribution. This highlights the fact that the sensitivity of narrow-band imaging surveys is limited both in flux and in equivalent width. The limit in flux determines whether the galaxy is detected at all in the various filters used. The limit in equivalent width determines whether the line emission can be detected over the continuum one. This is illustrated in Figure 2.4, which shows the color–magnitude diagram used by Shimasaku *et al.* (2004) to select Lyα emitters at $z \sim 4.79$. The narrow-band filter is centered at $\lambda = 7040$ Å, and the continuum image is obtained by summing the R-band and i-band images.

Let's consider a survey with one narrow-band filter and a broad-band one, a configuration often used in narrow-band surveys. Let's also assume that the sources are detected in the broad bandpass. The limit in equivalent width comes from the fact that, in order to detect a line emission, the fraction of the total flux detected through the narrow filter due to the line only must be larger than the total photometric relative error (from both the broad and the narrow filter). In other words, the equivalent width $E_{\rm w}$ of the line must be such that

$$\frac{E_{\rm w}}{{\rm FW_N}} \gtrsim 3\sqrt{\left(\frac{\Delta\phi}{\phi}\right)^2_{\rm Broad} + \left(\frac{\Delta\phi}{\phi}\right)^2_{\rm Narrow}}, \tag{2.1}$$

where $\phi_{\rm Broad}$ and $\phi_{\rm Narrow}$ are the fluxes collected in the broad and narrow filters, respectively, and ${\rm FW_N}$ is the FWHM of the narrow-band filter. This relation assumes that the contribution of the line emission to the broad-band magnitude is small, which is certainly

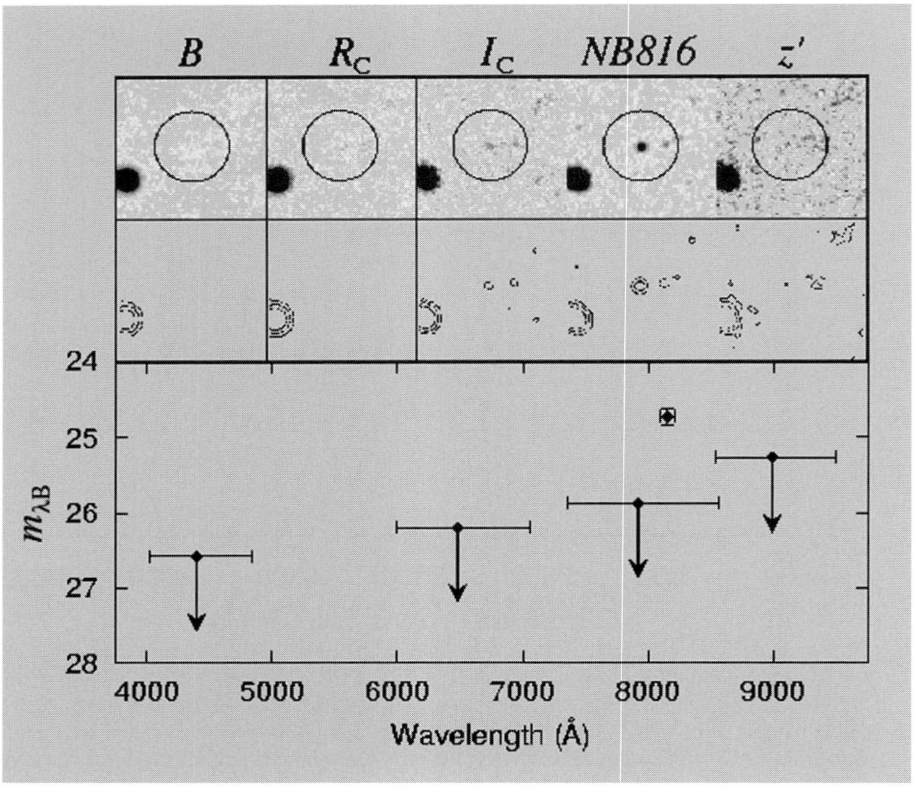

FIGURE 2.2. A narrow-band survey for Lyα-emitting galaxies at $z = 5.66$. The top panel shows a galaxy with strong Lyα emission, which is detected with high signal-to-noise ratio in the narrow-band filter, but undetected in the broad-band passbands, except in the I-band filter, where most of the observed flux is also coming from the line. The bottom panel shows the measured magnitudes, on the AB scale. From Taniguchi *et al.* (2003).

true when the equivalent width of the line is small and we are close to the threshold of detectability. We can express the relative error in each filter in terms of the flux limit, if we assume that the observations are such that the errors are dominated by the photon noise:

$$\frac{\Delta\phi}{\phi} = \frac{1}{3}\left(\frac{\phi_{\rm lim}}{\phi}\right)^{1/2}. \tag{2.2}$$

Let's make the further assumption that the source is characterized by a flat spectrum, which is a good approximation in the case of the UV SED of starburst galaxies with little dust obscuration (observed at optical wavelengths if the sources are at high redshifts). Then, we can rewrite Equation (4.2) in terms of the luminosity density detected in the narrow bandpass, the equivalent width of the Lyα emission, and the FWHM of the narrow-band filter,

$$f^{\rm c}\left(1 + \frac{E_{\rm w}}{\rm FW_N}\right) = f^{\rm c,N} \geq f^{\rm c,N}_{\rm Lim}\left[\frac{f^{\rm c,B}_{\rm Lim}}{f^{\rm c,N}_{\rm Lim}}\left(1 + \frac{E_{\rm w}}{\rm FW_N}\right) + 1\right]\left(\frac{\rm FW_N}{E_{\rm w}}\right)^2, \tag{2.3}$$

where $f^{\rm c,N}_{\rm Lim}$ and $f^{\rm c,B}_{\rm Lim}$ are the minimum continuum luminosity densities detectable in the two filters. This relation defines the minimum luminosity density which must be detected

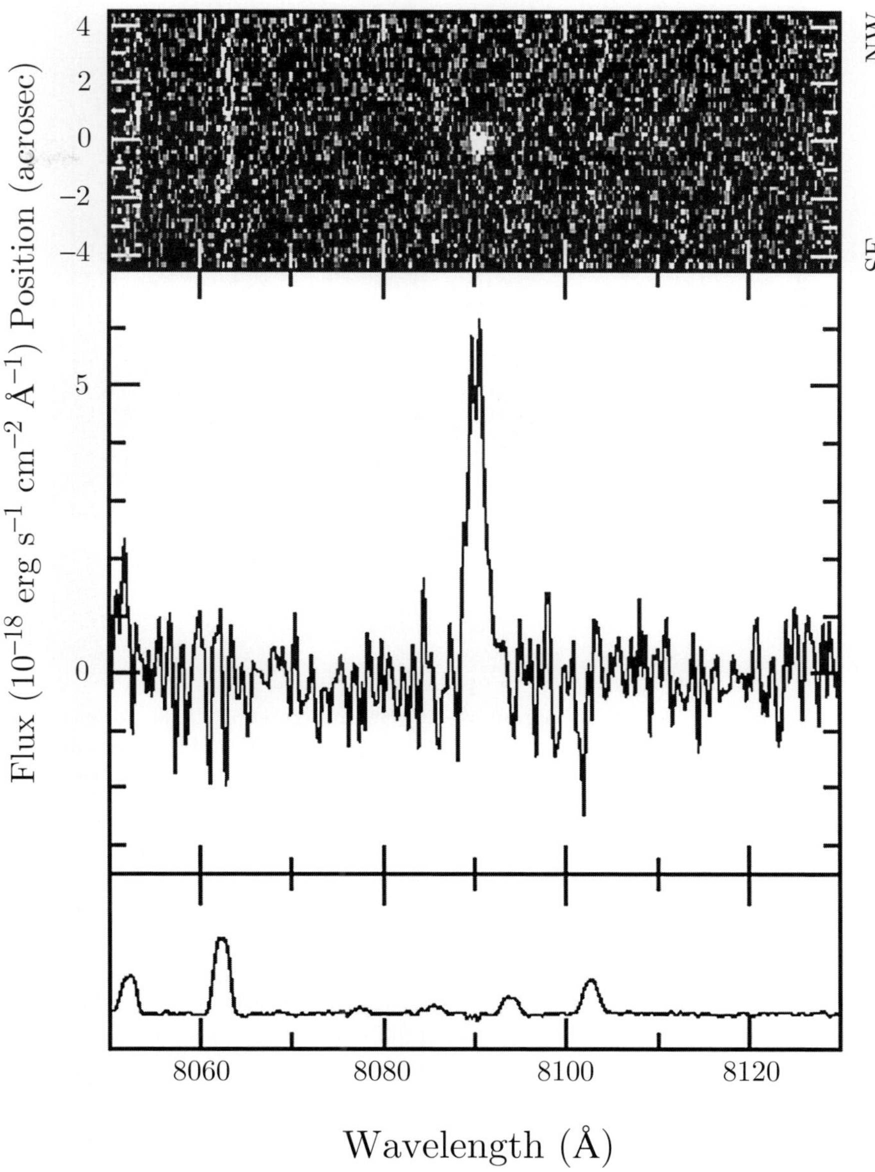

FIGURE 2.3. The two-dimensional spectrum (top) and extracted spectrum (middle) of the Lyα emitter at $z \sim 5.66$ shown in Figure 2.2. The night-sky spectrum is also plotted (bottom). Only the Lyα emission line is observed, with no continuum emission detected. Note the asymmetric profile of the Lyα line, an indicator of the very high redshift of the source, which is caused by intervening absorption by H I. From Taniguchi *et al.* (2003).

in order to reveal a line emission with equivalent width $E_{\rm w}$ using a narrow-band filter with FWHM equal to $\rm FW_N$.

2.3.2 *Spectroscopic surveys*

The main advantage of spectroscopy, which can be done with or without a slit, as a means to carry out emission-line surveys is that it allows one to probe a much larger spectral

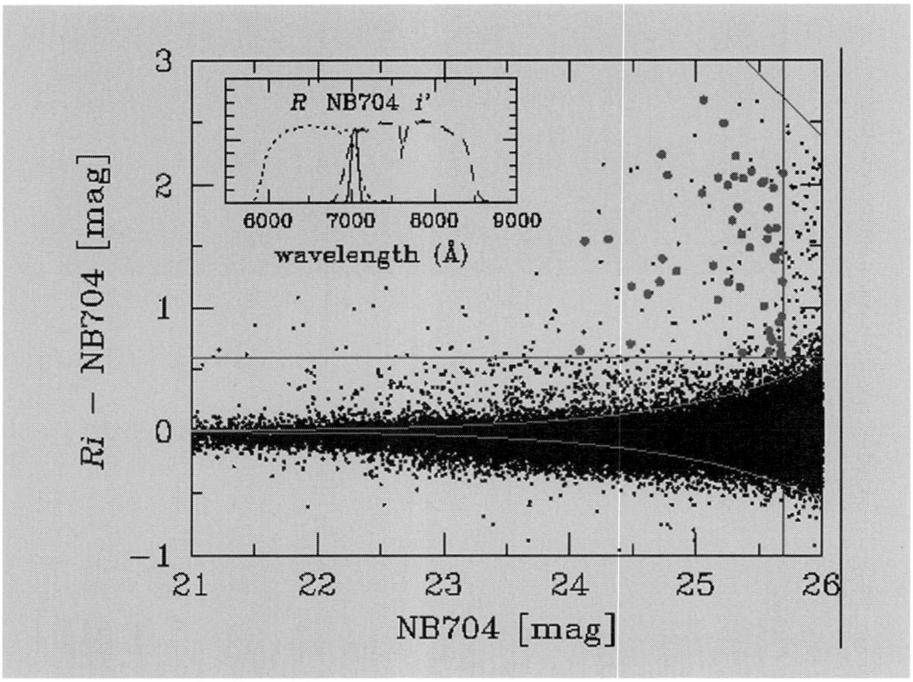

FIGURE 2.4. The color–magnitude diagram used by Shimasaku *et al.* (2004) to select Lyα emitters at $z \sim 4.79$. The narrow-band filter is centered at $\lambda = 7040$ Å, and the continuum image is obtained by summing the R-band and i-band images. The inset shows the transmittance of the filters. The lines indicate the 3σ error for flat-spectrum sources ($f_\nu \propto \nu^0$), the flux limit of the survey, and the color sensitivity threshold for line emitters to be detected.

range than narrow-band imaging, resulting in surveys that cover wider redshift and/or radial-velocity intervals. However, depending on the specific instrumental configuration of the survey, either survey area (and hence cosmic volume) or sensitivity, or both, can be significantly diminished compared with the case of narrow-band imaging.

In the case of long-slit surveys, or surveys done with masks capable of providing multislits, the sensitivity can actually be higher than in the case of narrow-band imaging, because when light is dispersed only the sky background at the same wavelength as the emission line contributes to the noise, whereas in the imaging case the background emission in the whole spectral range transmitted by the filter is a source of noise, while the line flux is the same in both cases (assuming a slit width sufficiently large that the whole of the line-emitting region is observed). However, the sky-area coverage of long-slit surveys is limited to those regions viewed through the slit or the multi-slit mask, which is often minimal compared with the case of imaging. Also, the observations are more involved than imaging, and the data reduction and analysis are of medium to high complexity, depending on whether multi-object masks are involved or not.

Slitless-objective-prism surveys or grism surveys cover much larger areas of the sky than can be covered by slit or multi-slit spectroscopy. However, their sensitivity is generally much lower than in the previous two cases, because the sky background is much higher. In particular, in this case every pixel of the detector array receives the full-sky background transmitted by the telescope–instrument combination. Furthermore, because these surveys are often carried out at very low dispersion, sensitivity is further reduced by continuum dilution of the line emission. Sky-area coverage is generally less than in

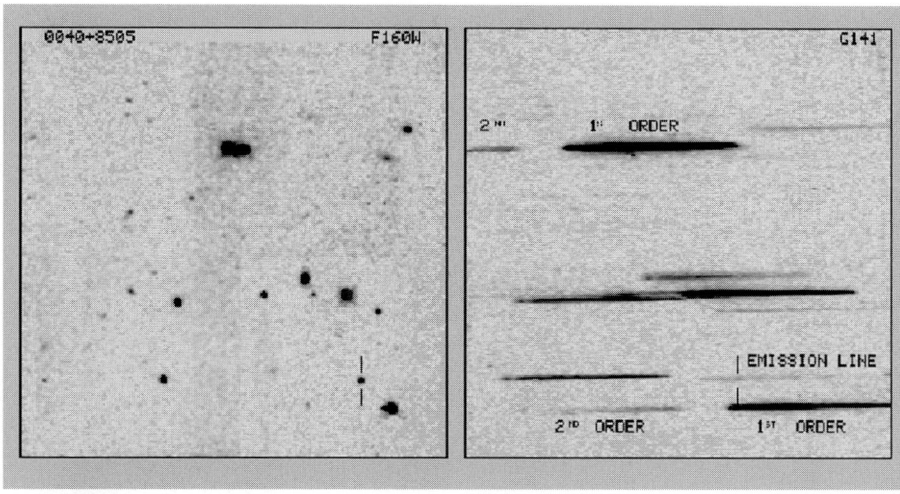

FIGURE 2.5. An example of dispersed images obtained with the HST and NICMOS during a blind slitless spectroscopic grism survey for Hα emitters at high redshift. Both undispersed (left) and dispersed (right) images are shown, as well as the location of the emission lines and the order of the spectra. From McCarthy *et al.* (1999).

imaging surveys, because the dispersed images occupy more detector area, and because confusion from adjacent overlapping spectra makes some portion of the dispersed image useless. Observations are generally easier than slit spectroscopy, but data reduction and especially analysis (calibration and extraction of the spectra) are usually complex. Figure 2.5 shows, as an example, the grism images taken as part of the Hα survey by McCarthy *et al.* (1999) with the Hubble Space Telescope (HST) and NICMOS. Both undispersed (left) and dispersed images (right) are shown. Figure 2.6 shows a gallery of grism spectra to illustrate the type of results and the diversity of spectra encountered in such surveys, while Figure 2.7 shows examples of extracted spectra.

2.3.3 *The sky background*

Finding distant galaxies is a contrast problem. The night sky, especially at the near-IR wavelengths to which the target emission lines are redshifted, is much brighter than the galaxies. Figure 2.8 shows the night-sky spectrum in the optical window, together with the wavelength ranges of traditional filters and the wavelengths of important emission lines, while Figure 2.9 shows a synthetic spectrum of the night sky in the near-IR windows (from Rousselot *et al.* 2000) with the relative intensity of the lines proportional to the photon flux.

Since, as the figures show, the atmospheric OH emission lines generally dominate this background in broad-band filters, as well as in narrow-band ones (unless they are very narrow and/or strategically positioned, as we are about to discuss), one way to make ground-based observations more sensitive is by adopting transmittance curves limited to wavelength intervals that fall between the molecular-line complexes. For example, narrow-band imaging through one of the larger OH windows at 8200 Å led to the first samples of Lyα-selected galaxies at redshift 5.7 (Hu *et al.* 2004; Rhoads *et al.* 2003; Taniguchi *et al.* 2003). Figure 2.1 illustrates this strategy in the implementation by Hu *et al.* (2004).

In principle, the sensitivity of observations over a relatively small wavelength interval can also be improved by dispersing the light. For example, if a final resolution of, say,

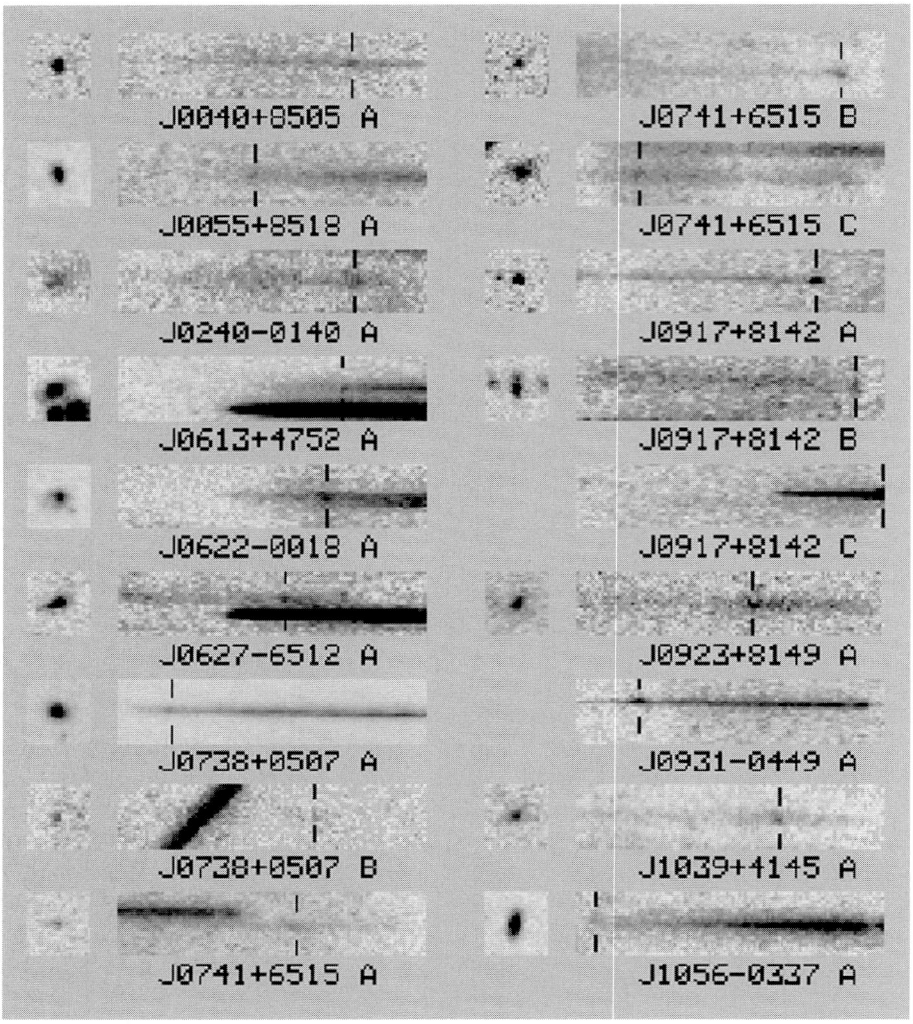

FIGURE 2.6. Examples of grism spectra obtained with the HST and NICMOS during a blind slitless spectroscopic grism survey for Hα emitters at high redshift by McCarthy *et al.* (1999).

3 Å is adequate for the emission lines targeted by the observations, the 150-Å-wide, OH-free window around 8200 Å discussed above can be dispersed with a spectrograph to improve the line-to-sky contrast by reducing the sky brightness by another factor of 50 (i.e. 150 Å/3 Å FWHM). Custom-designed, multi-slit masks can be employed in this case, when spectroscopic observations are made with a band-limiting, OH-suppression filter. This multi-slit-windows technique had been used previously to search for $z = 5.7$ and $z = 6.5$ Lyα emitters using 8-meter-class telescopes (Crampton 1999; Stockton 1999; Martin & Sawicki 2004; Tran *et al.* 2004). However, although some high-redshift Lyα emitters have been discovered serendipitously during deep blind spectroscopic searches, the volume probed in a typical long-slit observation is very small. The authors of the surveys mentioned above did not report detections of Lyα emitters, very likely because they did not probe large-enough volumes of space.

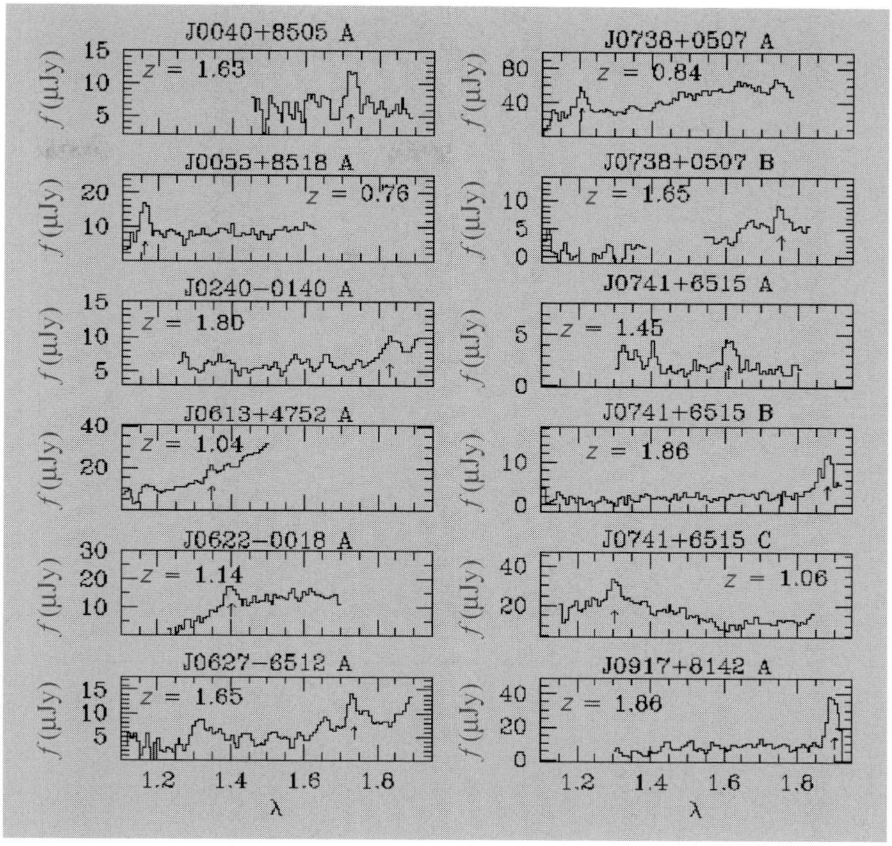

FIGURE 2.7. Extracted spectra from the survey by McCarthy *et al.* (1999) discussed above. The location of the emission lines is marked by an arrow.

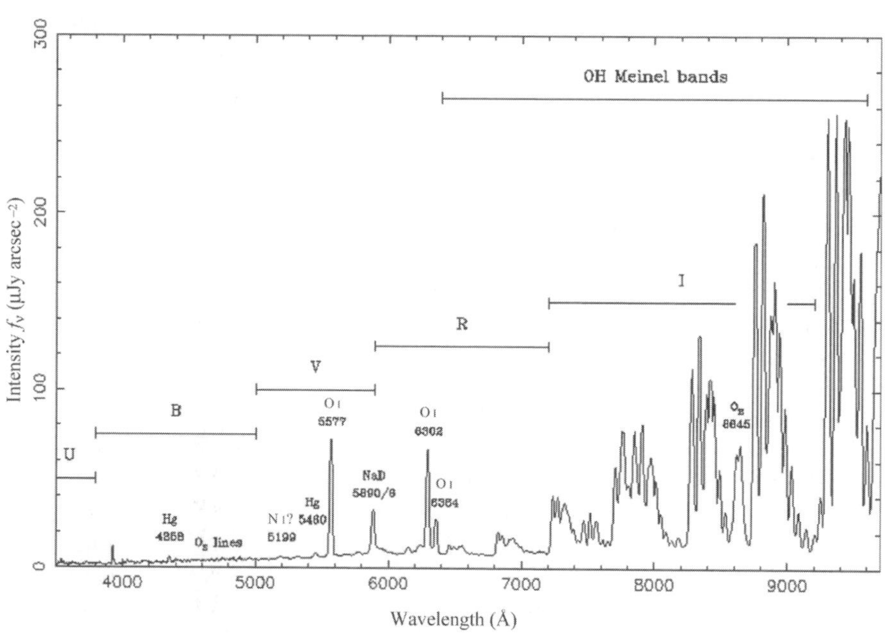

FIGURE 2.8. The spectrum of the night sky. Marked also are the wavelength ranges of traditional broad-band filters, as well as important emission lines of astronomical sources and of the sky.

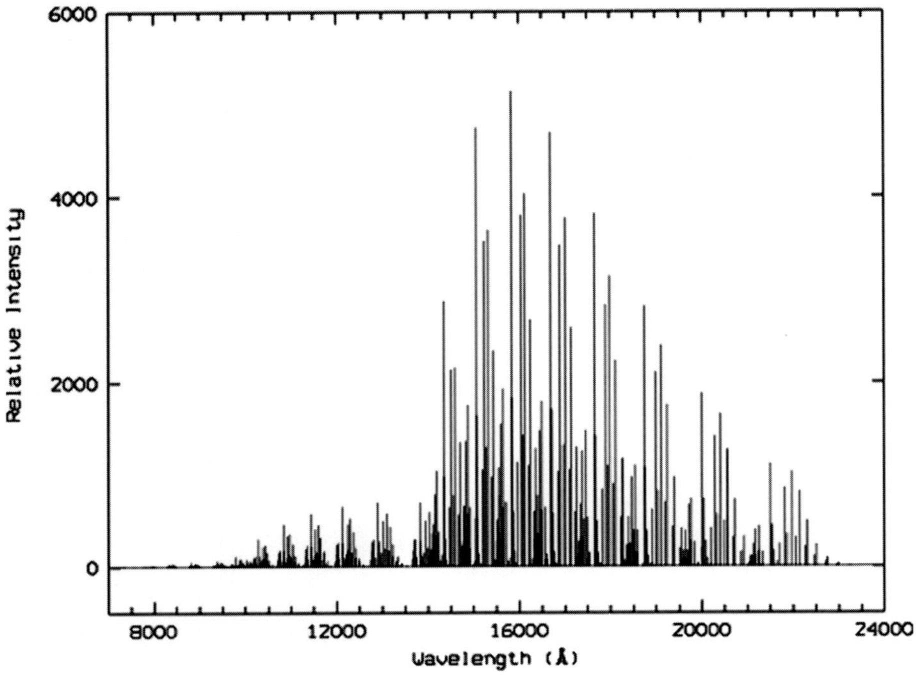

FIGURE 2.9. A synthetic spectrum of the night-sky OH emission in the near-IR wavelength range. From Rousselot *et al.* (2000). The relative intensity of the lines is proportional to the photon flux.

2.3.4 *Planning for a survey*

Regardless of the chosen methodology, emission-line surveys can be either targeted at sources whose redshifts are known to lie within a relatively small interval around a pre-established value, or blind searches for sources whose redshifts are expected to be distributed over an extended range. The difference between these two cases is important, because information on the sources' redshifts often drives or dictates the design of the experimental configuration, which in turn determines the sensitivity and/or the efficiency of the survey. As always, the design of the observations should be done only after the scientific questions to answer have been defined clearly and the trade-offs between often competing options thoroughly analyzed.

Examples of targeted surveys are searches for Lyα emission from damped Lyα-absorption systems observed in the spectra of distant quasi-stellar objects (QSOs) or searches for galaxies with strong emission lines, such as Lyα, [O II] or Hα, at the redshift of some known cluster or concentration of galaxies. In the first case a frequently chosen strategy is to obtain deep spectra covering the wavelength of the broad damped absorption feature with the spectrograph's slit centered on the QSO. This technique optimizes sensitivity, since it minimizes the background noise and typically fully covers the redshift interval whence the emission is expected to come. However, it covers only the volume of space subtended by the slit aperture, which includes the portion of the trough that causes the absorption observed along the line of sight to the QSO and the adjacent regions along the slit. Volume coverage can be increased by performing slitless spectroscopy, but the cost is a substantial loss in sensitivity (for a given exposure time). Since the redshift of the source is in a relatively small interval around a known value, narrow-band imaging is also an option. Unfortunately, the quasar prevents one from detecting sources that are closer to it than a few times the size of the point-spread function, depending on the apparent magnitude of the quasar.

Examples of blind surveys include the search for sources supposedly distributed over a large, unknown redshift range, such as distant galaxies at high redshift. Narrow-band imaging is often adopted in these cases, although both slit and slitless spectroscopy have been used as well. Imaging has the advantage of the simultaneous detection of a large number of sources, but covers only the limited redshift range allowed by the narrow-band filter. Spectroscopy generally allows one to probe a significantly larger redshift range but rather limited sky area, unless slitless spectroscopy is adopted (this, however, is penalized by rather low sensitivity).

2.4. Lyman-α surveys

2.4.1 *The search for galaxies at very high redshifts*

The Lyα emission line can be a strong feature in the far-UV spectrum of star-forming galaxies, when not obscured by dust, with equivalent width up to $E_{\rm w} \sim 150\,\text{Å}$ (see the discussion in Charlot & Fall 1993) or even larger in very young systems with top-heavy initial mass functions (IMFs) (e.g. Rhoads & Malhotra 2003). The Lyα emission in AGNs is typically stronger than that in star-forming galaxies, with equivalent width up to a few times 10^3 Å. In both cases, the line luminosities of both classes of sources, for redshifts up to $z \sim 7$, are within the sensitivity capabilities of modern telescopes and instrumentation. To illustrate this, let's consider the case of a star-forming galaxy with no dust obscuration, for which the expected luminosity density of the UV continuum at $\lambda = 1216$ Å is

$$L_{\rm UV} = \kappa \times 8.0 \times 10^{27} \times \frac{\rm SFR}{M_\odot\,{\rm yr}^{-1}}\ {\rm erg\,s^{-1}\,Hz^{-1}}$$
$$= \kappa \times 1.1 \times 10^{41} \times \frac{\rm SFR}{M_\odot\,{\rm yr}^{-1}}\ {\rm erg\,s^{-1}\,\text{Å}^{-1}},$$

using a Salpeter IMF (Madau *et al.* 1998), and a κ-correction $\kappa = 0.8$ to convert the luminosity density from 1500 Å to the wavelength of the Lyα line. Let's assume that the galaxy is forming stars at an average rate of $100{\rm M}_\odot\ {\rm yr}^{-1}$ and that there is no obscuration by dust. If such a rate is protracted over a period of time of the order of $\sim 10^8$ yr, the galaxy can assemble a Milky Way's worth of stars, of the order of $10^{10}\,{\rm M}_\odot$ (these numbers are in the range of what is observed for Lyman-break galaxies, see Giavalisco (2002)). Inserting these numbers into the formula above gives the luminosity density of the continuum, namely $L_{\rm UV} = 0.89 \times 10^{43}\,{\rm erg\,s^{-1}\,\text{Å}^{-1}}$. If the galaxy is placed at $z = 4$, the observed flux density (at the observed wavelength $\lambda_{\rm o} = (1+z)1216 = 6080\,\text{Å}$, approximately the central wavelength of the F606W filter in the ACS instrument on board the HST) is $f_{\rm UV} = 2.6 \times 10^{-29}\,{\rm erg\,s^{-1}\,cm^{-2}\,Hz^{-1}}$, corresponding to an AB magnitude of 22.9. Assuming that the Lyα line has an equivalent width $E_{\rm w} = 150$ Å, the line luminosity is $L_{\rm Ly\alpha} = 1.6 \times 10^{44}$ erg s^{-1}, which at the redshift of the galaxy corresponds to a flux $F_{\rm Ly\alpha} = 1.04 \times 10^{-15}$ erg s^{-1} cm^{-2}. Finally, assuming that the source is observed through a narrow-band filter with FWHM 100 Å centered on the line's redshifted wavelength (and that the filter bandpass is larger than the line width so that the whole flux is recorded), its narrow-band AB magnitude is ${\rm AB}_{\rm Ly\alpha} = 21.14$, namely 1.8 magnitudes brighter than the continuum.

Partridge & Peebles (1967a, b) were the first to discuss the expected far-UV spectral-energy distribution of star-forming galaxies and the general methodology of how to exploit these features, such as the 1216-Å Lyα line and the 912-Å Lyman limit, as observational flag poles to identify them at very high redshift. The first observational searches following these pioneering theoretical works were based on the continuum emission. Davis & Wilkinson (1974) and Partridge (1974) used photoelectric and photographic BVR photometry to look for red, extended sources with low surface brightness as primeval galaxy candidates (this was the first time that such a terminology was used), reporting no identifications. The methodology is qualitatively similar to the color selection used today for the Lyman-break technique (Giavalisco 2002). Indeed, both papers explicitly mention the possibility of a Lyman break in the spectrum of the galaxies caused by stellar atmosphere as a possible additional identifying feature. Koo & Kron (1980) used CCD and photographic slitless spectroscopy to identify sources at redshift $z \sim 5$ by means of the Lyman break or the Lyα emission. They also reported no detections. It is interesting that after these early, unsuccessful attempts to identify the "primeval galaxies," which essentially developed and implemented the idea behind the Lyman-break technique, the leading observational strategy to identify these sources focused mainly on the search for redshifted Lyα emission.

The first galaxies discovered at significant redshifts thanks to their Lyα emission lines were radio galaxies at $1.5 \lesssim z \lesssim 1.8$ (McCarthy 1993). Candidates were identified because of their strong radio luminosity and shape of the radio SED (steep radio sources). Redshift identifications were made with spectroscopy in the optical window, to which the UV is red-shifted (Spinrad & Djorgovski 1984a, b; Spinrad *et al.* 1985). Strictly speaking, therefore, these were both radio and Lyα surveys.

Djorgovski *et al.* (1985) were the first to use narrow-band imaging to search for Lyα at high redshifts. These authors targeted fields around known distant QSOs, looking for companion galaxies at the same redshift as the quasar. The idea behind this technique is that, since galaxies are clustered in space, the probability of detection is enhanced in the volumes nearby a known high-redshift galaxy, in this case the host to the QSO. These observations indeed yielded a successful detection of a Lyα-emitting source at $z = 3.215$ around the quasar PKS1614+051. Another detection of a Lyα emitter, at $z = 3.27$ around the gravitational lens MG 2016+112, was reported a year later by Schneider *et al.* (1986). In the following years a few more similar detections were reported by other groups (Steidel *et al.* 1991; Hu & Cowie 1987). However, authors of additional searches around 26 other QSOs (Djorgovski *et al.* 1987; Hu & Cowie 1987) reported negative results, showing that those early observations were sensitive only to the brightest, and thus rarest, objects of their type. The same strategy has also been adopted to look for companions of QSO damped-Lyα-absorption (DLA) systems at high redshifts, which are thought to be caused by protogalactic disks (Wolfe *et al.* 1993, 2000a, b), with both successful detections (Giavalisco *et al.* 1993; Moller & Warren 1993; Francis *et al.* 1996; Djorgovski *et al.* 1996; Warren & Moller 1996; Fynbo *et al.* 1999) and negative (Foltz *et al.* 1986; Smith *et al.* 1989; Wolfe 1986; Deharveng *et al.* 1990) or contradictory (Hunstead *et al.* 1990; Wolfe *et al.* 1992) results having been reported. It has recently been suggested that DLA systems are very likely not representative of the general population of star-forming galaxies at high redshifts (Wild *et al.* 2006).

The search for a population of Lyα-emitting star-forming galaxies at high redshifts in the field, namely conducted in inconspicuous "blank" regions of the sky as opposed to the "targeted" search described above, has been generally unsuccessful until relatively recently. Early surveys included searches in the redshift interval $2 \lesssim z \lesssim 6$ by means of narrow-band and long-slit spectroscopy (Pritchet and Hartwick 1990; Lowenthal *et al.* 1990; Djorgovski & Thompson 1992; De Propris *et al.* 1993; Thompson & Djorgovski 1995; Thompson *et al.* 1995), with essentially no detections.

Major progress in the identification of very distant galaxies by means of Lyα-emission-line surveys had to wait until the availability of the 8-meter-class telescopes and the sensitivity and large area of modern CCD arrays. Modern surveys started in the mid 1990s and continue today, with deep optical narrow-band imaging routinely returning Lyα emitters at high redshifts in large numbers, both around QSOs and in the field (Hu & McMahon 1996; Hu *et al.* 1996, 1998; Cowie & Hu 1998; Steidel *et al.* 2000; Kudritzki *et al.* 2000; Rhoads *et al.* 2000; Rhoads & Malhotra 2001; Malhotra & Rhoads 2002; Stiavelli *et al.* 2002; Rhoads, 2003; Taniguchi, 2003; Hu *et al.* 2004). The combination of the Subaru telescope with the large-area, prime-focus camera Supreme has been particularly successful at redshift $z > 4$ (Ouchi *et al.* 2003; Shimasaku *et al.* 2003, 2004; Nagao *et al.* 2004; Matsuda *et al.* 2005; Taniguchi *et al.* 2005; Kashikawa *et al.* 2006a, b; Shimasaku *et al.* 2006; Ajiki *et al.* 2006), with detections of Lyα-emitters up to redshifts $z \sim 7$, the current record holder being located at $z = 6.69$ (Iye *et al.* 2006).

Because of their double sensitivity thresholds of flux level and equivalent width (only galaxies with a relatively large equivalent width can be detected in these surveys, no matter how bright the continuum) and the very narrow redshift range that each filter can probe at a given time, these surveys are generally less efficient than surveys based on continuum emission at recovering the bright end of the luminosity function. However, they nicely complement the continuum surveys in that they can reach sources with much fainter continuum luminosity, which is particularly useful as a means to constrain the faint end of the luminosity function, which would not otherwise be accessible to continuum-based observations. They are also very effective for studying large concentrations of galaxies at the same redshifts, e.g. high-redshift clusters, as we will review in the next section.

One important caveat to keep in mind is that, while galaxies identified by means of their emission line are very useful for exploring fainter continuum flux levels than continuum-based observations – in fact, the method often provides the *only* way to effectively identify very faint galaxies, albeit with strong line emission – it is in general difficult to draw general conclusions based on such samples. This is because emission-line sources are a subset of the general populations, and in general one does not know to what degree they are representative of the population as a whole. An example is that of the faint end of the luminosity function of Lyman-break galaxies. While it is true that both Lyman-break galaxies and Lyα emitters at the same redshifts share similarities (they are both classes of star-forming galaxies characterized by relatively low dust obscuration), Lyα emitters are a subset of Lyman-break galaxies whose equivalent width is larger than $\approx$20 Å (at the current sensitivity level). It is not known whether the luminosity function of this subset is representative of the luminosity function of Lyman-break galaxies as a whole. It is important to study how the Lyα properties of Lyman-break galaxies depend on the continuum luminosity in order to identify statistical trends (e.g. Steidel *et al.* 2000; Shapley *et al.* 2003; Ando *et al.* 2006).

2.4.2 *Clusters and clustering from line-emission surveys*

One great advantage of selecting line-emitting galaxies from narrow-band images is that they are all distributed within a very narrow range of redshifts, opening up the possibility of studying spatial clustering and structures at very high redshifts. Of course, when carrying out these surveys, unless spectroscopic redshifts are obtained for every candidate, there is always the distinct possibility that the samples contain a significant number of sources at a redshift different from the targeted one. For example, narrow-band imaging targeting Lyα emitters at $z = 3.50$ can also select [O II] emitters at redshift $z = 0.47$ or Hβ emitters at $z = 0.13$. In practice, however, with a relatively modest amount of additional data (the acquisition of which is not as time-consuming as a complete spectroscopic follow-up) it is possible to minimize, or eliminate, contamination by those interlopers. For example, in the case of the searches for $z \sim 3.5$ Lyα emitters described above, U-band or B-band (depending on the exact redshift) imaging can be used to cull the high-redshift candidates from the low-redshift ones, because the former will be very faint or undetected in the U-band, which at $z = 3.5$ samples below the Lyman limit, while the latter will remain relatively bright. This strategy is illustrated in Figure 2.2.

Modern observations started after the discovery that the redshift distribution of star-forming galaxies at high redshifts selected from the colors of the UV continuum (Lyman-break galaxies) is characterized by large fluctuations over very narrow redshift intervals (Steidel *et al.* 1998). These under-dense and over-dense regions (spikes) are a manifestation of spatial clustering (Giavalisco *et al.* 1998; Adelberger *et al.* 1998) and contain information on the growth of structures, including the formation of clusters and superclusters, as well as on the topology of the large-scale structure.

The distribution of the Lyα's equivalent width of Lyman-break galaxies, which are selected regardless of their Lyα properties, is such that $\approx$30% of them can be detected at the current sensitivity (Shapley *et al.* 2001, 2003). Furthermore, the redshift range probed by the narrow-band filters commonly used in these surveys is well matched to the width in the redshift space of the observed structures, making emission-line surveys a very effective tool for studying the evolution of clusters and clustering at high redshift.

Steidel *et al.* (1999) used narrow-band imaging to map the distribution of Lyα emitters in a large over-density of Lyman-break galaxies at $z \sim 3.01$ to study its morphology, concluding that such a concentration is likely a protocluster. Hayashino *et al.* (2003) surveyed a much wider area around this structure, finding that the protocluster is part of a larger structure that extends up to a factor of $\sim$60 in linear size. This provides evidence of large-scale structure at $\sim$20% of the cosmic age with size and shape similar to that observed today. Matsuda *et al.* (2005) also reported evidence of filamentary structure within the same system. Shimasaku *et al.* (2003, 2004) and Ouchi *et al.* (2004) find evidence of strong spatial clustering of Lyα emitters at $z \sim 5$ and Ouchi *et al.* (2005) report similar evidence at redshift $z \sim 6$, namely at only $\sim$7% of the cosmic age.

Since they were recognized as the potential sites of protoclusters, the fields around high-redshift radio galaxies and radio-loud quasars have also been targeted by line-emission surveys looking for the redshifted Lyα (or Hα) emission lines either from the cluster members or from adjacent cosmic structures. Such surveys have successfully detected concentrations of galaxies at the same redshifts as the target AGNs and with properties consistent with their being the members of protoclusters (Miley *et al.* 2006; Overzier *et al.* 2006; van Breugel *et al.* 2006; Zirm *et al.* 2005; Venemans *et al.* 2002, 2003, 2004, 2005; Kurk *et al.* 2004a, b, c; Pentericci *et al.* 2000).

2.4.3 *Lyman-α surveys as probes of the cosmic reionization*

The evolution of the Lyα luminosity function of star-forming galaxies as a function of redshift, or the lack of it, can be effectively used to set constraints on the end of the cosmic reionization. This methodology is an interesting alternative to the traditional one based on distant quasars as probes for the presence of the Gunn–Peterson trough (Becker *et al.* 2001; Fan *et al.* 2002), since Lyα-emitting galaxies are now routinely identified at redshifts well in excess of that of the most distant quasars known. It also probes a different regime of the neutral fraction, ∼0.1 as opposed to ∼0.01, which is probed by the quasar absorption systems. The idea of the test is as follows (Malhotra & Rhoads 2004; Stern *et al.* 2005). Since Lyα is a resonance line, the Lyα photons are resonantly scattered by neutral hydrogen in the intergalactic medium (IGM). As a consequence, the Lyα photons emitted from a source embedded in a neutral IGM are expected to be strongly attenuated or even completely suppressed by the red damping wing of the Gunn–Peterson trough (Miralda-Escude & Rees 1998; Furlanetto *et al.* 2004). The transmitted fraction depends on the size of the H II region that surrounds the source, which in turn is a function of the luminosity and age of the source, and on the contribution from other adjacent, clustered sources (Santos 2004; Wyithe & Loeb 2004). Malhotra & Rhoads (2004) estimate an average attenuation by a factor of ≈3 in all Lyα fluxes if the sources are embedded in a neutral IGM compared with the case of an ionized one. This, in turn, would correspond to a rapid apparent evolution of the luminosity function of the Lyα emitters as the observations probe deeper and deeper into the end of the reionization epoch, if no evolution of the astrophysical properties of the sources occurs over the same time. Surveys conducted at $z = 5.7$ and $z = 6.5$ so far have not given definitive results. Malhotra & Rhoads (2004) found no measurable difference between the Lyα emitter luminosity functions at the two redshifts, concluding that the reionization was largely completed by $z \sim 6.5$. Kashikawa *et al.* (2006) found the luminosity function of Lyα emitters at $z \sim 5$ to be fainter, at the 2σ level, than that at the lower redshift, concluding that the cosmic reionization was likely still going on at $z \sim 6.5$. Clearly, large samples and higher redshifts need to be secured if we are to resolve this.

2.4.4 *Serendipitous discoveries*

Surveys for Lyα emission from high-redshift galaxies have also yielded serendipitous discoveries, such as groups or clusters of AGNs at high redshifts (Keel *et al.* 1999) and the Lyα "blobs." The latter, in particular, have received a great amount of attention recently, since their physical nature remains elusive. The Lyα "blobs" were discovered during searches for Lyα emitters associated either with QSO damped-Lyα absorbers (Fymbo *et al.* 1999) or protoclusters of Lyman-break galaxies (Steidel *et al.* 2000). They appear in deep narrow-band images as Lyα nebulae whose line emission is very diffuse and extends over a large spatial extent, $\gtrsim$100 kpc. The line luminosity is very bright, of the order of 10^{44}–10^{45} erg s^{-1}, about 20–40 times brighter than the Lyα luminosity of Lyman-break galaxies or even Lyα emitters at similar redshift. The Lyα equivalent width is also very large, well in excess of 100 Å. They bear a close resemblance to the extended Lyα emission often associated with powerful high-redshift radio galaxies, although the blobs' radio flux is at least about two orders of magnitude fainter than that of radiogalaxies at the same redshift. Further discoveries of blobs confirmed initial reports that they often have no obvious UV sources associated with them that could power the Lyα emission (Matsuda *et al.* 2004; Nilsson *et al.* 2006; Matsuda *et al.* 2006), although some seem to be associated with very red sources, namely very dusty starbursts, AGNs or old stellar systems (Steidel *et al.* 2000; Geach *et al.* 2007). Three physical mechanisms have been proposed to explain the observed phenomenology of the blobs, including photoionization from a buried AGN (Steidel *et al.* 2000; Haiman & Rees 2001; Chapman *et al.* 2004; Geach *et al.* 2007); superwinds associated with star formation (Taniguchi & Shioya 2000; Matsuda *et al.* 2006; Geach *et al.* 2007) and cold accretion of gas on dark-matter halos (Steidel *et al.* 2000; Haiman *et al.* 2000; Nilsson *et al.* 2006). While all of these mechanisms can certainly explain individual properties of the blobs, and in some cases even all of them, it is still not known which one is the most important.

2.5. Surveys for other emission lines

The Lyα emission line has a number of shortcomings that can preclude or limit its use in some types of surveys. Firstly, it is a UV line and, if not redshifted into the optical window or at near-IR wavelengths, it can be observed only from space, which imposes limits on the available telescope instrumentation. Secondly, it is a resonance line, since it is created by the transition from the first excited level to the ground level of the neutral hydrogen atom. This means that, once created, the Lyα photon can be absorbed and re-emitted multiple times by neutral hydrogen, effectively amplifying the optical path of the Lyα photons in the IGM. This resonant scattering can selectively destroy, or at least severely attenuate, the Lyα photons relative to the adjacent continuum if even modest amounts of gas are present in the IGM (Neufeld 1991; Charlot & Fall 1991, 1993; Laursen & Larsen 2006), although the geometry and clumpiness of the gas also plays a crucial role in determining the number of photons that can escape the source, and hence the observed Lyα equivalent width (see Laursen & Sommer-Larsen 2006 and references therein). This means that any statistical conclusion based on the properties of the Lyα emission will be biased in a way that is often difficult to model. Thirdly, because dust extinction is generally more severe at shorter wavelengths, the line is more easily obscured than are other lines at longer wavelengths, such as Hβ and Hα, and often is observed to be weaker than they are, even if its unobscured luminosity is higher.

Such problems do not affect, or are significantly less severe for, the Balmer lines and other optical nebular lines such as [O II] and [O III] that are typical of the spectra of star-forming galaxies and AGNs. Among the brightest of these features in star-forming galaxies is the Hα line, which has indeed been used in a variety of emission-line surveys, both in the local and in the distant Universe. The Hα line is a good tracer of star-formation activity and can be calibrated onto an absolute scale of star-formation rate, given some assumption on the IMF and some estimate of the dust obscuration (Kennicutt 1998).

In the local universe, where Hα is observed near its rest-frame wavelength $\lambda_{\mathrm{H}\alpha} = 6563\,\text{Å}$, the line has been used to identify a diverse class of sources, including ionized gas in the cores of globular clusters (Smith *et al.* 1976) and planetary nebulae both in our Galaxy (Rauch 1999) and in other ones (Okamura *et al.* 2002; Arnaboldi *et al.* 2003), using narrow-band imaging, and star-forming galaxies (Zamorano *et al.* 1994, 1996; Alonso *et al.* 1995; Gallego *et al.* 1995; Bennet & Moss 1998; McCarthy *et al.* 1999) with objective-prism spectroscopy.

At larger redshifts the line is shifted to spectral regions where the sky background becomes progressively brighter, making the observations more challenging. At $z \gtrsim 0.37$ the Hα line is observed at wavelengths longer than $\lambda \gtrsim 0.9\,\mu\text{m}$, requiring observations in the near-IR window (up to $z \sim 18$). Compared with optical observations, this means not only significantly lower sensitivity due to the much-brighter and less-stable sky, but also lower survey efficiency, because of the generally smaller area coverage of the imagers, since near-IR detector arrays are smaller than optical ones. Early near-IR searches for high-redshift Hα emitters did not provide a higher rate of detections than that of their optical, Lyα-based counterparts (Thompson *et al.* 1994, 1996; Pahre & Djorgovski, 1995; Teplitz *et al.* 1998), and only a small number of candidates has been selected, a few of which have been confirmed at high redshifts (Teplitz *et al.* 1999; Beckwith *et al.* 1998). Surveys with 8-meter-class telescopes have started very recently, and are currently being undertaken (e.g. Willis & Courbin 2005; Cuby *et al.* 2007). No detections have been reported yet, although the area surveyed so far is still small. Space-borne surveys with the HST and NICMOS grism have a significantly higher detection rate, because of the improved sensitivity afforded by the much lower sky background (McCarthy *et al.* 1999; Yan *et al.* 2000; Drozdovsky *et al.* 2005), but the area coverage of these surveys is relatively small (the available number of pixels is primarily used to take advantage of the high angular resolution of the HST).

The nebular lines of [O II] and [O III] have also been used to survey star formation at high redshifts, in the field and in clusters, with similar techniques, both from the ground and using the HST (e.g. Martin *et al.* 2000; Biretta *et al.* 2005; Drozdovsky *et al.* 2005).

The [O III] line has also been exploited to search for and/or study planetary nebulae (PNs), both those in the Milky Way and extragalactic ones, using both narrow-band imaging and spectroscopy, with and without slits (Jacoby *et al.* 1992; Mendez *et al.* 2001; Stanghellini *et al.* 2002, 2003; Okamura *et al.* 2002; Douglas *et al.* 2002; Arnaboldi *et al.* 2003; Gerhard *et al.* 2005; Merrett *et al.* 2006). While searching for PNs in the Virgo cluster with narrow-band imaging, Kudritzki *et al.* (2000) serendipitously discovered nine Lyα emitters at redshift $z \sim 3.1$, one of the first relatively large samples of such sources obtained with the 8-meter-class telescopes.

Finally, for completeness, we conclude this overview of technical and observational aspects and results of emission-line surveys with a brief mention of a number of works that exploited objective-prism imaging to identify a number of galactic and extragalactic types of sources. References can be found in these works for additional information on the observations and other, similar works.

Weaver & Babcock (2004) carried out a very deep survey for classical T Tauri stars using the CTIO Curtis–Schmidt objective prism, identifying 63 Hα-emitting stars. MacConnell (2003) identified a sample of southern cool carbon stars in near-IR objective-prism plates.

Rawlings *et al.* (2000) searched for highly reddened early-type stars with optical photometry and spectroscopy of stars in the Stephenson Objective Prism Survey. Stock *et al.* (1998) studied high-velocity stars with an objective-prism spectroscopic survey.

Iovino *et al.* (1996) identified a large sample of quasar candidates from an objective-prism survey. Jakobsen *et al.* (1993) studied a sample of 12 quasars at redshifts $z > 3$ in the far UV using the the objective prism on the Faint Object Camera on the HST to identify candidate HE II λ304 absorbers along the line of sight.

Ran & Chen (1993) conducted an objective-prism survey of QSOs. Bade *et al.* (1992) identified X-ray-luminous AGNs with objective-prism spectroscopy of ROSAT fields.

Hazard *et al.* (1986) identified high-redshift QSOs selected from IIIa-J objective-prism plates of the UK Schmidt Telescope. Webster (1982) identified a large sample of QSOs from an objective-prism survey in order to study their spatial clustering. Parker *et al.* (1987) reported the detection of features in the large-scale distribution of galaxies from wide-angle samples of objective-prism spectra.

REFERENCES

Adelberger, K. L., Steidel, C. C., Giavalisco, M., Dickinson, M., Pettini, M., Kellogg, M. 1998, ApJ 505, 18

Ajiki, M., Mobasher, B., Taniguchi, Y., Shioya, Y., Nagao, T., Murayama, T., Sasaki, S. 2006, ApJ, 638, 596

Alonso, O., Zamorano, J., Rego, M., Gallego, J. 1995, A&AS, 113, 399

Ando, M., Ohta, K., Iwata, I., Akiyama, M., Aoki, K., Tamura, N. 2006, ApJ, 645, 9

Arnaboldi, M. *et al.* 2003, AJ, 125, 514

Becker, R. *et al.* 2001, AJ, 122, 2850

Beckwith, S. V., Thompson, D., Mannucci, F., Djorgovski, S. G. 1998, AJ, 116, 1591

Bennett, S., Moss, C. 1998, A&AS, 132, 55

Biretta, J. A., Martel, A. R., McMaster, M., Sparks, W. B., Baum, S. A., Macchetto, F., McCarthy, P. J. 2002, NEWAR, 46, 181

Chapman, S. C., Scott, D., Windhorst, R. A., Frayer, D. T., Borys, C., Lewis, G. F., Ivison, R. J. 2004, ApJ, 606, 85

Cowie, L. L., Hu, E. M. 1998, AJ, 115, 1319

Crampton, D. 1999, JRASC, 93, 168

Cuby, J.-G., Hibon, P., Lidman, C., Le Fevre, O., Gilmozzi, R., Moorwood, A., van der Werf, P. 2007, A&A, 461, 911

Davis, M., Wilkinson, D. T. 1974, ApJ, 192, 251

Djorgovski, S., Strauss, M. A., Perley, R. A., Spinrad, H., McCarthy, P. 1985, AJ, 93, 1318

Djorgovski, S. G., Pahre, M. A., Bechtold J., Elston, R. 1996, Nature, 382, 234

Douglas, N. G. *et al.* 2002, PASP, 114, 1234

Drozdovsky, I., Yan, L., Chen, H.-W., Stern, D., Kennicutt, R., Spinrad, H., Dawson, S. 2005, AJ, 130, 1324

Fan, X. *et al.* 2002, AJ, 123, 1247

Francis, P. *et al.* 1996, ApJ, 457, 490

Furlanetto, S., Hernquist, L., Zaldarriaga, M. 2004, MNRAS, 354, 695

Fynbo, J. U., Moller, P., Warren, S. J. 1999, MNRAS, 305, 849

Gallego, J., Zamorano, J., Aragon-Salamanca, A., Rego, M. 1995, ApJ, 455, L1

Geach, J. E., Smail, I., Chapman, S. C., Alexander, D. M., Blain, A. W., Stott, J. P., Ivison, R. J 2007, ApJ, 655, 9

Gerhard, O., Arnaboldi, M., Freeman, K., Kashikawa, N., Okamura, S., Yasuda, N. 2005, ApJ, 621, 93

Giavalisco, M. 2002, ARA&A, 40, 579

Giavalisco, M., Macchetto, F. D., Sparks, W. B. 1994, A&A, 288, 103

Giavalisco, M., Steidel, C. C., Adelberger, K. L., Dickinson, M. E., Pettini, M., Kellogg, M. 1998, ApJ, 503, 543

Haiman, Z., Rees, M. 2001, ApJ, 556, 87

Hazard, C., Morton, D., McMahon, R., Sargent, W., Terlevich, R. 1986, MNRAS, 223, 87

Heiman, Z., Sppans, M., Quataert, E. 2000, ApJ, 537, 5

Hu, E., Cowie, L. L. 1987, ApJ, 317, L7

Hu, E. M., Cowie, L. L., Capak, P., McMahon, R. G., Hayashino, T., Komiyama, Y. 2004, AJ, 127, 563

Hu, E. M., Cowie, L. L., McMahon, R. G. 1998, ApJ, 502, 99

Hu, E. M., McMahon, R. G. Egami, E. 1996, ApJ, 459, 53

Hu, E. M., McMahon, R. G. 1996, Nature, 382, 281

Iovino, A., Clowes, R., Shaver, P. 1996, A&AS, 119, 265

Iye, M., Ota, K., Kashikawa, N., Furusawa, H., Hashimoto, T., Hattori, T., Matsuda, Y., Morokuma, T., Ouchi, M., Shimasaku, K. 2006, Nature, 443, 186

Jacoby, G., Branch, D., Ciardullo, R., Davies, R., Harris, W., Pierce, M., Pritchet, C., Tonry, J., Welch, D. 1992, PASP, 104, 599

Jakobsen, P. *et al.* 1993, ApJ, 417, 528

Kashikawa, N. *et al.* 2006a, ApJ, 648, 7

Kashikawa, N. *et al.* 2006b, ApJ, 637, 631

Keel, W. C., Cohen, S. H., Windhorst, R. A., Waddington, I. 1999, AJ, 118, 2547

Koo, D. C., Kron, R. G. 1980, PASP, 545, 537

Kudritzki, R. P., Mendez, R. H., Feldmeier, J. J, Ciardulo, R., Jacoby, G. H., Freeman, K. C., Arnaboldi, M., Capaccioli, M., Gerhard, O., Ford, H. C. 2000, ApJ, 536, 19

Kurk, J. D., Pentericci, L., Overzier, R. A., Röttgering, H. J. A., Miley, G. K. 2004a, A&A, 428, 81

Kurk, J. D., Pentericci, L., Röttgering, H. J. A., Miley, G. K. 2004b, A&A, 428, 793

Kurk, J. D., Cimatti, A., di Serego Alighieri, S., Vernet, J., Daddi, E., Ferrara, A., Ciardi, B. 2004c, A&A, 422, 13

Larson, R. B. 1974 MNRAS, 166, 585

Loveday, J., Maddox, S. J., Efstathiou, G., Peterson, B. A. 1995, ApJ, 442, 457

Lowenthal, J. D., Koo, D. C., Guzman, R., Gallego, J., Phillips, A. C., Faber, S. M., Vogt, N. P., Illingworth, G. D., Gronwall, C. 1997, ApJ, 481, 673

MacConnell, J. 2003, PASP, 115, 351

Madau, P., Pozzetti, L., Dickinson, M. E. 1998, ApJ, 498, 106

Malhotra, S., Rhoads, J. 2004, ApJ, 617, 5

Martin, C., Sawicki, M. 2004, ApJ, 603, 41

Martin, C., Lotz, J., Ferguson, H. 2000, ApJ, 543, 97

Matsuda, Y., Yamada, T., Hayashino, T., Yamauchi, R., Nakamura, Y. 2006, ApJ, 640, 123

Matsuda, Y. *et al.* 2004, AJ, 128, 569

Matsuda, Y. *et al.* 2005, ApJ, 634, 125

McCarthy, P. *et al.* 1999, ApJ, 520, 548

McCarthy, P.J. 1993, ARA&A, 31, 639

Méndez, R. H., Riffeser, A., Kudritzki, R.-P., Matthias, M., Freeman, K. C., Arnaboldi, M., Capaccioli, M., Gerhard, O. E. 2001, ApJ, 563, 135

Merrett, H. R. *et al.* 2006, MNRAS, 369, 120

Miley, G. *et al.* 2006, ApJ, 650, 29

Miralda-Escude, J., Rees, M. 1998, ApJ, 497, 21

Nagao, T. *et al.* 2004, ApJ, 613, 9

Nilsson, K., Fynbo, J., Moller, P., Sommer-Larsen, J., Ledoux, C. 2006, A&A, 452, 23

Okamura, S. *et al.* 2002, PASJ, 54, 8830

Oke, J. B., Gunn, J. E. 1983, ApJ, 266, 713

Ouchi, M. *et al.* 2003, ApJ, 582, 600

Overzier, R. *et al.* 2006, ApJ, 637, 580

Parker, Q. A., Beard, S. M., MacGillivray, H. T., 1987, A&A, 173, 5

Pentericci, L., Kurk, J. D., Röttgering, H. J. A., Miley, G. K., van Breugel, W., Carilli, C. L., Ford, H., Heckman, T., McCarthy, P., Moorwood, A. 2000, A&A, 361, 25

Ran, H., Chen, H. S. 1993, Ap&SS, 200, 279

Rawlings, M.G. *et al.* 2000, ApJS, 131, 531

Rhoads, J., Dey, A., Malhotra, S., Stern, D., Spinrad, H., Jannuzi, B., Dawson, S., Brown, Michael J., Landes, E. 2003, AJ, 125, 1006

Rhoads, J., Malhotra, S. 2001, ApJ, 563

Rhoads, J., Malhotra, S., Dey, A., Stern, D., Spinrad, H., Jannuzi, B. T. 2000, ApJ, 545, 85

Rousselot, P., Lidman, C., Cuby, J.-G., Moreels, G., Monnet, G. 2000, A&A, 354, 1134

Santos, M. 2004, MNRAS, 349, 1137

Sawicki, M., Yee, H. K. C. 1998, AJ, 115, 1329

Schneider, D., Gunn, J., Turner, E., Lawrence, C., Hewitt, J., Schmidt, M., Burke, B. 1986, AJ, 91, 991

Shapley, A. E., Steidel, C. C., Adelberger, K. L., Dickinson, M., Giavalisco, M., Pettini, M. 2001, ApJ, 562, 95

Shapley, A. E., Steidel, C. C., Pettini, M., Adelberger, K. L. 2003, ApJ, 588, 65

Shimasaku, K. *et al.* 2003, ApJ, 586, 111

Shimasaku, K. *et al.* 2006, PASJ, 58, 313

Shimasaku, K., Hayashino, T., Matsuda, Y., Ouchi, M., Ohta, K., Okamura, S., Tamura, H., Yamada, T., Yamauchi, R. 2004, ApJ, 605, 93

Spinrad, H., Djorgovski, S. 1984

Spinrad, H., Filippenko, A. V., Wyckoff, S., Stocke, J. T., Wagner, M. R., Lawrie, D. G. 1985, ApJ, 299, L7

Stanghellini, L., Shaw, R., Balick, B., Mutchler, M., Blades, J. C., Villaver, E. 2003, ApJ, 596, 997

Stanghellini, L., Shaw, R., Mutchler, M., Palen, S., Balick, B., Blades, J. C. 2002, ApJ, 575, 178

Steidel, C., Dickinson, M., Sargent, W. 1991, AJ, 101, 1187

Steidel, C. C., Adelberger, K. L., Dickinson, M. E., Giavalisco, M., Pettini, M., Kellogg, M. 1998, ApJ, 492, 428

Steidel, C. C., Adelberger, K. L., Giavalisco, M., Dickinson, M. E., Giavalisco, M. Pettini, M. 1999, ApJ, 519, 1

Steidel, C. C., Adelberger, K. L., Shapley, A. E., Pettini, M., Dickinson, M, Giavalisco, M. 2000, ApJ, 532, 170

Stern, D., Yost, S., Eckart, M., Harrison, F., Helfand, D., Djorgovski, S., Malhotra, S., Rhoads, J. 2005, ApJ, 619, 12

Stiavelli, M., Scarlatta, C. Panagia, N. Treu, T., Bertin, G., Bertola, F. 2001, ApJ, 561, 37

Stock, J., Rose, J., Agostinho, R., 1998, PASP, 110, 1434

Stockton, A. 1999, Ap&SS, 269, 209

Taniguchi, Y. *et al.* 2003, ApJ, 585, 97

Taniguchi, Y. *et al.* 2005, PASJ, 57, 165

Taniguchi, Y., Shioya, Y. 2000, ApJ, 532, 13

Teplitz, H. I., Malkan, M. A., McLean, I. S. 1999, ApJ, 514, 33

Teplitz, H. I., Malkan, M. A., McLean, I. S. 1998, ApJ, 506, 519

Thompson, D., Djorgovski, S. G. 1995, AJ, 110, 982

Thompson, D., Djorgovski, S., Beckwith, S. V. 1994, AJ, 107, 1

Thompson, D., Djorgovski, S., Trauger, J. 1995, AJ, 110, 963

Thompson, D., Mannucci, F., Beckwith, S. V. 1996, AJ, 112, 1794

Tran, K.-V., Lilly, S., Crampton, D., Brodwin, M. 2004, ApJ, 612, 89

Turnshek, D. A., Wolfe, A. M., Lanzetta, K. M., Briggs, F. H., Cohen, R. D., Foltz, C. B., Smith, H. E., Wilkes, B. J. 1989, ApJ, 344, 567

van Breugel, W., de Vries, W., Croft, S., De Breuck, C., Dopita, M., Miley, G., Reuland, M., Röttgering, H. 2006, AN, 327, 175

van Starkenburg, L., van der Werf, P. P., Yan, L., Moorwood, A. F. M. 2006, A&A, 450, 25

Venemans, B. P. *et al.* 2002a, A&A, 424, 17

Venemans, B. P., Kurk, J. D., Miley, G. K., Röttgering, H. J. A., van Breugel, W., Carilli, C. L., De Breuck, C., Ford, H., Heckman, T., McCarthy, P., Pentericci, L. 2002b, ApJ, 569, 11

Venemans, B. P. *et al.* 2005, A&A, 431, 793

Walsh, J. R. 1995, *Proceedings of the ESO/ST-ECF workshop "Calibrating and Understanding HST and ESO Instruments"*, 25–28 April 1995, Garching, ed. Piero Benvenuti, p. 27.

Warren, S. J., Moller, P. 1996, A&A, 311, 25

Weaver, W., Babcock, A. 2004, PASP, 116, 1035,

Webster, A. 1982, MNRAS, 199, 683

Weymann, R., Stern, D., Bunker, A., Spinrad, H., Chaffee, F., Thompson, R. I., Storrie-Lombardi, L. 1998, ApJ, 505, 95

Willis, J., Courbin, F. 2005, MNRAS, 357, 1348

Wyithe, S., Loeb, A. 2003, ApJ, 588, 69

Wolfe, A. M., Prochaska, J. X. 2000a, ApJ, 545, 603

Wolfe, A. M., Prochaska, J. X. 2000b, ApJ, 545, 591

Wolfe, A., Turnshek, D., Lanzetta, K., Lu, L. 1993, ApJ, 404, 480

Wolfe, A., Turnshek, D., Smith, H., Cohen, R. 1986, ApJS, 61, 249

Yan, L., McCarthy, P., Weymann, R., Malkan, M., Teplitz, H., Storrie-Lombardi, L., Smith, M., Dressler, A. 2000, AJ, 120, 575

Zamorano, J., Gallego, J., Rego, M., Vitores, A. G., Alonso, O. 1996, ApJS, 105, 343

Zamorano, J., Rego, M., Gallego, J. G., Vitores, A. G., Gonzalez-Riestra, R., Rodriguez-Caderot, G. 1994, ApJS, 95, 387

Zirm, A. *et al.* 2005, ApJ, 630, 68

3. The astrophysics of early galaxy formation

PIERO MADAU

3.1. Preamble

Hydrogen in the Universe recombined about half a million years after the Big Bang, and cooled down to a temperature of a few kelvins until the first non-linearities developed, and evolved into stars, galaxies, and black holes that lit up the Universe again. In currently popular cold-dark-matter flat cosmologies (ΛCDM), some time beyond a redshift of 10, the gas within halos with virial temperatures $T_{\rm vir} \geq 10^4$ K – or, equivalently, with masses $M \geq 10^8[(1+z)/10]^{-3/2} M_\odot$ – cooled rapidly due to the excitation of hydrogen Lyα and fragmented. Massive stars formed with some initial mass function (IMF), synthesized heavy elements, and exploded as Type II supernovae after a few times 10^7 yr, enriching the surrounding medium: these subgalactic stellar systems, aided perhaps by an early population of accreting black holes in their nuclei, generated the ultraviolet radiation and mechanical energy that contributed to the reheating and reionization of the cosmos. It is widely believed that collisional excitation of molecular hydrogen may have allowed gas in even smaller systems – with virial temperatures of 1000 K, corresponding to masses around $5 \times 10^5[(1+z)/10]^{-3/2} M_\odot$ – to cool and form stars at even earlier times (Couchman & Rees 1986; Haiman *et al.* 1996; Tegmark *et al.* 1997; Abel *et al.* 2000, 2002; Fuller & Couchman 2000; Bromm *et al.* 2002; Reed *et al.* 2005; Kuhlen & Madau 2005). Throughout the epoch of structure formation, the all-pervading intergalactic medium (IGM), which contains most of the ordinary baryonic material left over from the Big Bang, becomes clumpy under the influence of gravity, and acts as a sink for the gas that gets accreted, cools, and forms stars within subgalactic fragments and as a source for the metal-enriched material, energy, and radiation which they eject. The well-established existence of heavy elements like carbon and silicon in the Lyα-forest clouds at z =2–6 (Songaila 2001; Pettini *et al.* 2003; Ryan-Weber *et al.* 2006) may be indirect evidence for such an early episode of pregalactic star formation. The recently released Wilkinson Microwave Anisotropy Probe (WMAP) 3-year data require the Universe to be fully reionized by redshift $z_{\rm r} = 11 \pm 2.5$ (Spergel *et al.* 2006), another indication that significant star-formation activity started at very early cosmic times.

The last decade has witnessed great advances in our understanding of the high-redshift Universe, thanks to breakthroughs achieved with satellites, 8–10-m-class telescopes, and experiments on the cosmic microwave background (CMB). Large surveys such as the Sloan Digital Sky Survey (SDSS), together with the use of novel instruments and observational techniques, have led to the discovery of galaxies and quasars at redshifts in excess of 6. At the time of writing, nine quasars with $z > 6$ have already been found (Fan 2006), and one actively star-forming galaxy has been confirmed spectroscopically at $z = 6.96$ (Iye *et al.* 2006). These sources probe an epoch when the Universe was <7% of its current age. Keck and VLT observations of redshifted H I Lyα ("forest") absorption have been shown to be a sensitive probe of the distribution of gaseous matter in the Universe (Rauch 1998). Gamma-ray bursts have recently displayed their potential to replace quasars as the preferred probe of early star formation and chemical enrichment: GRB050904, the most-distant event known to date, is at $z = 6.39$ (Haislip *et al.* 2006).

The Emission-Line Universe, ed. J. Cepa. Published by Cambridge University Press.

The underlying goal of all these efforts is to understand the growth of cosmic structures, the properties of galaxies and their evolution, and, ultimately, to map the transition from the cosmic "dark age" to an ionized Universe populated with luminous sources.

Progress has been equally significant on the theoretical side. The key idea of currently popular cosmological scenarios, that primordial density fluctuations grow by gravitational instability driven by collisionless CDM, has been elaborated upon and explored in detail through large-scale numerical simulations on supercomputers, leading to a hierarchical ("bottom-up") scenario of structure formation. In this model, the first objects to form are on subgalactic scales, and merge to make progressively bigger structures ("hierarchical clustering"). Ordinary matter in the Universe follows the dynamics dictated by the dark matter until radiative, hydrodynamic, and star-formation processes take over. According to these calculations, a truly intergalactic and protogalactic medium (the main repository of baryons at high redshift) collapses under the influence of dark-matter gravity into flattened and filamentary structures, which are seen in absorption against background quasi-stellar objects (QSOs). Gas condensation in the first baryonic objects is possible through the formation of H_2 molecules, which cool via roto-vibrational transitions down to temperatures of a few hundred kelvins. In the absence of an ultraviolet (UV) photodissociating flux and of ionizing X-ray radiation, three-dimensional simulations of early structure formation show that the fraction of cold, dense gas available for accretion onto seed black holes or star formation exceeds 20% for halos more massive than $10^6 M_\odot$ already at redshifts of 20 (Machacek *et al.* 2003; Yoshida *et al.* 2003).

In spite of some significant achievements in our understanding of the formation of cosmic structures, there are still many challenges facing hierarchical-clustering theories. While quite successful at matching the observed large-scale density distribution (e.g. the properties of galaxy clusters, galaxy clustering, and the statistics of the Lyα forest), CDM simulations appear to produce halos that are too centrally concentrated compared with the mass distribution inferred from the rotation curves of (dark-matter-dominated) dwarf galaxies, and to predict too many dark-matter subhalos compared with the number of dwarf satellites observed within the Local Group (Moore *et al.* 1999; Klypin *et al.* 1999). Another perceived difficulty, which is arguably connected with the "missing-satellites problem", e.g. Bullock *et al.* (2000), is our inability to predict when and how the Universe was reheated and reionized, i.e. to understand the initial conditions of the galaxy-formation process and the basic building blocks of today's massive baryonic structures. We know that at least some galaxies and quasars had already formed when the Universe was less than 10^9 yr old. But when did the first luminous clumps form, was star formation efficient in baryonic objects below the atomic cooling mass, and what was the impact of these early systems on the surrounding intergalactic gas? The crucial processes of star formation and "feedback" (e.g. the effect of the energy input from the earliest generations of sources on later ones) in the nuclei of galaxies are still poorly understood. Accreting black holes can release large amounts of energy to their surroundings, and may play a role in regulating the thermodynamics of the interstellar, intracluster, and intergalactic medium. The detailed astrophysics of these processes is, however, unknown. Although we may have a sketchy history of the production of the chemical elements in the Universe, we know little about how and where exactly they were produced and how they are distributed in the IGM and in the intracluster medium. Finally, where are the first stars and their remnants now, and why are the hundreds of massive satellites predicted to survive today in the Milky Way halo dark?

Here I will describe some of the basic tools that have been developed to study the dawn of galaxies, state the lessons learned, and summarize our current understanding of the birth of the earliest astrophysical objects.

3.2. The dark ages

3.2.1 *Cosmological preliminaries*

Recent CMB experiments, in conjunction with new measurements of the large-scale structure in the present-day Universe, the SN Ia Hubble diagram, and other observations, have led to a substantial reduction in the uncertainties in the parameters describing the ΛCDM concordance cosmology. We appear to be living in a flat ($\Omega_{\rm m} + \Omega_\Lambda = 1$) universe dominated by a cosmological constant and seeded with an approximately scale-invariant primordial spectrum of Gaussian density fluctuations. A ΛCDM cosmology with $\Omega_{\rm m} = 0.24$, $\Omega_\Lambda = 0.76$, $\Omega_{\rm b} = 0.042$, $h = H_0/100\ {\rm km\,s^{-1}\,Mpc^{-1}} = 0.73$, $n = 0.95$, and $\sigma_8 = 0.75$ is consistent with the best-fit parameters from the WMAP 3-year data release (Spergel *et al.* 2006). Here $\Omega_{\rm m} = \rho^0_{\rm m}/\rho^0_{\rm c}$† is the present-day matter density (including cold dark matter as well as a contribution $\Omega_{\rm b}$ from baryons) relative to the critical density $\rho^0_{\rm c} = 3H_0^2/(8\pi G)$, Ω_Λ is the vacuum energy contribution, H_0 is the Hubble constant today, n is the spectral index of the matter power spectrum at inflation, and σ_8 normalizes the power spectrum: it is the root-mean-square amplitude of mass fluctuations in a sphere of radius $8h^{-1}$ Mpc. The lower values of σ_8 and n compared with WMAP 1-year results (Spergel *et al.* 2003) have the effect of delaying structure formation and reducing small-scale power. In this cosmology, and during the epochs of interest here, the expansion rate H evolves according to the Friedmann equation

$$H = \frac{1}{a}\frac{{\rm d}a}{{\rm d}t} = H_0[\Omega_{\rm m}(1+z)^3 + \Omega_\Lambda]^{1/2}, \tag{3.1}$$

where z is the redshift. Light emitted by a source at time t is observed today (at time t_0) with a redshift $z \equiv 1/a(t) - 1$, where a is the cosmic scale factor; $a(t_0) \equiv 1$. The age of the (flat) Universe today is

$$t_0 = \int_0^\infty \frac{{\rm d}z'}{(1+z')H(z')} = \frac{2}{3H_0\sqrt{\Omega_\Lambda}} \ln\left(\frac{\sqrt{\Omega_\Lambda}+1}{\sqrt{\Omega_{\rm m}}}\right) = 1.025\,H_0^{-1} = 13.7\ {\rm Gyr} \tag{3.2}$$

(Ryden 2003). At high redshift, the Universe approaches the Einstein–de Sitter behaviour, $a \propto t^{2/3}$, and its age is given by

$$t(z) \approx \frac{2}{3H_0\sqrt{\Omega_{\rm m}}}(1+z)^{-3/2}. \tag{3.3}$$

The average baryon density today is $n^0_{\rm b} = 2.5 \times 10^{-7}\ {\rm cm^{-3}}$, and the hydrogen density is $n_{\rm H} = (1 - Y_{\rm p})n_{\rm b} = 0.75 n_{\rm b}$, where $Y_{\rm p}$ is the primordial mass fraction of helium. The photon density is $n^0_\gamma = (2.4/\pi^2)[k_{\rm B}T_0/(\hbar c)]^3 = 400\,{\rm cm^{-3}}$, where $T_0 = 2.728 \pm 0.004$K (Fixsen *et al.* 1996), and the cosmic baryon-to-photon ratio is then $\eta = n_{\rm b}/n_\gamma = 6.5 \times 10^{-10}$. The electron-scattering optical depth to redshift $z_{\rm r}$ is given by

$$\tau_{\rm e}(z_{\rm r}) = \int_0^{z_{\rm r}} {\rm d}z\, n_{\rm e}\sigma_{\rm T} c \frac{{\rm d}t}{{\rm d}z} = \int_0^{z_{\rm r}} {\rm d}z\, \frac{n_{\rm e}(z)\sigma_{\rm T} c}{(1+z)H(z)} \tag{3.4}$$

where $n_{\rm e}$ is the electron density at redshift z and $\sigma_{\rm T}$ the Thomson cross-section. Assuming a constant electron fraction $x_{\rm e} \equiv n_{\rm e}/n_{\rm H}$ with redshift, and neglecting the vacuum energy contribution in $H(z)$, this can be rewritten as

$$\tau_{\rm e}(z_{\rm r}) = \frac{x_{\rm e} n^0_{\rm H} \sigma_{\rm T} c}{H_0\sqrt{\Omega_{\rm m}}} \int_0^{z_{\rm r}} {\rm d}z\,\sqrt{1+z} = 0.0022[(1+z_{\rm r})^{3/2} - 1]. \tag{3.5}$$

† Hereafter densities measured at the present epoch will be indicated by either superscript or subscript "0."

Ignoring helium, the 3-year WMAP polarization data requiring $\tau_e = 0.09 \pm 0.03$ are consistent with a universe in which x_e changes from essentially close to zero to unity at $z_r = 11 \pm 2.5$, and $x_e \approx 1$ thereafter.

Given a population of objects with proper number density $n(z)$ and geometric cross-section sigma $\Sigma(z)$, the incremental probability dP that a line of sight will intersect one of the objects in the redshift interval dz at redshift z is

$$\mathrm{d}P = n(z)\Sigma(z)c\frac{\mathrm{d}t}{\mathrm{d}z}\,\mathrm{d}z = n(z)\Sigma(z)c\frac{\mathrm{d}z}{(1+z)H(z)}. \tag{3.6}$$

3.2.2 *Physics of recombination*

Recombination marks the end of the plasma era, and the beginning of the era of neutral matter. Atomic hydrogen has an ionization potential of $I = 13.6$ eV: it takes 10.2 eV to raise an electron from the ground state to the first excited state, from which a further 3.4 eV will free it,

$$\gamma + \mathrm{H} \longleftrightarrow \mathrm{p} + \mathrm{e}^-. \tag{3.7}$$

According to Boltzmann statistics, the number-density distribution of non-relativistic particles of mass m_i in thermal equilibrium is given by

$$n_i = g_i\left(\frac{m_i k_\mathrm{B} T}{2\pi\hbar^2}\right)^{3/2} \exp\left(\frac{\mu_i - m_i c^2}{k_\mathrm{B} T}\right), \tag{3.8}$$

where g_i and μ_i are the statistical weight and chemical potential of the species. When photoionization equilibrium also holds, then $\mu_\mathrm{e} + \mu_\mathrm{p} = \mu_\mathrm{H}$. Recalling that $g_\mathrm{e} = g_\mathrm{p} = 2$ and $g_\mathrm{H} = 4$, one then obtains the Saha equation:

$$\frac{n_\mathrm{e} n_\mathrm{p}}{n_\mathrm{H} n_{\mathrm{H\,I}}} = \frac{x_\mathrm{e}^2}{1 - x_\mathrm{e}} = \frac{(2\pi m_\mathrm{e} k_\mathrm{B} T)^{3/2}}{n_\mathrm{H}(2\pi\hbar)^3} \exp\left(-\frac{I}{k_\mathrm{B} T}\right), \tag{3.9}$$

which can be rewritten as

$$\ln\left(\frac{x_\mathrm{e}^2}{1 - x_\mathrm{e}}\right) = 52.4 - 1.5\ln(1+z) - 58\,000(1+z)^{-1}. \tag{3.10}$$

The ionization fraction goes from 0.91 to 0.005 as the redshift decreases from 1500 to 1100, and the temperature drops from 4100 K to 3000 K. The time elapsed is less than 200 000 yr.

Although the Saha equation describes reasonably well the initial phases of the departure from complete ionization, the assumption of equilibrium rapidly ceases to be valid in an expanding universe, and recombination freezes out. The residual electron fraction can be estimated as follows. The rate at which electrons recombine with protons is

$$\frac{\mathrm{d}n_\mathrm{e}}{\mathrm{d}t} = -\alpha_\mathrm{B} n_\mathrm{e} n_\mathrm{p} \equiv -\frac{n_\mathrm{p}}{t_\mathrm{rec}}, \tag{3.11}$$

where t_rec is the characteristic time for recombination and α_B is the radiative recombination coefficient. This is the product of the electron-capture cross-section σ_n and the electron velocity v_e, averaged over a thermal distribution and summed over all excited states $n \geq 2$ of the hydrogen atom, $\alpha_\mathrm{B} = \sum_n \langle \sigma_n v_\mathrm{e} \rangle$. The radiative-recombination coefficient is well approximated by the fitting formula

$$\alpha_\mathrm{B} = 6.8 \times 10^{-13}\, T_3^{-0.8}\,\mathrm{cm}^3\,\mathrm{s}^{-1} = 1.85 \times 10^{-10}(1+z)^{-0.8}\,\mathrm{cm}^3\,\mathrm{s}^{-1}, \tag{3.12}$$

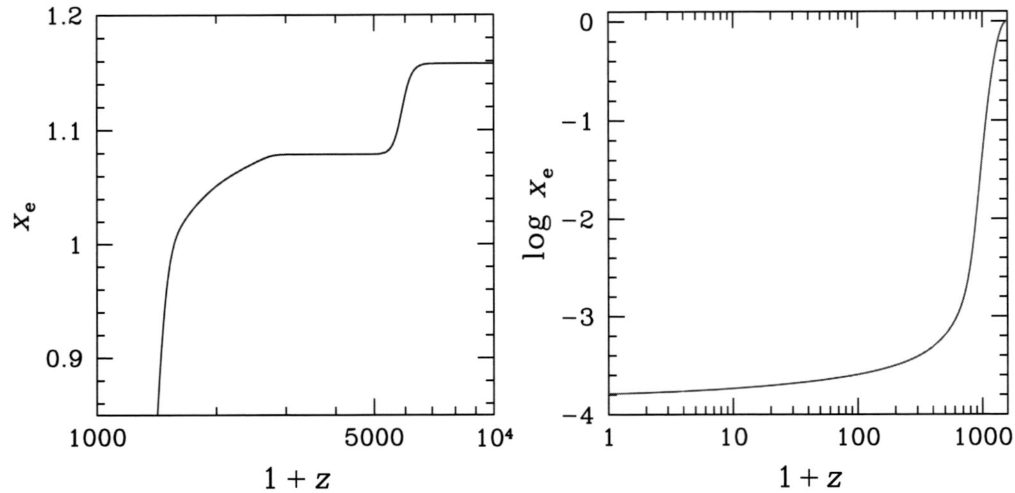

FIGURE 3.1. Helium and hydrogen recombination for a WMAP 3-year cosmology. The step at earlier times in the left panel is due to the recombination of He III into He II. We used the code RECFAST (Seager *et al.* 1999) to compute the electron fraction $x_{\rm e}$. Note that the residual electron fraction (determined by the condition $t_{\rm rec}H = 1$) scales with cosmological parameters as $x_{\rm e} \propto \sqrt{\Omega_{\rm m}}/(\Omega_{\rm b}h)$.

where $T_3 \equiv T/3000$ K and in the second equality I used $T = T_0(1+z)$. When the recombination rate falls below the expansion rate, i.e. when $t_{\rm rec} > 1/H$, the formation of neutral atoms ceases and the remaining electrons and protons have negligible probability of combining with each other:

$$t_{\rm rec}H = \frac{H_0\sqrt{\Omega_{\rm m}}}{x_{\rm e}n_{\rm H}^0\alpha_{\rm B}}(1+z)^{-3/2} = \frac{0.0335}{x_{\rm e}}(1+z)^{-0.7} = \frac{2.5\times10^{-4}}{x_{\rm e}}, \qquad (3.13)$$

where I assumed $z = 1100$ in the third equality.

The recombination of hydrogen in an expanding universe is actually delayed by a number of subtle effects that are not taken into account in the above formulation. An e^- captured to the ground state of atomic hydrogen produces a photon that immediately ionizes another atom, leaving no net change. When an e^- is instead captured to an excited state, the allowed decay to the ground state produces a resonant Lyman-series photon. These photons have large capture cross-sections and put atoms into a high energy state that is easily photoionized again, thereby annulling the effect. That leaves two main routes to the production of atomic hydrogen: (1) two-photon decay from the 2s level to the ground state; and (2) loss of Lyα resonance photons by virtue of the cosmological redshift. The resulting recombination history was derived by Peebles (1968) and Zel'dovich *et al.* (1969).

In the redshift range $800 < z < 1200$, the fractional ionization varies rapidly and is given approximately by the fitting formula

$$x_{\rm e} = 0.042\left(\frac{z}{1000}\right)^{11.25}. \qquad (3.14)$$

This is a fit to a numerical output from the code RECFAST – an improved calculation of the recombination of H I, He I, and He II in the early Universe involving a line-by-line treatment of each atomic level (Seager *et al.* 1999) (see Figure 3.1). Using this expression, we can again compute the optical depth of the Universe for Thomson scattering by free

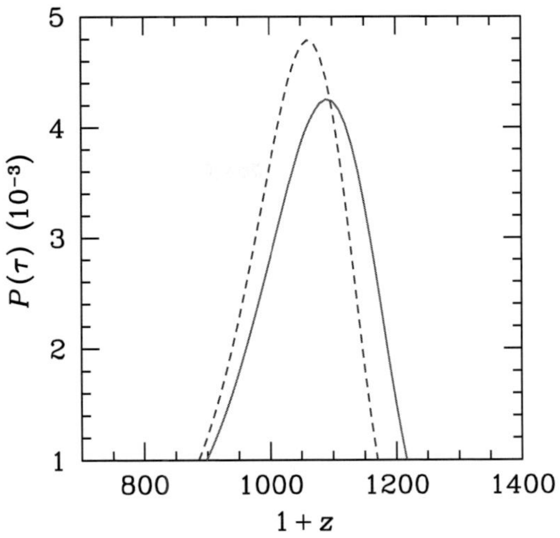

FIGURE 3.2. The visibility function: solid line, RECFAST with WMAP 3-year cosmological parameters; dashed line, Jones & Wyse (1985).

electrons:

$$\tau_e(z) = \int_0^z \mathrm{d}z \, \frac{x_e n_H \sigma_T c}{(1+z)H(z)} \approx 0.3\left(\frac{z}{1000}\right)^{12.75}. \tag{3.15}$$

This optical depth is unity at $z_{dec} = 1100$. From the optical depth we can compute the visibility function $P(\tau)$, the probability that a photon was last scattered in the interval $(z, z + \mathrm{d}z)$. This is given by

$$P(\tau) = \mathrm{e}^{-\tau}\,\frac{\mathrm{d}\tau}{\mathrm{d}z} = 0.0038\left(\frac{z}{1000}\right)^{11.75} \mathrm{e}^{-0.3(z/1000)^{12.75}}, \tag{3.16}$$

and has a sharp maximum at $z = z_{dec}$ and a width of $\Delta z \approx 100$ (see Figure 3.2). The finite thickness of the last scattering surface has important observational consequences for CMB temperature fluctuations on small scales.

3.2.3 *Coupling of gas and radiation*

The CMB plays a key role in the early evolution of structures. First, it sets the epoch of decoupling, when baryonic matter becomes free to move through the radiation field to form the first generation of gravitationally bound systems. Second, it fixes the matter temperature that in turn determines the Jeans scale of the minimum size of the first bound objects.

Following Peebles (1993), consider an electron e^- moving at non-relativistic speed $v \ll c$ through the CMB. In the e^- rest-frame, the CMB temperature measured at angle θ from the direction of motion is

$$T(\theta) = T\left(1 + \frac{v}{c}\cos\theta\right). \tag{3.17}$$

The radiation energy density per unit volume per unit solid angle $\mathrm{d}\Omega = \mathrm{d}\phi\,\mathrm{d}\cos\theta$ around θ is

$$\mathrm{d}u = [a_B T^4(\theta)]\frac{\mathrm{d}\Omega}{4\pi} \tag{3.18}$$

(here $a_{\rm B}$ is the radiation constant), and the net drag force (i.e. the component of the momentum transfer along the direction of motion, integrated over all directions of the radiation) felt by the electron is

$$F = \int_{4\pi} \sigma_{\rm T}\, {\rm d}u \cos\theta = \int \sigma_{\rm T}(a_{\rm B}T^4)\left(1+\frac{v}{c}\cos\theta\right)^4 \cos\theta \,\frac{{\rm d}\Omega}{4\pi}$$
$$= \frac{4}{3}\frac{\sigma_{\rm T} a_{\rm B} T^4}{c} v. \qquad (3.19)$$

This force will be communicated to the protons through electrostatic coupling. The formation of the first gravitationally bound systems is limited by radiation drag, as the drag force per baryon is $x_{\rm e}F$, where $x_{\rm e}$ is the fractional ionization. The mean force divided by the mass $m_{\rm p}$ of a hydrogen atom gives the deceleration time of the streaming motion:

$$t_{\rm s}^{-1} = \frac{1}{v}\frac{{\rm d}v}{{\rm d}t} = \frac{4}{3}\frac{\sigma_{\rm T} a_{\rm B} T_0^4 (1+z)^4}{m_{\rm p} c} x_{\rm e}. \qquad (3.20)$$

The product of the expansion rate H and the velocity-dissipation time $t_{\rm s}$ is

$$t_{\rm s} H = 7.6 \times 10^5 h x_{\rm e}^{-1} (1+z)^{-5/2}. \qquad (3.21)$$

Prior to decoupling $t_{\rm s}H \ll 1$. Since the characteristic time for the gravitational growth of mass-density fluctuations is of the order of the expansion timescale, baryonic density fluctuations become free to grow only after decoupling.

We can use the above results to find the rate of relaxation of the matter temperature $T_{\rm e}$ to that of the radiation. The mean energy per electron in the plasma is $E = 3k_{\rm B}T_{\rm e}/2 = m_{\rm e}\langle v^2\rangle$. The rate at which an electron is doing work against the radiation drag force is Fv, so the plasma transfers energy to the radiation at the mean rate, per electron, of

$$-\frac{{\rm d}E}{{\rm d}t} = \langle Fv \rangle = \frac{4}{3}\frac{\sigma_{\rm T} a_{\rm B} T^4}{c}\langle v^2\rangle \propto T_{\rm e}. \qquad (3.22)$$

At thermal equilibrium $T_{\rm e} = T$, and this rate must be balanced by the rate at which photons scattering off electrons increase the matter energy:

$$\frac{{\rm d}E}{{\rm d}t} = \frac{4}{3}\frac{\sigma_{\rm T} a_{\rm B} T^4}{c}\frac{3k_{\rm B}}{m_{\rm e}}(T - T_{\rm e}). \qquad (3.23)$$

Thus the rate of change of the matter temperature is

$$\frac{{\rm d}T_{\rm e}}{{\rm d}t} = \frac{2}{3k_{\rm B}}\frac{{\rm d}E}{{\rm d}t} = \frac{x_{\rm e}}{1+x_{\rm e}}\frac{8\sigma_{\rm T} a_{\rm B} T^4}{3m_{\rm e}c}(T - T_{\rm e}). \qquad (3.24)$$

The factor $x_{\rm e}/(1+x_{\rm e})$ accounts for the fact that the plasma energy-loss rate per unit volume is $-n_{\rm e}\,{\rm d}E/{\rm d}t = -x_{\rm e}n_{\rm H}\,{\rm d}E/{\rm d}t$, while the total plasma energy density is $(n_{\rm e} + n_{\rm p} + n_{\rm H\,I})3k_{\rm B}T_{\rm e}/2 = n_{\rm H}(1+x_{\rm e})3k_{\rm B}T_{\rm e}/2$. The expression above can be rewritten as

$$\frac{{\rm d}T_{\rm e}}{{\rm d}t} = \frac{T - T_{\rm e}}{t_{\rm C}}, \qquad (3.25)$$

where the "Compton cooling timescale" is

$$t_{\rm C} = \frac{3m_{\rm e}c}{4\sigma_{\rm T} a_{\rm B} T^4}\frac{1+x_{\rm e}}{2x_{\rm e}} = \frac{7.4\times 10^{19}\ {\rm s}}{(1+z)^4}\left(\frac{1+x_{\rm e}}{2x_{\rm e}}\right). \qquad (3.26)$$

The characteristic condition for thermal coupling is then

$$t_{\rm C} H = \frac{240 h\sqrt{\Omega_{\rm m}}}{(1+z)^{5/2}}\frac{1+x_{\rm e}}{2x_{\rm e}}\left(\frac{T}{T_{\rm e}} - 1\right)^{-1} < 1. \qquad (3.27)$$

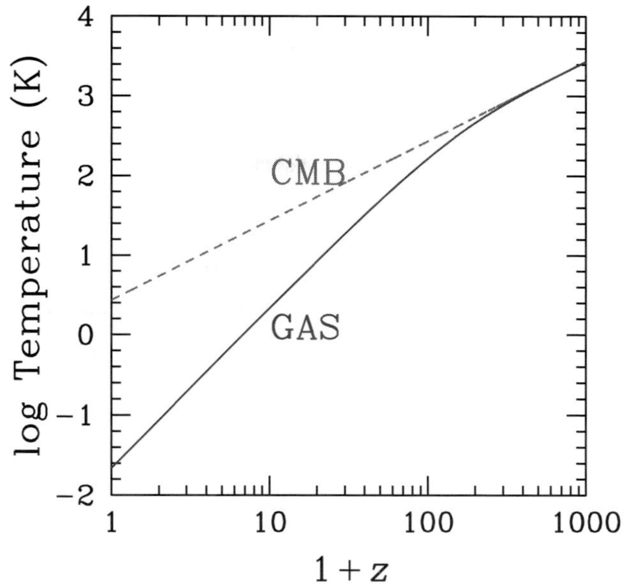

FIGURE 3.3. Evolution of the radiation (dashed line, labeled CMB) and matter (solid line, labeled GAS) temperatures after recombination, in the absence of any reheating mechanism.

For fully ionized gas $x_e = 1$ and the Compton cooling timescale is shorter than the expansion time at all redshifts $z > z_c = 5$. With increasing redshift above 5, it becomes increasingly difficult to keep optically thin ionized plasma hotter than the CMB (Figure 3.3). At redshift $z > 100$, well before the first energy sources (stars, accreting black holes) turn on, hydrogen will have the residual ionization $x_e = 2.5 \times 10^{-4}$. With this ionization, the characteristic relation for the relaxation of the matter temperature is

$$t_C H = \frac{1.7 \times 10^5}{(1+z)^{5/2}} \left(\frac{T}{T_e} - 1 \right)^{-1} . \tag{3.28}$$

The coefficient of the fractional temperature difference reaches unity at the "thermalization redshift" $z_{th} \approx 130$. That is, the residual ionization is enough to keep the matter in temperature equilibrium with the CMB well after decoupling. At redshift lower than z_{th} the temperature of intergalactic gas falls adiabatically faster than that of the radiation, $T_e \propto a^{-2}$.

From the analysis above, the rate of change of the radiation energy density due to Compton scattering can be written as

$$\frac{du}{dt} = \frac{4}{3} \frac{\sigma_T a_B T^4}{c} \frac{3 k_B n_e}{m_e} (T_e - T), \tag{3.29}$$

or

$$\frac{du}{u} = 4\, dy, \qquad dy \equiv (n_e \sigma_T c\, dt) \frac{k_B (T_e - T)}{m_e c^2} = d\tau_e \frac{k_B (T_e - T)}{m_e c^2}. \tag{3.30}$$

Compton scattering causes a distortion of the CMB spectrum, depopulating the Rayleigh–Jeans regime in favor of photons in the Wien tail. The "Compton parameter"

$$y = \int_0^z \frac{k_B T_e}{m_e c^2} \frac{d\tau_e}{dz} dz \tag{3.31}$$

is a dimensionless measure of this distortion, and is proportional to the pressure of the electron gas $n_e k_B T_e$. The COBE satellite has shown the CMB to be thermal to high accuracy, setting a limit $y \leq 1.5 \times 10^{-5}$ (Fixsen *et al.* 1996). This can be shown to imply

$$\langle x_e T_e \rangle [(1+z)^{3/2} - 1] < 4 \times 10^7 \text{ K}. \tag{3.32}$$

A Universe that was reionized and reheated at $z = 20$ to $(x_e, T_e) = (1, >4 \times 10^5 \text{ K})$, for example, would violate the COBE y-limit.

3.2.4 *Hydrogen molecules in the early Universe*

In the absence of heavy metals or dust grains, the processes of radiative cooling, cloud collapse, and star formation in the earliest astrophysical objects were undoubtedly quite different from those today. Saslaw & Zipoy (1967) and Peebles & Dicke (1968) were the first to realize the importance of gas-phase reactions for the formation of the simplest molecule, H_2, in the post-recombination epoch. The presence of even a trace abundance of H_2 is of direct relevance for the cooling properties of primordial gas. In the absence of molecules this would be an extremely poor radiator: cooling by Lyα photons is ineffective at temperatures $\lesssim$8000 K, which is well above the matter and radiation temperature in the post-recombination era. It is the ability of dust-free gas to cool down to low temperatures that controls the formation of the first stars in subgalactic systems.

The formation of H_2 in the intergalactic medium is catalyzed by the residual free electrons and ions, through the reactions

$$\text{H} + \text{e}^- \rightarrow \text{H}^- + \gamma, \tag{3.33}$$

$$\text{H}^- + \text{H} \rightarrow \text{H}_2 + \text{e}^-, \tag{3.34}$$

and

$$\text{H}^+ + \text{H} \rightarrow \text{H}_2^+ + \gamma, \tag{3.35}$$

$$\text{H}_2^+ + \text{H} \rightarrow \text{H}_2 + \text{H}^+. \tag{3.36}$$

The second sequence produces molecular hydrogen at a higher redshift than the first, since the dissociation-energy thresholds of the intermediate species H_2^+ and H^- are 2.64 and 0.75 eV, respectively. The direct radiative association of $\text{H} + \text{H} \rightarrow \text{H}_2 + \gamma$ is highly forbidden, owing to the negligible dipole moment of the homonuclear H_2 molecule. The amount of H_2 that forms depends on the evolution of the gas density. Galli & Palla (1998) have analyzed in detail the case in which the production of hydrogen follows the general expansion of the Universe. Table 3.1 shows the reaction rates for hydrogen and helium species that enter into what they call the *minimal model*, i.e. the reduced set of processes that reproduces with excellent accuracy the full chemistry of H_2 molecules. The first column gives the reaction, the second the rate coefficient (in cm^3 s^{-1} for collisional processes; in s^{-1} for photo-processes), and the third the temperature range of validity of the rate coefficients and remarks on the rate (see Galli & Palla 1998). The gas and radiation temperatures are indicated by T_e and T, respectively.

The evolution of the abundance of H_2 follows the well-known behaviour (e.g. Lepp & Shull 1984) whereby an initial steep rise at $z \sim 400$ is determined by the H_2^+ channel, followed by a small contribution from H^- at $z \sim 100$. The freeze-out primordial fraction of H_2 is $[H_2/H] \sim 10^{-6}$. This is too small to trigger runaway collapse, fragmentation, and star formation. The fate of a subgalactic system at high redshifts depends therefore on its

TABLE 3.1. A reaction network for primordial H_2 chemistry

Reaction	Rate ($cm^3\ s^{-1}$ or s^{-1})	Notes
$H^+ + e \to H + \gamma$	$R_{c2} = 8.76 \times 10^{-11}(1+z)^{-0.58}$	
$H + \gamma \to H^+ + e$	$2.41 \times 10^{15} T^{1.5} \exp(-39\,472/T) R_{c2}$	
$H + e \to H^- + \gamma$	$1.4 \times 10^{-18} T_e^{0.928} \exp(-T_e/16\,200)$	
$H^- + \gamma \to H + e$	$1.1 \times 10^{-1} T^{2.13} \exp(-8823/T)$	
$H^- + H \to H_2 + e$	$\{ 1.5 \times 10^{-9}$	$T_e \leq 300$
	$4.0 \times 10^{-9} T_e^{-0.17}$	$T_e > 300$
$H^- + H^+ \to 2H$	$5.7 \times 10^{-6} T_e^{-0.5} + 6.3 \times 10^{-8}$	
	$-\,9.2 \times 10^{-11} T_e^{0.5} + 4.4 \times 10^{-13} T_e$	
$H + H^+ \to H_2^+ + \gamma$	$\mathrm{dex}[-19.38 - 1.523 \log T_e$	
	$+\,1.118(\log T_e)^2 - 0.1269(\log T_e)^3]$	$T_e \leq 10^{4.5}$
$H_2^+ + \gamma \to H + H^+$	$\{ 2.0 \times 10^1 T^{1.59} \exp(-82\,000/T)$	$v = 0$
	$1.63 \times 10^7 \exp(-32\,400/T)$	LTE
$H_2^+ + H \to H_2 + H^+$	6.4×10^{-10}	
$H_2 + H^+ \to H_2^+ + H$	$\{ 3.0 \times 10^{-10} \exp(-21\,050/T_e)$	$T_e \leq 10^4$
	$1.5 \times 10^{-10} \exp(-14\,000/T_e)$	$T_e > 10^4$
$He + H^+ \to HeH^+ + \gamma$	$\{ 7.6 \times 10^{-18} T_e^{-0.5}$	$T_e \leq 10^3$
	$3.45 \times 10^{-16} T_e^{-1.06}$	$T_e > 10^3$
$HeH^+ + H \to He + H_2^+$	9.1×10^{-10}	
$HeH^+ + \gamma \to He + H^+$	$6.8 \times 10^{-1} T^{1.5} \exp(-22\,750/T)$	

LTE, local thermodynamic equilibrium.

ability to rapidly increase its H_2 content following virialization. In the low-density limit ($n_H < 0.1\ cm^{-3}$), the H_2 cooling function Λ_{H_2} (in erg $cm^3\ s^{-1}$) is well approximated by the expression

$$\log \Lambda_{H_2}(n_H \to 0) = -103.0 + 97.59 \log T_e - 48.05(\log T_e)^2 + 10.80(\log T_e)^3 - 0.9032(\log T_e)^4, \tag{3.37}$$

over the range $10\,K \leq T_e \leq 10^4$ K (Galli & Palla 1998).

3.3. The emergence of cosmic structure

3.3.1 *Linear theory*

As shown above, it is only after hydrogen recombination that baryons can start falling into the already growing dark-matter perturbations. Gas pressure can resist the force of gravity, and small-scale perturbations in the baryonic fluid will not grow in amplitude. However, at sufficiently large scales, gravity can overpower pressure gradients, thereby allowing the perturbation to grow.

The linear evolution of sub-horizon density perturbations in the dark-matter–baryon fluid is governed during the matter-dominated era by two second-order differential equations:

$$\ddot{\delta}_{dm} + 2H\dot{\delta}_{dm} = \frac{3}{2} H^2 \Omega_m^z (f_{dm}\delta_{dm} + f_b\delta_b), \tag{3.38}$$

for the dark matter, and

$$\ddot{\delta}_{\rm b} + 2H\dot{\delta}_{\rm b} = \frac{3}{2}H^2\Omega_{\rm m}^z(f_{\rm dm}\delta_{\rm dm} + f_{\rm b}\delta_{\rm b}) - \frac{c_{\rm s}^2}{a^2}k^2\delta_{\rm b}, \tag{3.39}$$

for the baryons, where $\delta_{\rm dm}(k)$ and $\delta_{\rm b}(k)$ are the Fourier components of the density fluctuations in the dark matter and baryons,† $f_{\rm dm}$ and $f_{\rm b}$ are the corresponding mass fractions, $c_{\rm s}$ is the gas sound speed, k the (comoving) wavenumber, and the derivatives are taken with respect to cosmic time. Here $\Omega_{\rm m}^z \equiv 8\pi G\rho(t)/(3H^2) = \Omega_{\rm m}(1+z)^3/[\Omega_{\rm m}(1+z)^3 + \Omega_\Lambda]$ is the time-dependent matter-density parameter, and $\rho(t)$ is the total background matter density. Because there is five times more dark matter than baryons, it is the former that defines the pattern of gravitational wells in which structure formation occurs. In the case that $f_{\rm b} \simeq 0$ and the Universe is static ($H = 0$), Equation (3.38) becomes

$$\ddot{\delta}_{\rm dm} = 4\pi G\rho\delta_{\rm dm} \equiv \frac{\delta_{\rm dm}}{t_{\rm dyn}^2}, \tag{3.40}$$

where $t_{\rm dyn}$ denotes the dynamical timescale. This equation admits the solution $\delta_{\rm dm} = A_1\exp(t/t_{\rm dyn}) + A_2\exp(-t/t_{\rm dyn})$. After a few dynamical times, only the exponentially growing term is significant: gravity tends to make small density fluctuations in a static pressureless medium grow exponentially with time. The additional term $\propto H\dot{\delta}_{\rm dm}$ present in an expanding universe can be thought of as a “Hubble-friction” term that acts to slow down the growth of density perturbations. Equation (3.38) admits the general solution for the growing mode

$$\delta_{\rm dm}(a) = \frac{5\Omega_{\rm m}}{2}H_0^2 H\int_0^a \frac{da'}{(\dot{a}')^3}, \tag{3.41}$$

where the constants have been chosen so that an Einstein–de Sitter universe ($\Omega_{\rm m} = 1$, $\Omega_\Lambda = 0$) gives the familiar scaling $\delta_{\rm dm}(a) = a$ with coefficient unity. The right-hand side of Equation (3.41) is called the linear growth factor $D(a)$. Different values of $\Omega_{\rm m}$, and Ω_Λ lead to different linear growth factors: growing modes actually decrease in density, but not as fast as the average universe does. Note how, in contrast to the exponential growth found in the static case, the growth of perturbations even in the case of an Einstein–de Sitter universe is just algebraic. A remarkable approximation formula to the growth factor in a flat universe follows from Lahav *et al.* (1991),

$$\delta_{\rm dm}(a) = D(a) \simeq \frac{5\Omega_{\rm m}^z}{2(1+z)}\left((\Omega_{\rm m}^z)^{4/7} - \frac{(\Omega_{\rm m}^z)^2}{140} + \frac{209}{140}\Omega_{\rm m}^z + \frac{1}{70}\right)^{-1}. \tag{3.42}$$

This is good to within a few percent in regions of plausible $\Omega_{\rm m}$ and Ω_Λ.

Equation (3.39) shows that, on large scales (i.e. small k), pressure forces can be neglected and baryons simply track the dark-matter fluctuations. Conversely, on small scales (i.e. large k), pressure dominates and baryon fluctuations will be suppressed relative to the dark matter. Gravity and pressure are equal at the characteristic Jeans scale:

$$k_{\rm J} = \frac{a}{c_{\rm s}}\sqrt{4\pi G\rho}. \tag{3.43}$$

Corresponding to this critical wavenumber $k_{\rm J}$ there is a critical cosmological Jeans mass $M_{\rm J}$, defined as the total mass (gas plus dark matter) enclosed within the sphere of

† For each fluid component ($i = {\rm b, dm}$) the real-space fluctuation in the density field, $\delta_i(\mathbf{x}) \equiv \delta\rho_i(\mathbf{x})/\rho_i$, can be written as a sum over Fourier modes, $\delta_i(\mathbf{x}) = \int d^3\mathbf{k}(2\pi)^{-3}\delta_i(\mathbf{k})\exp(i\mathbf{k}\cdot\mathbf{x})$.

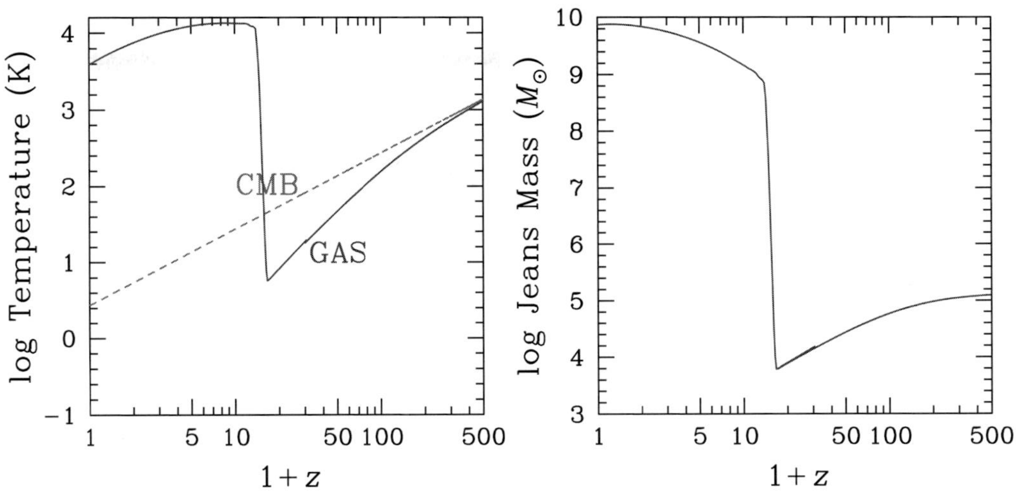

FIGURE 3.4. Left: the same as Figure 3.3 but for a Universe reionized by ultraviolet radiation at $z_r = 11$. Right: the cosmological (gas plus dark matter) Jeans mass.

physical radius equal to $\pi a/k_J$,

$$M_J = \frac{4\pi}{3}\rho\left(\frac{\pi a}{k_J}\right)^3 = \frac{4\pi}{3}\rho\left(\frac{5\pi k_B T_e}{12 G\rho m_p \mu}\right)^{3/2} \approx 8.8\times 10^4 M_\odot \left(\frac{aT_e}{\mu}\right)^{3/2}, \qquad (3.44)$$

where μ is the mean relative molecular mass. The evolution of M_J is shown in Figure 3.4. In the post-recombination Universe, the baryon–electron gas is thermally coupled to the CMB, $T_e \propto a^{-1}$, and the Jeans mass is independent of redshift and comparable to the mass of globular clusters, $M_J \approx 10^5 M_\odot$. For $z < z_{th}$, the gas temperature drops as $T_e \propto a^{-2}$, and the Jeans mass decreases with time, $M_J \propto a^{-3/2}$. This trend is reversed by the reheating of the IGM. The energy released by the first collapsed objects drives the Jeans mass up to galaxy scales (Figure 3.4): baryonic-density perturbations stop growing as their mass drops below the new Jeans mass. In particular, photo-ionization by the UV radiation from the first stars and quasars would heat the IGM to temperatures of $\approx 10^4$ K (corresponding to a Jeans mass $M_J \sim 10^{10} M_\odot$ at $z_r = 11$), suppressing gas infall into low-mass systems and preventing new (dwarf) galaxies from forming.†

3.3.2 *Statistics of density fields*

The observed uniformity of the CMB guarantees that density fluctuations must have been quite small at decoupling, implying that the evolution of the density contrast can be studied at $z \leq z_{dec}$ using linear theory, and that each mode $\delta(k)$ evolves independently. The inflationary model predicts a scale-invariant primordial power spectrum of density fluctuations $P(k) \equiv \langle|\delta(k)|^2\rangle \propto k^n$, with $n = 1$ (the so-called Harrison–Zel'dovich spectrum). It is the index n that governs the balance between large- and small-scale power. In the case of a Gaussian random field with zero mean, the power spectrum contains the complete statistical information about the density inhomogeneity. It is often more convenient to use the dimensionless quantity $\Delta_k^2 \equiv [k^3 P(k)/(2\pi^2)]$, which is the power per logarithmic interval in wavenumber k. In the matter-dominated epoch, this quantity retains its initial

† When the Jeans mass itself varies with time, linear gas fluctuations tend to be smoothed on a (filtering) scale that depends on the full thermal history of the gas instead of the instantaneous value of the sound speed (Gnedin & Hui 1998).

primordial shape ($\Delta_k^2 \propto k^{n+3}$) only on very large scales. Small-wavelength modes enter the horizon earlier on and their growth is suppressed more severely during the radiation-dominated epoch: on small scales the amplitude of Δ_k^2 is essentially suppressed by four powers of k (from k^{n+3} to k^{n-1}). If $n = 1$, then small scales will have nearly the same power except for a weak, logarithmic dependence. Departures from the initially scale-free form are described by the transfer function $T(k)$, defined such that $T(0) = 1$:

$$P(k, z) = Ak^n \left(\frac{D(z)}{D(0)}\right)^2 T^2(k), \tag{3.45}$$

where A is the normalization. An accurate fitting function for $T(k)$ in a ΛCDM universe is

$$T_k = \frac{\ln(1 + 2.34q)}{2.34q} \left[1 + 3.89q + (16.1q)^2 + (5.46q)^3 + (6.71q)^4\right]^{-1/4}, \tag{3.46}$$

where

$$q \equiv \frac{k/\mathrm{Mpc}^{-1}}{\Omega_\mathrm{m} h^2 \exp(-\Omega_\mathrm{b} - \Omega_\mathrm{b}/\Omega_\mathrm{m})}. \tag{3.47}$$

This is the fit given by Bardeen *et al.* (1986) modified to account for the effects of baryon density following Sugiyama (1995).

Another useful measurement of inhomogeneity is the quantity

$$\sigma^2(M) = \frac{1}{2\pi^2} \int_0^\infty \mathrm{d}k\, k^2 P(k) |W(kR)|^2. \tag{3.48}$$

This is the mass variance of the density field smoothed by a window function $W(kR)$ over a spherical volume of comoving radius R and average mass $M = H_0^2 \Omega_\mathrm{m} R^3/(2G)$. If the window is a top-hat in real space, then its Fourier transform is $W(kR) = 3(kR)^{-3}[\sin(kR) - kR\cos(kR)]$. The significance of the window function is the following: the dominant contribution to $\sigma(M)$ comes from perturbation components with wavelengths $\lambda = 2\pi/k > R$, because those with higher frequencies tend to be averaged out within the window volume. Since the fluctuation spectrum is falling with decreasing k, waves with much larger λ contribute only a small amount. Hence, in terms of a mass $M \propto \lambda^3 \propto k^{-3}$, we have

$$\sigma(M) \propto M^{-(3+n)/n}. \tag{3.49}$$

Since $P(k)$ is not a strict power law, n should be thought of as an approximate local value $\mathrm{dln}\, P/\mathrm{dln}\, k$ in the relevant range. For $n > -3$, the variance $\sigma(M)$ decreases with increasing M; this implies that, on average, smaller masses condense out earlier than larger masses. Structures grow by the gradual separation and collapse of progressively larger units. Each parent unit will in general be made up of a number of smaller progenitor clumps that had collapsed earlier. This leads to a hierarchical pattern of clustering. In CDM, $n \geq -3$ at small M, increases with increasing M, and reaches the asymptotic value $n = 1$ for $M \geq 10^{15} M_\odot$. The power spectrum on galactic scales can be approximated by $n \approx -2$.

3.3.3 *Spherical collapse*

At late times the density contrast at a given wavelength becomes comparable to unity, linear perturbation theory fails at this wavelength, and a different model must be used to follow the collapse of bound dark-matter systems ("halos"). Small scales are the first to become non-linear. Consider, at some initial time t_i, a spherical region of size r_i and mass M that has a slight constant overdensity δ_i relative to the background ρ_b,

$$\rho(t_i) = \rho_b(t_i)(1 + \delta_i). \tag{3.50}$$

As the Universe expands, the over-dense region will expand more slowly than does the background, reach a maximum radius, and eventually collapse under its own gravity to form a bound virialized system. Such a simple model is called a spherical top-hat. As long as radial shells do not cross each other during the evolution, the motion of a test particle at (physical) radius r is governed by the equation (ignoring the vacuum-energy component)

$$\frac{d^2 r}{dt^2} = -\frac{GM}{r^2}, \tag{3.51}$$

where $M = (4\pi/3)r_i^3 \rho_b(t_i)(1 + \delta_i) = \text{constant}$. On integrating we obtain

$$\frac{1}{2}\left(\frac{dr}{dt}\right)^2 - \frac{GM}{r} = E, \tag{3.52}$$

where E is a constant of integration. If $E > 0$ then $\dot{r}^2$ will never become zero and the shell will expand forever. If $E < 0$ instead, then as r increases $\dot{r}$ will eventually become zero and later negative, implying a contraction and a collapse. Let's choose t_i to be the time at which δ_i is so small that the over-dense region is expanding along with the Hubble flow. Then $\dot{r}_i = (\dot{a}/a)r_i = H(t_i)r_i \equiv H_i r_i$ at time t_i, and the initial kinetic energy will be

$$K_i \equiv \left(\frac{\dot{r}^2}{2}\right)_{t=t_i} = \frac{H_i^2 r_i^2}{2}. \tag{3.53}$$

The potential energy at $t = t_i$ is

$$|U| = \left(\frac{GM}{r}\right)_{t=t_i} = G\frac{4\pi}{3}\rho_b(t_i)(1+\delta_i)r_i^2 = \frac{1}{2}H_i^2 r_i^2 \Omega_i(1+\delta_i) = K_i\Omega_i(1+\delta_i), \tag{3.54}$$

with Ω_i denoting the initial value of the matter-density parameter of the smooth background Universe. The total energy of the shell is therefore

$$E = K_i - K_i\Omega_i(1+\delta_i) = K_i\Omega_i(\Omega_i^{-1} - 1 - \delta_i). \tag{3.55}$$

The condition $E < 0$ for the shell to eventually collapse becomes $(1 + \delta_i) > \Omega_i^{-1}$, or

$$\delta_i > \Omega_i^{-1} - 1. \tag{3.56}$$

In an Einstein–de Sitter universe at early times ($\Omega_i = 1$), this condition is satisfied by any over-dense region with $\delta_i > 0$. In this case the patch will always collapse.

Consider now a shell with $E < 0$ in a background Einstein–de Sitter universe that expands to a maximum radius r_{max} ("turnaround", $\dot{r} = 0$) and then collapses. The solution to the equation of motion can be written as the parametric equation for a cycloid

$$r = \frac{r_{max}}{2}(1 - \cos\theta), \qquad t = t_{max}\frac{\theta - \sin\theta}{\pi}. \tag{3.57}$$

The background density evolves as $\rho_b(t) = (6\pi G t^2)^{-1}$, and the density contrast becomes

$$\delta = \frac{\rho(t)}{\rho_b(t)} - 1 = \frac{9}{2}\frac{(\theta - \sin\theta)^2}{(1-\cos\theta)^3} - 1. \tag{3.58}$$

For comparison, linear theory (in the limit of small t) gives

$$\delta_{\rm L} = \frac{3}{20}\left(\frac{6\pi t}{t_{\rm max}}\right)^{2/3} = \frac{3}{20}[6(\theta - \sin\theta)]^{2/3}. \tag{3.59}$$

This all agrees with what we knew already: at early times the sphere expands with the $a \propto t^{2/3}$ Hubble flow and density perturbations grow proportionally to a. We can now see how linear theory breaks down as the perturbation evolves. There are three interesting epochs in the final stage of its development, which we can read directly from the above solutions.

Turnaround. The sphere breaks away from the Hubble expansion and reaches a maximum radius at $\theta = \pi$, $t = t_{\rm max}$. At this point the true density enhancement with respect to the background is

$$\delta = \frac{9}{2}\frac{\pi^2}{2^3} - 1 = \frac{9}{16}\pi^2 - 1 = 4.55, \tag{3.60}$$

which is definitely in the non-linear regime. By comparison, linear theory predicts

$$\delta_{\rm L} = \frac{3}{20}(6\pi)^{2/3} = 1.062. \tag{3.61}$$

Collapse. If gravity alone operates, then the sphere will collapse to a singularity at $\theta = 2\pi$, $t = 2t_{\rm max}$. This occurs when

$$\delta_{\rm L} = \frac{3}{20}(12\pi)^{2/3} = 1.686. \tag{3.62}$$

Thus from Equation (3.42) we see that a top-hat collapses at redshift z if its linear over-density *extrapolated to the present day* is

$$\delta_{\rm c}(z) = \frac{\delta_{\rm L}}{D(z)} = 1.686(1+z), \tag{3.63}$$

where the second equality holds in a flat Einstein–de Sitter universe. This is termed the critical over-density for collapse. An object of mass M collapsing at redshift z has an over-density that is $\nu_{\rm c}$ times the linearly extrapolated density contrast today, $\sigma_0(M)$, on that scale,

$$\nu_{\rm c} = \delta_{\rm c}(z)/\sigma_0(M). \tag{3.64}$$

The mass scale associated with typical $\nu_{\rm c} = 1\sigma$ non-linear fluctuations in the density field decreases from $10^{13} M_\odot$ today to about $10^7 M_\odot$ at redshift 5. At $z = 10$, halos of $10^{10} M_\odot$ collapse from much rarer $\nu_{\rm c} = 3\sigma$ fluctuations.

Virialization. Collapse to a point at $\theta = 2\pi$ ($t = 2t_{\rm max} \equiv t_{\rm vir}$) will never occur in practice because, before this happens, the approximation that matter is distributed in spherical shells and that the random velocities of the particles are small will break down. The dark matter will reach virial equilibrium by a process known as "violent relaxation": since the gravitational potential is changing with time, individual particles do not follow orbits that conserve energy, and the net effect is to widen the range of energies available to them. Thus a time-varying potential can provide a relaxation mechanism that operates on the dynamical timescale rather than on that of the much longer two-body relaxation time. This process will convert the kinetic energy of collapse into random motions.

At $t = t_{\max}$ all the energy is in the form of gravitational potential energy, and $E = U = -GM/r_{\max}$. At virialization $U = -2K$ (the virial theorem) and $E = U + K = U/2 = -GM/(2r_{\rm vir})$. Hence $r_{\rm vir} = r_{\max}/2$. The mean density of the virialized object is then $\rho_{\rm vir} = 2^3\rho_{\max}$, where $\rho_{\max}$ is the density of the shell at turnaround. From Equation (3.60) we have $\rho_{\max} = (9/16)\pi^2\rho_{\rm b}(t_{\max})$, and $\rho_{\rm b}(t_{\max}) = \rho_{\rm b}(t_{\rm vir})(t_{\rm vir}/t_{\max})^2 = 4\,\rho_{\rm b}(t_{\rm vir})$. On combining these relations we get

$$\rho_{\rm vir} = 2^3\rho_{\max} = 2^3(9/16)\pi^2\rho_{\rm b}(t_{\max}) = 2^3(9/16)\pi^2 4\,\rho_{\rm b}(t_{\rm vir}) = 18\pi^2\rho_{\rm b}(t_{\rm vir}). \quad (3.65)$$

Therefore the density contrast at virialization in an Einstein–de Sitter universe is

$$\Delta_{\rm c} = 178. \quad (3.66)$$

In a universe with a cosmological constant, the collapse of a top-hat spherical perturbation is described by

$$\frac{{\rm d}^2 r}{{\rm d}t^2} = -\frac{GM}{r^2} + \frac{\Lambda}{3}c. \quad (3.67)$$

In a flat $\Omega_{\rm m} + \Omega_\Lambda = 1$ cosmology, the final over-density *relative to the critical density* gets modified according to the fitting formula (Bryan & Norman 1998)

$$\Delta_{\rm c} = 18\,\pi^2 + 82d - 39d^2, \quad (3.68)$$

where $d \equiv \Omega_{\rm m}^z - 1$ is evaluated at the collapse redshift. A spherical top-hat collapsing today in a universe with $\Omega_{\rm m} = 0.24$ has a density contrast at virialization of $\Delta_{\rm c} = 93$. This corresponds to an over-density relative to the background matter density ($=\Omega_{\rm m}\rho_{\rm c}^0$) of $\Delta_{\rm c}/\Omega_{\rm m} = 388$: the faster expansion of a low-density universe means that the perturbation turns around and collapses when a larger density contrast has been produced. For practical reasons a density contrast of 200 relative to the background is often used to define the radius, r_{200}, that marks the boundary of a virialized region.

A halo of mass M collapsing at redshift z can be described in terms of its virial radius $r_{\rm vir}$, circular velocity $V_{\rm c}$, and virial temperature $T_{\rm vir}$ (Barkana & Loeb 2001):

$$r_{\rm vir} = \left(\frac{2GM}{\Delta_{\rm c}H^2}\right)^{1/3} = 1.23\,{\rm kpc}\left(\frac{M}{10^8 M_\odot}\right)^{1/3} f^{-1/3}\left(\frac{1+z}{10}\right)^{-1}, \quad (3.69)$$

$$V_{\rm c} = \left(\frac{GM}{r_{\rm vir}}\right)^{1/2} = 21.1\,{\rm km\,s^{-1}}\left(\frac{M}{10^8 M_\odot}\right)^{1/3} f^{1/6}\left(\frac{1+z}{10}\right)^{1/2}, \quad (3.70)$$

$$T_{\rm vir} = \frac{\mu m_{\rm p} V_{\rm c}^2}{2k_{\rm B}} = 1.6\times 10^4\,{\rm K}\left(\frac{M}{10^8 M_\odot}\right)^{2/3} f^{1/3}\left(\frac{1+z}{10}\right), \quad (3.71)$$

where $f \equiv (\Omega_{\rm m}/\Omega_{\rm m}^z)[\Delta_{\rm c}/(18\pi^2)]$ and $\mu = 0.59$ is the mean relative molecular mass for fully ionized primordial gas ($\mu = 1.23$ for neutral primordial gas). The binding energy of the halo is approximately

$$E_{\rm b} = \frac{1}{2}\frac{GM^2}{r_{\rm vir}} = 4.42\times 10^{53}\,{\rm erg}\left(\frac{M}{10^8 M_\odot}\right)^{5/3} f^{1/3}\left(\frac{1+z}{10}\right), \quad (3.72)$$

where the coefficient of 1/2 is exact for a singular isothermal sphere. The binding energy of the baryons is smaller by a factor $\Omega_{\rm m}/\Omega_{\rm b} \simeq 5.7$. The energy deposition by supernovae in the shallow potential wells of subgalactic systems may then lift out metal-enriched material from the host (dwarf) halos, causing the pollution of the IGM at early times (Madau *et al.* 2001).

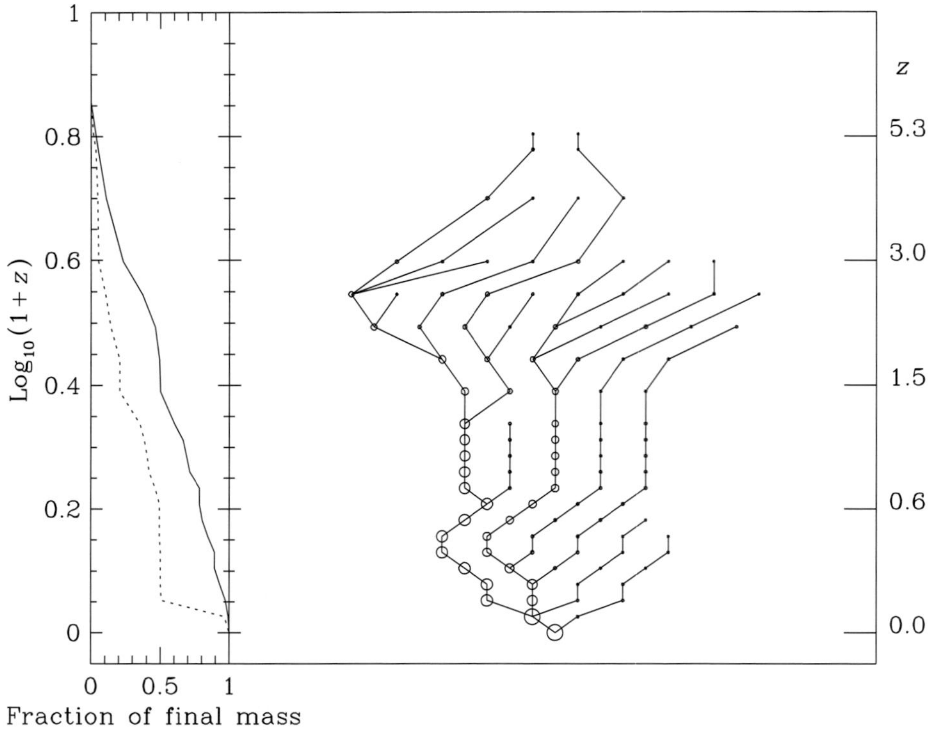

FIGURE 3.5. An example of a merger tree obtained from an N-body simulation of a $9 \times 10^{12}h^{-1}M_\odot$ halo at redshift $z = 0$. Each circle represents a dark-matter halo identified in the simulation, the area of the circle being proportional to halo mass. The vertical position of each halo on the plot is determined by the redshift z at which the halo is identified; the horizontal positioning is arbitrary. The solid lines connect halos to their progenitors. The solid line in the panel on the left-hand side shows the fraction of the final mass contained in resolved progenitors as a function of redshift. The dotted line shows the fraction of the final mass contained in the largest progenitor as a function of redshift. From Helly *et al.* (2003.)

3.3.4 *Dark-halo mergers*

The assumption that virialized objects form from smooth spherical collapse, while providing a useful framework for thinking about the formation histories of gravitationally bound dark-matter halos, does not capture the real nature of structure formation in CDM theories. In these models galaxies are assembled hierarchically through the merging of many smaller subunits that formed in a similar manner at higher redshift (see Figure 3.5). Galaxy halos experience multiple mergers during their lifetime, with those between comparable-mass systems ("major mergers") expected to result in the formation of elliptical galaxies (e.g. Barnes 1988; Hernquist 1992). Figure 3.6 shows the number of major mergers (defined as mergers for which the mass ratio of the progenitors is >0.3) per unit redshift bin experienced by halos of various masses. For galaxy-sized halos this quantity happens to peak in the redshift range 2–4, corresponding to the epoch when the observed space density of optically selected quasars also reaches a maximum.

The merger between a large parent halo and a smaller satellite system will evolve under two dynamical processes: dynamical friction, that causes the orbit of the satellite to decay toward the central regions; and tidal stripping, that removes material from the satellite and adds it to the diffuse mass of the parent. Since clustering is hierarchical, the satellite will typically form at earlier times and have a higher characteristic density

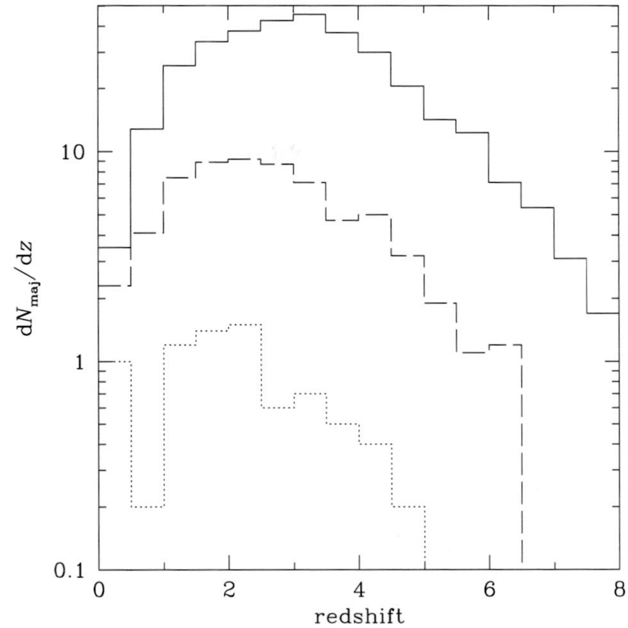

FIGURE 3.6. Mean numbers of major mergers experienced per unit redshift by halos with masses $>10^{10}M_\odot$. Solid line: progenitors of an $M_0 = 2\times10^{13}M_\odot$ halo at $z = 0$. Dashed line: the same for $M_0 = 3\times10^{12}M_\odot$. Dotted line: the same for $M_0 = 10^{11}M_\odot$. From Volonteri *et al.* (2003.)

and a smaller characteristic radius. The study of the assembly of dark-matter halos by repeated mergers is particularly well suited to the *N*-body methods that have been developed in the past two decades. Numerical simulations of structure formation by dissipationless hierarchical clustering from Gaussian initial conditions indicate a roughly universal spherically averaged density profile for the resulting halos (Navarro *et al.* 1997, hereafter NFW):

$$\rho_{\rm NFW}(r) = \frac{\rho_{\rm s}}{cx(1+cx)^2}, \qquad (3.73)$$

where $x \equiv r/r_{\rm vir}$ and the characteristic density $\rho_{\rm s}$ is related to the concentration parameter c by

$$\rho_{\rm s} = \frac{3H^2}{8\pi G}\frac{\Delta_{\rm c}}{3}\frac{c^3}{\ln(1+c) - c/(1+c)}. \qquad (3.74)$$

This function fits the numerical data presented by NFW over a radius range of about two orders of magnitude. Equally good fits are obtained for high-mass (rich galaxy cluster) and low-mass (dwarf) halos. Power-law fits to this profile over a restricted radial range have slopes that steepen from -1 near the halo center to -3 at large $cr/r_{\rm vir}$. Bullock *et al.* (2001) found that the concentration parameter follows a log-normal distribution such that the median depends on the halo mass and redshift,

$$c_{\rm med}(M,z) = \frac{c_*}{1+z}\left(\frac{M}{M_*}\right)^\alpha, \qquad (3.75)$$

where M_* is the mass of a typical halo collapsing today. The halos in the simulations by Bullock *et al.* were best described by $\alpha = -0.13$ and $c_* = 9.0$, with a scatter around the median of $\sigma_{\log c} = 0.14$ dex.

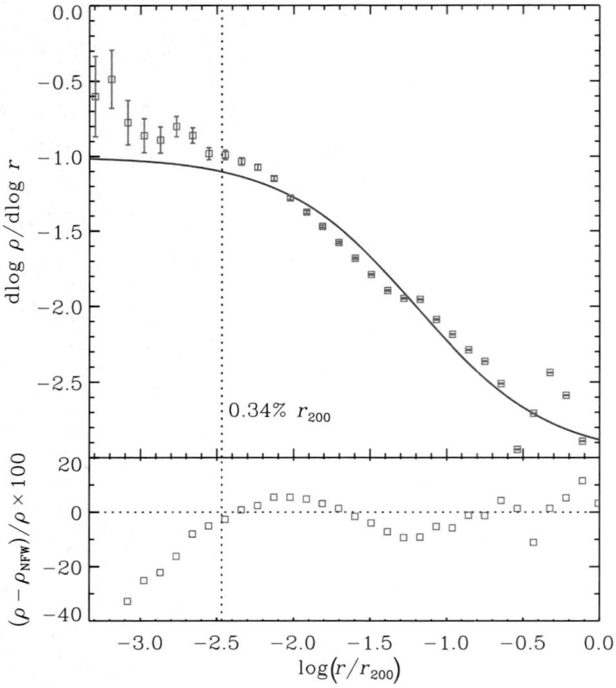

FIGURE 3.7. Top: the logarithmic slope of the density profile of the "Via Lactea" halo, as a function of radius. Densities were computed in 50 radial logarithmic bins, and the local slope was determined by a finite-difference approximation using one neighboring bin on each side. The thin line shows the slope of the best-fit NFW profile with concentration $c = 12$. The vertical dotted line indicates the estimated convergence radius: local densities (but not necessarily the logarithmic slopes) should be correct to within 10% outside of this radius. Bottom: the residuals in percent between the density profile and the best-fit NFW profile, as a function of radius. Here r_{200} is the radius within which the enclosed average density is 200 times the background value, $r_{200} = 1.35 r_{\rm vir}$ in the adopted cosmology. From Diemand *et al.* (2007a).

"Via Lactea," the highest-resolution N-body simulation to date of the formation of a Milky Way-sized halo (Diemand *et al.* 2007a, b), shows that the fitting formula proposed by NFW with concentration $c = 12$ provides a reasonable approximation to the density profile down a convergence radius of $r_{\rm conv} = 1.3\,$kpc (Figure 3.7). Within the region of convergence, deviations from the best-fit NFW matter density are typically less than 10%. From 10 kpc down to $r_{\rm conv}$ Via Lactea is actually denser than predicted by the NFW formula. Near $r_{\rm conv}$ the density approaches the NFW value again while the logarithmic slope is shallower (-1.0 at $r_{\rm conv}$) than predicted by the NFW fit.

3.3.5 *Assembly history of a Milky Way halo*

The simple spherical top-hat collapse ignores shell crossing and mixing, accretion of self-bound clumps, triaxiality, angular momentum, random velocities, and large-scale tidal forces. It is interesting at this stage to use "Via Lactea" and study in more details, starting from realistic initial conditions, the formation history of a Milky Way-sized halo in a ΛCDM cosmology. The Via Lactea simulation was performed with the PKDGRAV tree-code (Stadel 2001) using the best-fit cosmological parameters from the WMAP 3-year data release. The galaxy-forming region was sampled with 234 million particles of mass $2.1 \times 10^4 M_\odot$, evolved from redshift 49 to the present with a force resolution of 90 pc and adaptive time-steps as short as 68 500 yr, and centered on an isolated halo that had

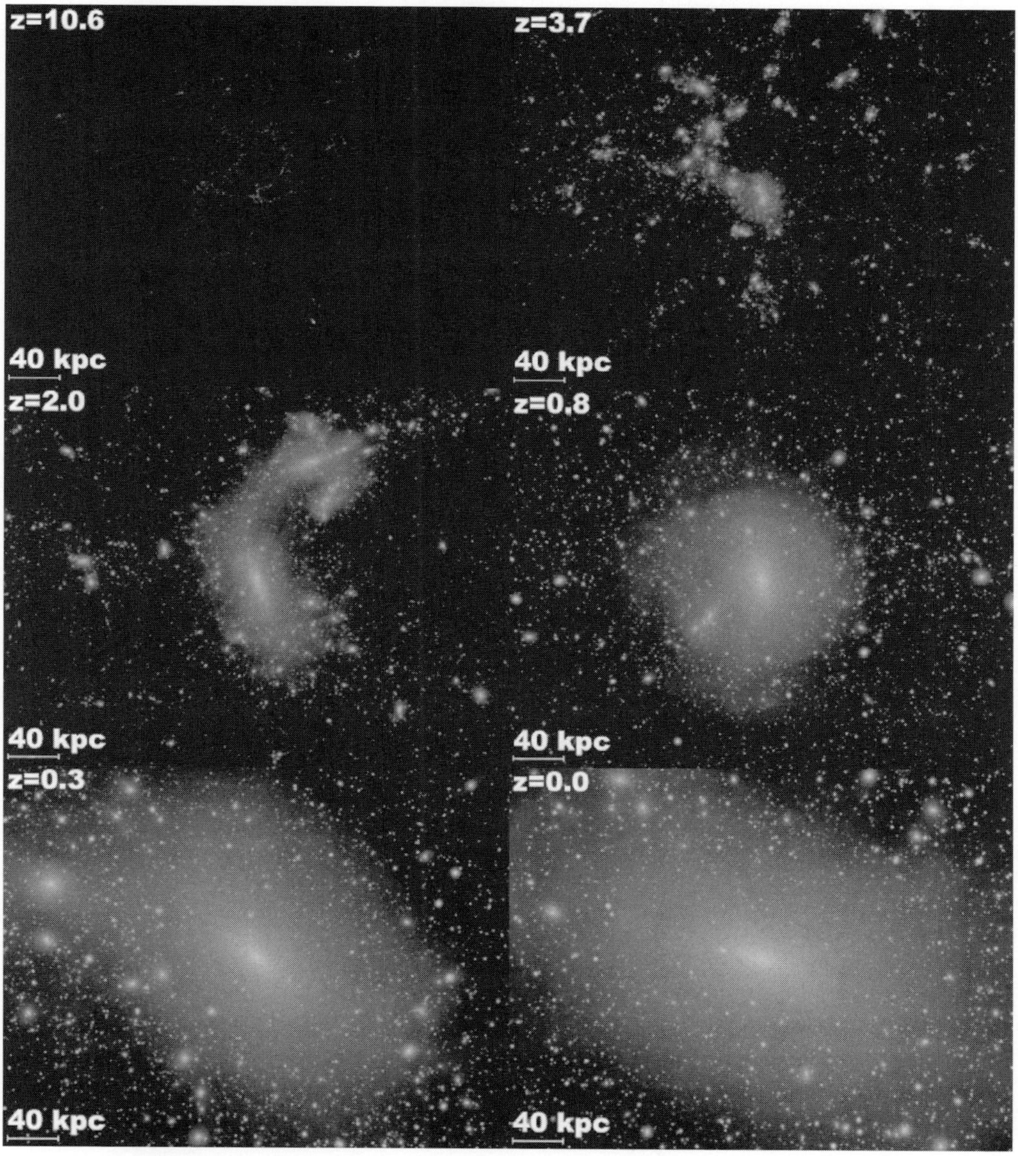

FIGURE 3.8. Projected dark-matter density-squared map of our simulated Milky Way-halo region ("Via Lactea") at various redshifts, from $z = 10.6$ to the present. The image covers an area of 400×300 physical kiloparsecs.

no major merger after $z = 1.7$, making it a suitable host for a Milky Way-like disk galaxy (see Figure 3.8). The number of particles is an order of magnitude larger than had been used in previous simulations. The run was completed in 320 000 CPU hours on NASA's Project Columbia supercomputer, currently one of the fastest machines available. (More details about the Via Lactea run are given in Diemand *et al.* (2007a, b). Movies, images, and data are available at http://www.ucolick.org/~diemand/vl.) The host halo mass at $z = 0$ is $M_{200} = 1.8 \times 10^{12} M_\odot$ within a radius of $r_{200} = 389$ kpc.

Following the spherical top-hat model, the common procedure used to describe the assembly of a dark-matter halo is to define at each epoch a virial radius $r_{\rm vir}$ (or,

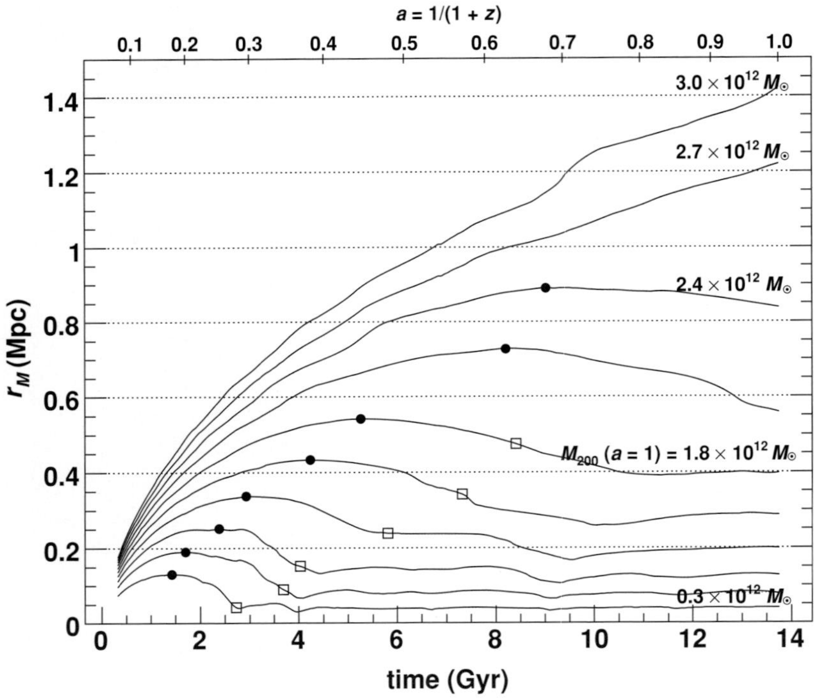

FIGURE 3.9. Evolution of radii r_M enclosing a fixed mass versus cosmic time or scale factor a. The enclosed mass grows in constant amounts of $0.3 \times 10^{12} M_\odot$ from bottom to top. Initially all spheres are growing in the physical (non-comoving) units used here. Inner shells turn around, collapse, and stabilize, whereas the outermost shells are still expanding today. Solid circles: points of maximum expansion at the turnaround time $t_{\rm max}$. Open squares: time after turnaround when r_M first contracts within 20% of the final value. These mark the approximate epoch of "stabilization."

equivalently, r_{200}), which depends on the cosmic background density at the time. As the latter decreases with the Hubble expansion, formal virial radii and masses grow with cosmic time even for stationary halos. Studying the transformation of halo properties within $r_{\rm vir}$ (or some fraction of it) mixes real physical change with apparent evolutionary effects caused by the growing radial window, and makes it hard to disentangle the two. Figure 3.9 shows the formation of Via Lactea where radial shells enclosing a fixed mass, r_M, have been used instead. Unlike $r_{\rm vir}$, r_M stops growing as soon as the mass distribution of the host halo becomes stationary on the corresponding scale. Note that *mass and substructure are constantly exchanged between these shells*, since r_M is not a Lagrangian radius enclosing the same material at all times, just the same amount of it. The fraction of material belonging to a given shell in the past that still remains within the same shell today is shown in Figure 3.10. The mixing is larger before stabilization, presumably because of shell crossing during collapse, and smaller near the center, where most of the mass is in a dynamically cold, concentrated old component (Diemand *et al.* 2005b). Outer shells numbers 9 and 10, for example, retain today less than 25% of the particles that originally belonged to them at $a < 0.4$.

Note that the collapse times also appear to differ from the expectations of spherical top-hat behavior. Shell number 5, for example, encloses a mean density of about $100\rho_{\rm c}^0$ today, has a virial mass of $1.5 \times 10^{12} M_\odot$, and should have virialized just now according to the spherical top-hat model. It did so instead much earlier, at $a = 0.6$. Even the

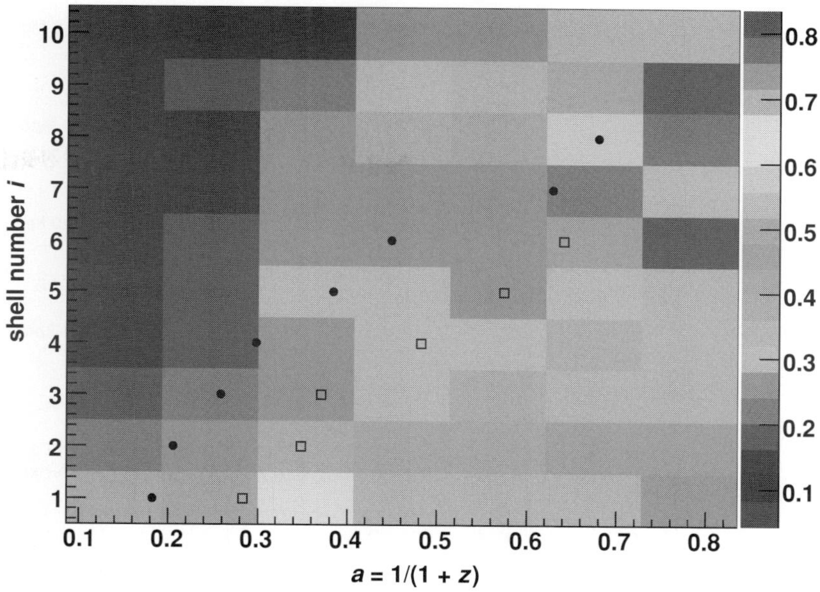

FIGURE 3.10. The fraction of material belonging to shell i at epoch a that remains in the same shell today. Shells are the same as in Figure 3.9, numbered from 1 (inner) to 10 (outer). Solid circles: time of maximum expansion. Open squares: stabilization epoch. Mass mixing generally decreases with time and toward the halo center.

next-larger shell with $1.8 \times 10^{12} M_\odot$ stabilized before $a = 0.8$. It appears that the spherical top-hat model provides only a crude approximation to the virialized regions of simulated galaxy halos.

To understand the mass-accretion history of the Via Lactea halo it is useful to analyze the evolution of mass within fixed physical radii. Figure 3.11 shows that the mass within all radii from the resolution limit of $\simeq$1 kpc up to 100 kpc grows during a series of major mergers before $a = 0.4$. After this phase of active merging and growth by accretion the halo mass distribution remains almost perfectly stationary at all radii. Only the outer regions ($\sim$400 kpc) experience a small amount of net mass accretion after the last major merger. The mass within 400 kpc increases only mildly, by a factor of 1.2 from $z = 1$ to the present. During the same time the mass within radii of 100 kpc and smaller, the peak circular velocity, and the radius at which this is reached all remain constant to within 10%. The fact that mass definitions inspired by the spherical top-hat model fail to accurately describe the real assembly of galaxy halos is clearly seen in Figure 3.11, where M_{200} is shown to increase at late times even when the halo's physical mass remains the same. This is just an artificial effect caused by the growing radial windows $r_{\rm vir}$ and r_{200} as the background density decreases. For Via Lactea M_{200} increases by a factor of 1.8 from $z = 1$ to the present, while the real physical mass within a 400-kpc sphere grows by only a factor of 1.2 during the same time interval, and by an even smaller factor at smaller radii.

3.3.6 *Smallest SUSY–CDM microhalos*

As already mentioned above, the key idea of the standard cosmological paradigm for the formation of structure in the Universe, namely that primordial density fluctuations grow by gravitational instability driven by cold, collisionless dark matter, is constantly being elaborated upon and explored in detail through supercomputer simulations, and tested

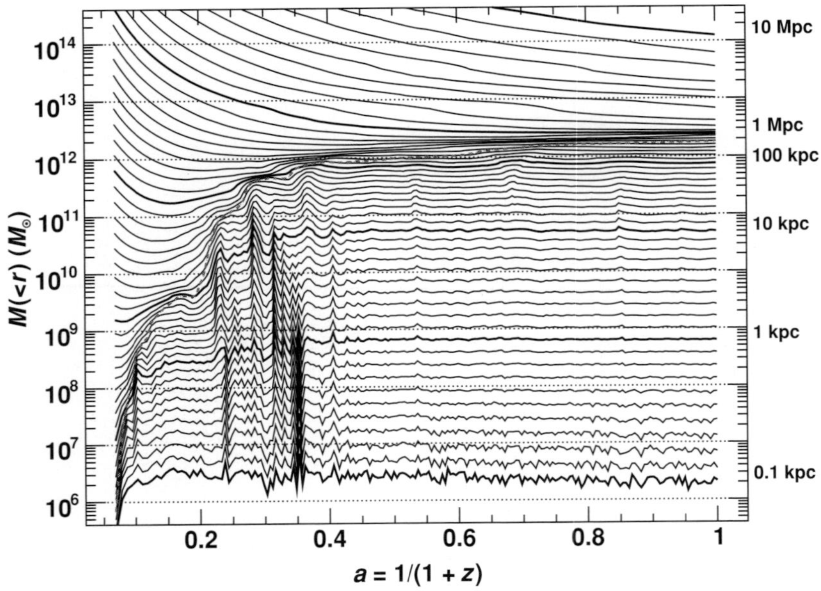

FIGURE 3.11. The mass-accretion history of Via Lactea. Masses within spheres of fixed physical radii centered on the main progenitor are plotted against the cosmological expansion factor a. The thick solid lines correspond to spheres with radii given by the labels on the left. The thin solid lines correspond to nine spheres of intermediate radii that are 1.3, 1.6, 2.0, 2.5, 3.2, 4.0, 5.0, 6.3, and 7.9 times larger than the next-smaller labeled radius. Dashed line: M_{200}. The halo is assembled during a phase of active merging before $a \simeq 0.37$ ($z \simeq 1.7$) and its net mass content remains practically stationary at later times.

against a variety of astrophysical observations. The leading candidate for dark matter is the neutralino, a weakly interacting massive particle predicted by the supersymmetry (SUSY) theory of particle physics. While in a SUSY–CDM scenario the mass of bound dark-matter halos may span about twenty orders of magnitude, from the most-massive galaxy clusters down to Earth-mass clumps (Green *et al.* 2004), it is the smallest microhalos that collapse first, and only these smallest scales are affected by the nature of the relic dark-matter candidate.

Recent numerical simulations of the collapse of the earliest and smallest gravitationally bound CDM clumps (Diemand *et al.* 2005a; Gao *et al.* 2005) have shown that tiny virialized microhalos form at redshifts above 50 with internal density profiles that are quite similar to those of present-day galaxy clusters. At these epochs a significant fraction of neutralinos has already been assembled into non-linear Earth-mass over-densities. If this first generation of dark objects were to survive gravitational disruption during the early hierarchical merger and accretion process – as well as late tidal disruption from stellar encounters (Zhao *et al.* 2007) – then over 10^{15} such clumps may populate the halo of the Milky Way. The nearest microhalos may be among the brightest sources of γ-rays from neutralino annihilation. Since the annihilation rate increases quadratically with the matter density, small-scale clumpiness may enhance the total γ-ray flux from nearby extragalactic systems (like M31), making them detectable by the forthcoming GLAST satellite or the next-generation of air Čerenkov telescopes.

The possibility of observing the fingerprints of the smallest-scale structure of CDM in direct and indirect dark-matter searches hinges on the ability of microhalos to survive the hierarchical clustering process as substructure within the larger halos that form at

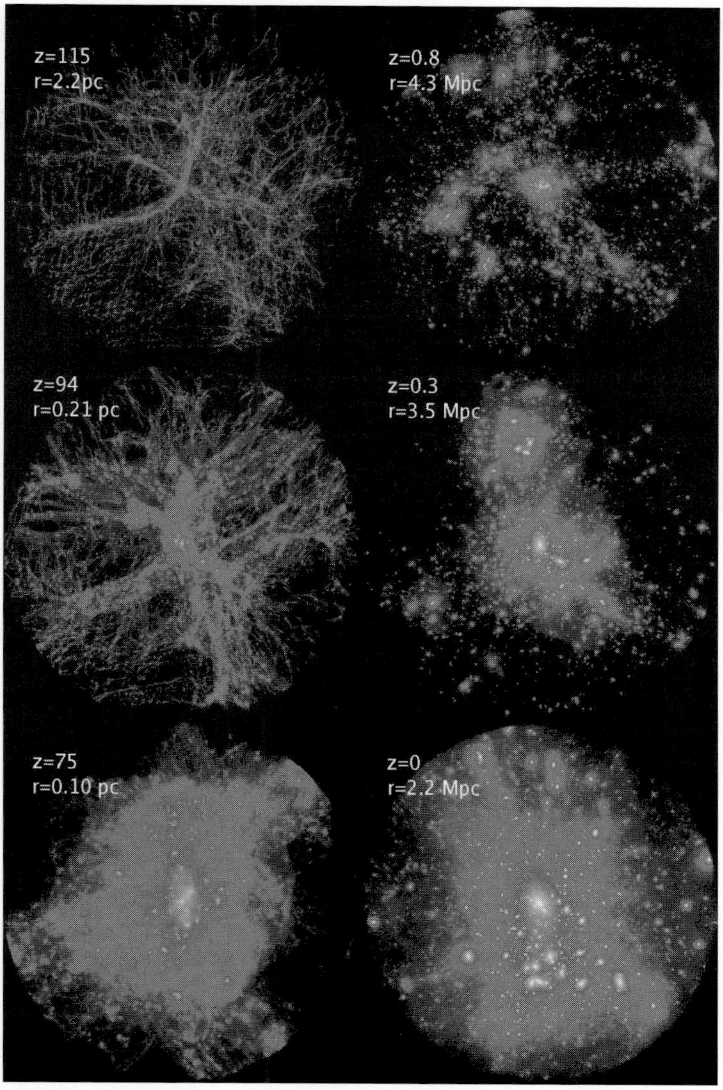

FIGURE 3.12. Local dark-matter-density maps. The left panels illustrate the almost simultaneous structure formation in the SUSY run at different epochs, within a sphere of physical radius r including a mass of $0.014 M_\odot$. A galaxy-cluster halo (right panels) forms in the standard hierarchical fashion: the dark-matter distribution is shown within a sphere of radius r including a mass of $5.9 \times 10^{14} M_\odot$. The SUSY and cluster halos have concentration parameters for an NFW profile of $c = 3.7$ and $c = 3.5$, respectively. In each image the logarithmic color scale ranges from 10 to 10^6 times $\rho_c(z)$.

later times. In recent years high-resolution N-body simulations have enabled the study of gravitationally bound subhalos with $M_{\rm sub}/M \geq 10^{-6}$ on galaxy (and galaxy-cluster) scales (e.g. Moore *et al.* 1999; Klypin *et al.* 1999; Stoehr *et al.* 2003). The main differences between these subhalos – the surviving cores of objects that fell together during the hierarchical assembly of galaxy-sized systems – and the tiny sub-microhalos discussed here is that on the smallest CDM scale the effective index of the linear power spectrum of mass-density fluctuations is close to -3. In this regime typical halo-formation times

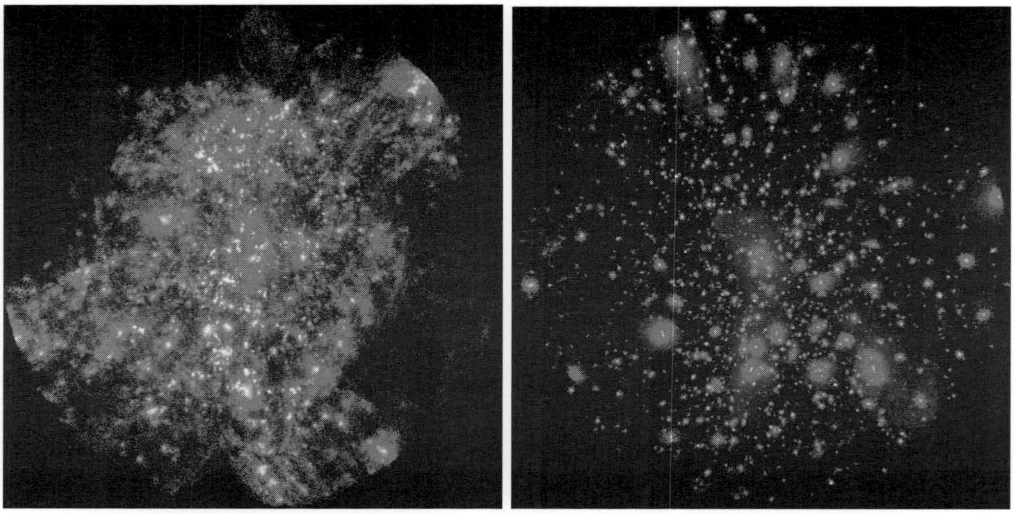

FIGURE 3.13. A phase-space density (ρ/σ^3, where σ is the one-dimensional velocity dispersion) map for a $z = 75$ SUSY halo (left) and a $z = 0$ galaxy cluster (right). Note the different color scales: relative to the average phase-space density, the logarithmic color scale ranges from 10 to 10^5 in the SUSY halo and from 10 to 10^7 in the cluster halo.

depend only weakly on halo mass, the capture of small clumps by larger ones is very rapid, and sub-microhalos may be more easily disrupted.

In Diemand *et al.* (2006) we presented a large N-body simulation of early substructure in a SUSY–CDM scenario characterized by an exponential cutoff in the power spectrum at $10^{-6} M_\odot$. The simulation resolves a $0.014 M_\odot$ parent "SUSY" halo at $z = 75$ with 14 million particles. Compared with a $z = 0$ galaxy cluster, substructure within the SUSY host is less evident both in phase space and in physical space (see Figures 3.12 and 3.13), and it is less resistant against tidal disruption. As the Universe expands by a factor of 1.3, between 20 and 40 percent of well-resolved SUSY substructure is destroyed, compared with only ~1 percent in the low-redshift cluster. Nevertheless SUSY substructure is just as abundant as in $z = 0$ galaxy clusters, i.e. the normalized mass and circular velocity functions are very similar.

3.4. The dawn of galaxies

3.4.1 *Uncertainties in the power spectrum*

As mentioned in the introduction, some shortcomings on galactic and sub-galactic scales of the currently favored model of hierarchical galaxy formation in a universe dominated by CDM have recently appeared. The significance of these discrepancies is still being debated, and "gastrophysical" solutions involving feedback mechanisms may offer a possible way out. Other models have been constructed in an attempt to solve the apparent small-scale problems of CDM at a more fundamental level, i.e. by reducing small-scale power. Although the "standard" ΛCDM model for structure formation assumes a scale-invariant initial power spectrum of density fluctuations with index $n = 1$, the recent WMAP data reveal strong evidence for a departure from scale invariance, with a best-fit value $n = 0.951^{+0.015}_{-0.019}$. Furthermore, the 1-year WMAP results favored a slowly varying spectral index, $dn/d\ln k = -0.031^{+0.016}_{-0.018}$, i.e. a model in which the spectral index varies

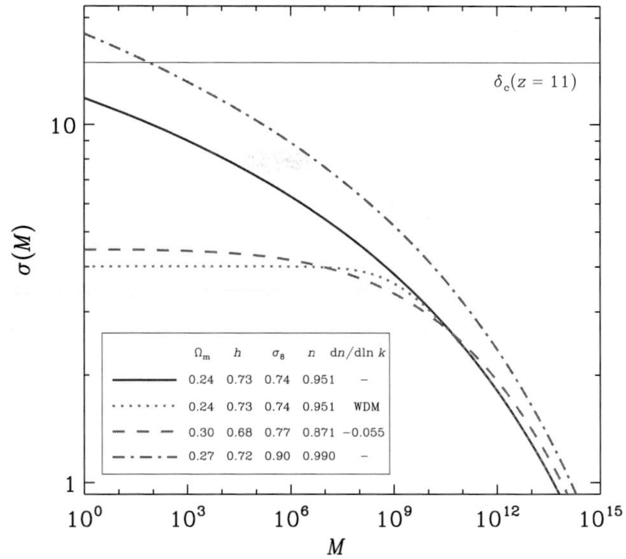

FIGURE 3.14. The variance of the matter-density field versus mass M, for several different cosmologies, all based on WMAP results. Solid curve: 3-year-WMAP-only best-fit model. Dotted curve: ΛWDM with a particle mass $m_X = 2$ keV, otherwise the same as before. Dashed curve: 3-year-WMAP best-fit running spectral index model. Dash–dotted curve: 1-year-WMAP-only best-fit tilted model. Here n refers to the spectral index at $k = 0.05\,\mathrm{Mpc}^{-1}$. The horizontal line at the top of the figure shows the value of the extrapolated critical collapse over-density $\delta_c(z)$ at the reionization redshift $z = 11$.

as a function of wavenumber k (Spergel *et al.* 2003). In the 3-year WMAP data such a "running spectral index" leads to a marginal improvement in the fit. Models with either $n < 1$ or $dn/d\ln k < 0$ predict a significantly lower amplitude of fluctuations on small scales than does standard ΛCDM. The suppression of small-scale power has the advantage of reducing the amount of substructure in galactic halos and makes small halos form later (when the Universe was less dense) (Zentner & Bullock 2002).

Figure 3.14 shows the linearly extrapolated (to $z = 0$) variance of the mass-density field for a range of cosmological parameters. Note that the new WMAP results prefer a low value for σ_8, the rms mass fluctuation in an $8h^{-1}$-Mpc sphere. This is consistent with a normalization by the $z = 0$ X-ray-cluster abundance (Reiprich & Böhringer 2002). For comparison we have also included a model with a higher normalization, $\sigma_8 = 0.9$, from the best-fit model to 1-year WMAP data. In the CDM paradigm structure formation proceeds bottom-up, so it then follows that the loss of small-scale power modifies structure formation most severely at the highest redshifts, significantly reducing the number of self-gravitating objects then. This, of course, will make it more difficult to reionize the Universe early enough.

It has been argued, for example, that one popular modification of the CDM paradigm, warm dark matter (WDM), has so little structure at high redshift that it is unable to explain the WMAP observations of an early epoch of reionization (Barkana *et al.* 2001). Yet the WMAP running-index model may suffer from a similar problem. A look at Figure 3.14 shows that $10^6 M_\odot$ halos will collapse at $z = 11$ from 2.4σ fluctuations in a tilted ΛCDM model with $n = 0.951$ and $\sigma_8 = 0.74$ (best-fit 3-year WMAP model), but from much rarer 3.7σ and 3.6σ fluctuations in the WDM and running-index models, respectively. The problem is that scenarios with increasingly rarer halos at early

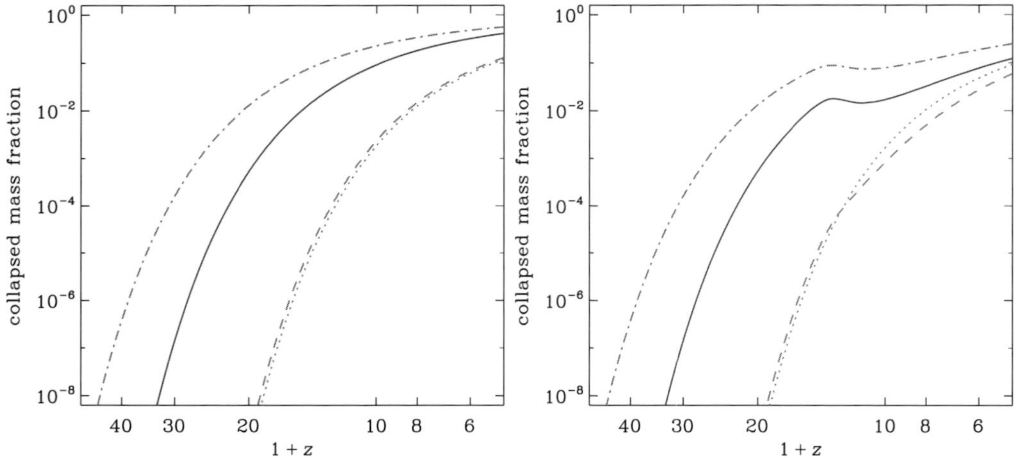

FIGURE 3.15. The mass fraction in all collapsed halos above the cosmological filtering (Jeans) mass as a function of redshift, for various power spectra. Curves are the same as in Figure 3.14. Left panel: filtering mass computed in the absence of reionization. Right panel: filtering mass computed assuming that the Universe is reionized and reheated to $T_e = 10^4$ K by UV radiation at $z \simeq 11$.

times require even-more-extreme assumptions (i.e. higher star-formation efficiencies and UV-photon-production rates) in order to be able to reionize the Universe suitably early (e.g. Somerville *et al.* 2003; Wyithe & Loeb 2003; Ciardi *et al.* 2003; Cen 2003). Figure 3.15 depicts the mass fraction in all collapsed halos with masses above the cosmological filtering mass for a case without reionization and one with reionization occurring at $z \simeq 11$. At early epochs this quantity appears to vary by orders of magnitude among models.

3.4.2 *First baryonic objects*

The study of the non-linear regime for the baryons is far more complicated than that of the dark matter because of the need to take into account pressure gradients and radiative processes. As a dark-matter halo grows and virializes above the cosmological Jeans mass through merging and accretion, baryonic material will be shock-heated to the effective virial temperature of the host and compressed to the same fractional over-density as the dark matter. The subsequent behavior of gas in a dark-matter halo depends on the efficiency with which it can cool. It is useful here to identify two mass scales for the host halos: (1) a *molecular cooling mass* M_{H_2} above which gas can cool via roto-vibrational levels of H_2 and contract, $M_{H_2} \approx 10^5[(1+z)/10]^{-3/2} M_\odot$ (virial temperature above 200 K); and (2) *an atomic cooling mass* M_H above which gas can cool efficiently and fragment via excitation of hydrogen Lyα, $M_H \approx 10^8[(1+z)/10]^{-3/2} M_\odot$ (virial temperature above 10^4 K). Figure 3.16 shows the cooling mechanisms at various temperatures for primordial gas. Figure 3.17 shows the fraction of the total mass in the Universe that is in collapsed dark-matter halos with masses greater than M_{H_2} and M_H at various epochs.

High-resolution hydrodynamics simulations of early structure formation are a powerful tool with which to track in detail the thermal and ionization history of a clumpy IGM and guide studies of primordial star formation and reheating. Such simulations performed in the context of ΛCDM cosmologies have shown that the first stars (the so-called "Population III") in the Universe formed out of metal-free gas in dark-matter minihalos of mass above a few times $10^5 M_\odot$ (Abel *et al.* 2000; Fuller & Couchman 2000; Yoshida *et al.* 2003; Reed *et al.* 2005) condensing from the rare high-ν_c peaks of the primordial density

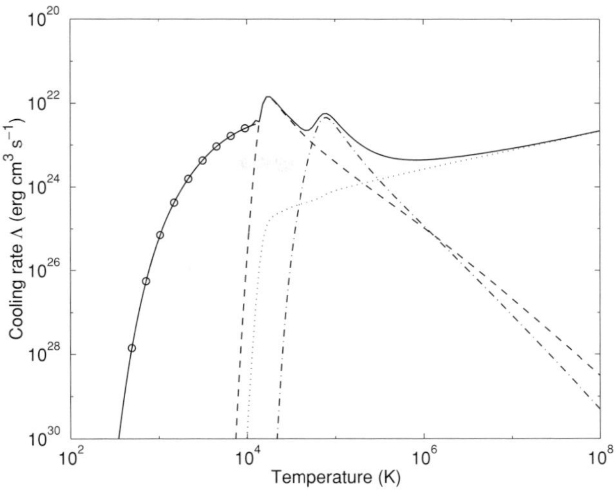

FIGURE 3.16. Cooling rates for *Bremsstrahlung* (dotted line), H (dashed line), and He (dash–dotted line) line cooling, and H_2 (circles) cooling. The e^-, H II, He II, and He III abundances were computed assuming collisional equilibrium, and the H_2 fractional abundance was fixed at 3×10^{-4}, which is typical for early objects. At temperatures between 100 and 10 000 K, the H_2 molecule is the most-effective coolant. From Fuller & Couchman (2000).

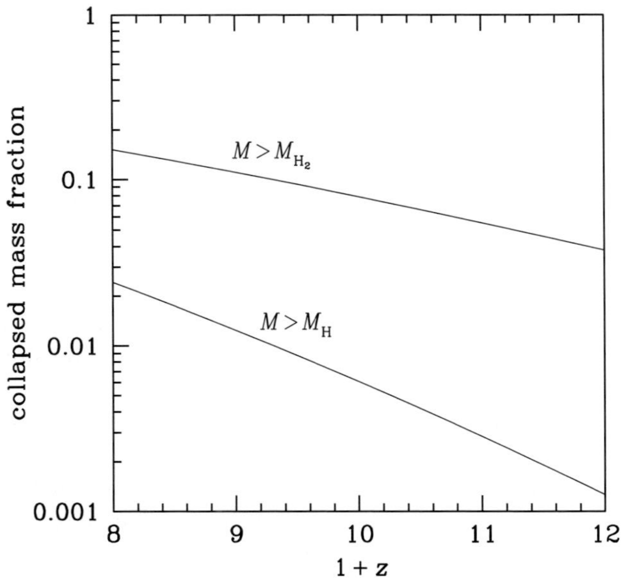

FIGURE 3.17. Total mass fraction in all collapsed dark-matter halos above the molecular cooling and the atomic cooling masses, M_{H_2} and M_H, as a function of redshift.

fluctuation field at $z > 20$, and were likely very massive (e.g. Abel *et al.* 2002; Bromm *et al.* 2002); see Bromm & Larson (2004) and Ciardi & Ferrara (2005) for recent reviews. In Kuhlen & Madau (2005) we used a modified version of ENZO, an adaptive-mesh-refinement (AMR), grid-based hybrid (hydrodynamic plus N-body) code developed by Bryan & Norman (see http://cosmos.ucsd.edu/enzo/) to solve the cosmological hydrodynamics equations and study the cooling and collapse of primordial gas in the first baryonic structures. The simulation samples the dark-matter density field in a 0.5-Mpc box with a mass resolution of $2000 M_\odot$ to ensure that halos above the cosmological Jeans

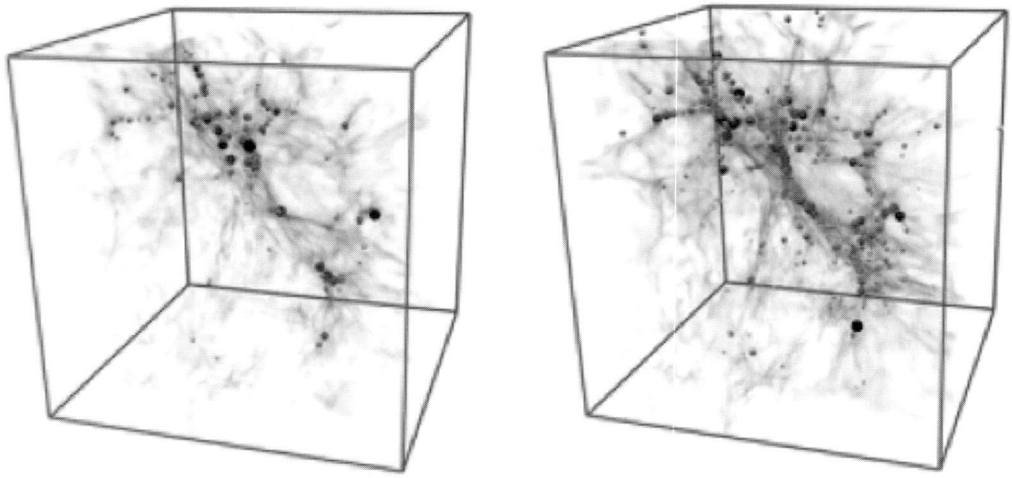

FIGURE 3.18. A three-dimensional volume rendering of the IGM in the inner 0.5-Mpc simulated box at $z = 21$ (left panel) and $z = 15.5$ (right panel). Only gas with over-density $4 < \delta_b < 10$ is shown: the locations of dark-matter minihalos are marked by spheres with sizes proportional to halo mass. At these epochs, the halo-finder algorithm identifies 55 ($z = 21$) and 262 ($z = 15.5$) bound clumps within the simulated volume. From Kuhlen & Madau (2005).

mass are well resolved at all redshifts $z < 20$. The AMR technique allows us to home in, with progressively finer resolution, on the densest parts of the "cosmic web." During the evolution from $z = 99$ to $z = 15$, refined (child) grids with twice the spatial resolution of the coarser (parent) grid are introduced when a cell reaches a dark-matter over-density (baryonic overdensity) of 2.0 (4.0). Dense regions are allowed to dynamically refine to a maximum resolution of 30 pc (comoving). The code evolves the non-equilibrium rate equations for nine species (H, H^+, H^-, e, He, He^+, He^{++}, H_2, and H_2^+) in primordial self-gravitating gas, including radiative losses from atomic- and molecular-line cooling, and Compton cooling by the cosmic background radiation.

The clustered structure around the most-massive peaks of the density field is clearly seen in Figure 3.18, a three-dimensional volume rendering of the simulated volume at redshifts 21 and 15.5. This figure shows gas at $4 < \delta_b < 10$, with the locations of dark-matter minihalos marked by spheres colored and sized according to their mass (the spheres are only markers; the actual shape of the halos is typically non-spherical). Several interleaving filaments are visible, at the intersections of which minihalos are typically found. At $z = 21$, 55 bound halos are identified in the simulated volume: by $z = 17.5$ this number has grown to 149, and by $z = 15.5$ to 262 halos. At this epoch, only four halos have reached the critical virial temperature for atomic cooling, $T_{\rm vir} = 10^4$ K.

The primordial fractional abundance of H_2 in the low-density IGM is small, $x_{H_2} \equiv [H_2/H] \simeq 2 \times 10^{-6}$, since at $z > 100$ H_2 formation is inhibited because the required intermediaries, either H_2^+ or H^-, are destroyed by CMB photons. Most of the gas in the simulation therefore cools by adiabatic expansion. Within collapsing minihalos, however, gas densities and temperatures are large enough that H_2 formation is catalyzed by H^- ions through the associative detachment reaction $H + H^- \longrightarrow H_2 + e^-$, and the molecular fraction increases at the rate $dx_{H_2}/dt \propto x_e n_{\rm H\,I} T_{\rm vir}^{0.88}$, where x_e is the number of electrons per hydrogen atom. For $I_{\rm vir}$ less than about a few thousand kelvins the virialization shock is not ionizing, the free electrons left over from recombination are depleted in the denser regions, and the production of H_2 stalls at a temperature-dependent

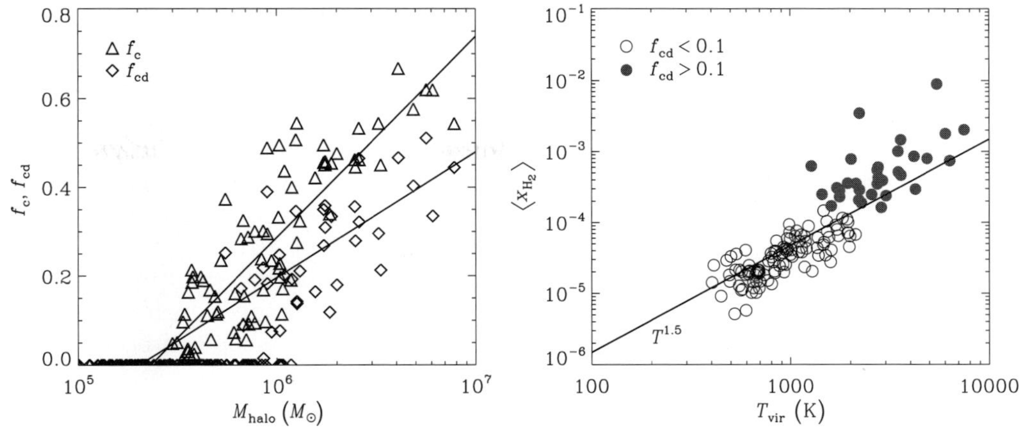

FIGURE 3.19. Left: the fraction of cold and cold plus dense gas within the virial radius of all halos identified at $z = 17.5$ with $T_{vir} > 400$ K, as a function of halo mass. Triangles: f_c, the fraction of halo gas with $T < 0.5\,T_{vir}$ and over-density $\delta_b > 1000$ that cooled via roto-vibrational transitions of H_2. Diamonds: the f_{cd}, fraction of gas with $T < 0.5T_{vir}$ and $\rho > 10^{19} M_\odot\,\mathrm{Mpc}^{-3}$ that is available for star formation. The straight lines represent mean regression analyses of f_c and f_{cd} with the logarithm of halo mass. Right: the mass-weighted mean H_2 fraction as a function of virial temperature for all halos at $z = 17.5$ with $T_{vir} > 400$ K and $f_{cd} < 0.1$ (empty circles) or $f_{cd} > 0.1$ (filled circles). The straight line marks the scaling of the temperature-dependent asymptotic molecular fraction.

asymptotic molecular fraction $x_{H_2} \approx 10^{-8}\,T_{vir}^{1.5}\,\ln(1 + t/t_{rec})$, where t_{rec} is the hydrogen-recombination time-scale (Tegmark *et al.* 1997). A typical H_2 fraction in excess of 200 times the primordial value is therefore produced after the collapse of structures with virial temperatures of order 10^3 K. This is large enough to efficiently cool the gas and allow it to collapse within a Hubble time unless significant heating occurs during this phase (Abel *et al.* 2000; Yoshida *et al.* 2003).

Figure 3.19 (left panel) shows the fraction of cold gas within the virial radius as a function of halo mass for all the halos identified at redshift 17.5. Following Machacek *et al.* (2001), we define f_c as the fraction of gas with temperature $<0.5T_{vir}$ and density more than 1000 times the background (this is the halo gas that is able to cool below the virial temperature because of H_2) and f_{cd} as the fraction of gas with temperature $<0.5T_{vir}$ and (physical) density $>10^{19} M_\odot$ Mpc^{-3} (this is the self-gravitating gas available for star formation). As in Machacek *et al.* (2001), we find that both f_c and f_{cd} are correlated with halo mass. The threshold for significant baryonic condensation (non-zero f_{cd}) is approximately $5 \times 10^5 M_\odot$ at these redshifts (Haiman *et al.* 1996). Also depicted in Figure 3.19 (right panel) is the mass-weighted mean molecular fraction of all halos with $T_{vir} > 400$ K. Filled circles represent halos with $f_{cd} > 0.1$, while open circles represent the others. The straight line marks the scaling of the asymptotic molecular fraction in the electron-depletion transition regime. The maximum gas density reached at redshift 15 in the most-refined region of our simulation is $4 \times 10^5\,\mathrm{cm}^{-3}$ (corresponding to an overdensity of 3×10^8): within this cold pocket the excited states of H_2 are in LTE and the cooling time is nearly independent of density.

3.4.3 *21-cm signatures of the neutral IGM*

It has long been known that neutral hydrogen in the diffuse IGM and in gravitationally collapsed structures may be directly detectable at frequencies corresponding to the red-shifted 21-cm line of hydrogen (Field 1959; Sunyaev & Zel'dovich 1975; Hogan & Rees

1979); see Furlanetto *et al.* (2006) for a recent review. The emission or absorption of 21-cm photons from neutral gas is governed by the spin temperature T_S, defined as

$$n_1/n_0 = 3\exp(-T_*/T_S). \tag{3.76}$$

Here n_0 and n_1 are the number densities of hydrogen atoms in the singlet and triplet $n = 1$ hyperfine levels, and $T_* = 0.068\,\mathrm{K}$ is the temperature corresponding to the energy difference between the levels. To produce an absorption or emission signature against the CMB, the spin temperature must differ from the temperature of the CMB, $T = 2.73(1+z)\,\mathrm{K}$. At $30 \le z \le 200$, prior to the appearance of non-linear baryonic objects, the IGM cools adiabatically faster than the CMB, spin-exchange collisions between hydrogen atoms couple T_S to the kinetic temperature T_e of the cold gas, and cosmic hydrogen can be observed in absorption (Scott & Rees 1990; Loeb & Zaldarriaga 2004). At lower redshifts, the Hubble expansion rarefies the gas and makes collisions inefficient: the spin states go into equilibrium with the radiation, and as T_S approaches T the 21-cm signal diminishes. It is the first luminous sources that make uncollapsed gas in the Universe shine again in the 21-cm line, by mixing the spin states either via Lyα scattering or via an increase in free-electron–atom collisions (e.g. Madau *et al.* 1997; Tozzi *et al.* 2000; Nusser 2005).

While the atomic physics of the 21-cm transition is well understood in the cosmological context, exact calculations of the radio signal expected during the era between the collapse of the first baryonic structures and the epoch of complete reionization have been difficult to do, since the result depends on the spin temperature, gas over-density, hydrogen neutral fraction, and line-of-sight peculiar velocities. When $T_S = T_e$, the visibility of the IGM at 21 cm revolves around the quantity $(T_e - T)/T_e$. If $T_e < T$, the IGM will appear in absorption against the CMB; in the opposite case it will appear in emission. To determine the kinetic temperature of the IGM during the formation of the first sources, one needs a careful treatment of the relevant heating mechanisms such as photoionization and shock-heating. In addition to the signal produced by the cosmic web, minihalos with virial temperatures of a few thousand kelvins form in abundance at high redshift, and are sufficiently hot and dense to emit collisionally excited 21-cm radiation (Iliev *et al.* 2002).

The H I spin temperature is a weighted mean of T_e and T,

$$T_S = \frac{T_* + T + yT_e}{1+y}, \tag{3.77}$$

where the coupling efficiency y is the sum of three terms,

$$y = \frac{T_*}{AT_e}(C_H + C_e + C_p). \tag{3.78}$$

Here $A = 2.85 \times 10^{-15}\,\mathrm{s}^{-1}$ is the spontaneous emission rate and C_H, C_e, and C_p are the de-excitation rates of the triplet due to collisions with neutral atoms, electrons, and protons. A fourth term must be added in the presence of ambient Lyα radiation, since intermediate transitions to the 2p level can mix the spin states and couple T_S to T_e, the "Wouthuysen–Field" effect (Wouthuysen 1952; Field 1958; Hirata 2006). In the absence of Lyα photons, to unlock the spin temperature of a neutral medium with (say) $T_e = 500\,\mathrm{K}$ from the CMB requires a collision rate $C_H > AT/T_*$, corresponding to a baryon over-density $\delta_b > 5[(1+z)/20]^{-2}$. Not only dense gas within virialized minihalos but also intergalactic filamentary structure heated by adiabatic compression or shock-heating may then be observable in 21-cm emission. A population of X-ray sources ("miniquasars") turning on at early stages (e.g. Madau *et al.* 2004; Ricotti *et al.* 2005) may make even the

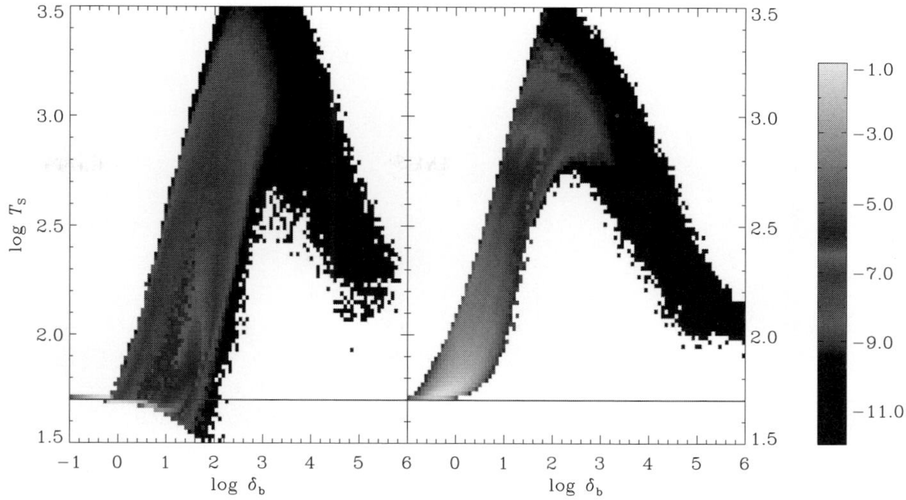

FIGURE 3.20. The two-dimensional distribution of spin temperature versus baryonic over-density at $z = 17.5$. The shading indicates the fraction of the simulated volume at a given (δ_b, T_S). Left: the NS run. The volume- and mass-averaged spin temperatures are 48.6 and 67.5 K, respectively. Right: MQ run. Only gas with neutral fraction >90% is shown in the figure. The volume- and mass-averaged spin temperatures are 82.6 and 138.0 K, respectively. The horizontal line marks the temperature of the CMB at that redshift.

low-density IGM visible in 21-cm emission as structures develop in the pre-reionization era.

In Kuhlen *et al.* (2006) we used the same hydrodynamic simulations of early structure formation in a ΛCDM universe as discussed in Section 1.4.2 to investigate the spin temperature and 21-cm brightness of the diffuse IGM prior to the epoch of cosmic reionization, at $10 < z < 20$. The two-dimensional distribution of gas over-density and spin temperature at $z = 17.5$ is shown in Figure 3.20 for a simulation with no radiation sources ("NS") and one ("MQ") in which a miniquasar powered by a $150 M_\odot$ black hole turns on at redshift 21 within a host halo of mass $2 \times 10^6 M_\odot$. The miniquasar shines at the Eddington rate and emits X-ray radiation with a power-law energy spectrum $\propto E^{-1}$ in the range 0.2–10 keV. The color coding in this phase diagram indicates the fraction of the simulated volume at a given (δ_b, T_S). In both runs we have assumed that there is no Lyα mixing, so that *the visibility of hydrogen at 21 cm is entirely determined by collisions.* Only gas with neutral fraction >90% is shown in the figure. The low-density IGM in the NS run lies on the yellow $T_S = T = 50.4$-K line: this is gas cooled by the Hubble expansion to $T_e \ll T$ that cannot unlock its spin states from the CMB and therefore remains invisible. At over-densities between a few and ~200, H–H collisions become efficient and adiabatic compression and shocks from structure formation heat up the medium to well above the radiation temperature. The coupling coefficient during this epoch is $y \sim \delta_b T_e^{-0.64}$: gas in this regime has $T < T_S \sim y T_e < T_e$ and appears in *emission* against the CMB (red and green swath). Some residual hydrogen with over-density up to a few tens, however, is still colder than the CMB, and is detectable in *absorption*. At higher densities, $y \gg 1$ and $T_S \to T_e$: the blue cooling branch follows the evolutionary tracks in the kinetic temperature–density plane of gas shock-heated to virial values, $T_e = 2000$–10^4 K, which is subsequently cooling down to ~100 K because of H_2 line emission.

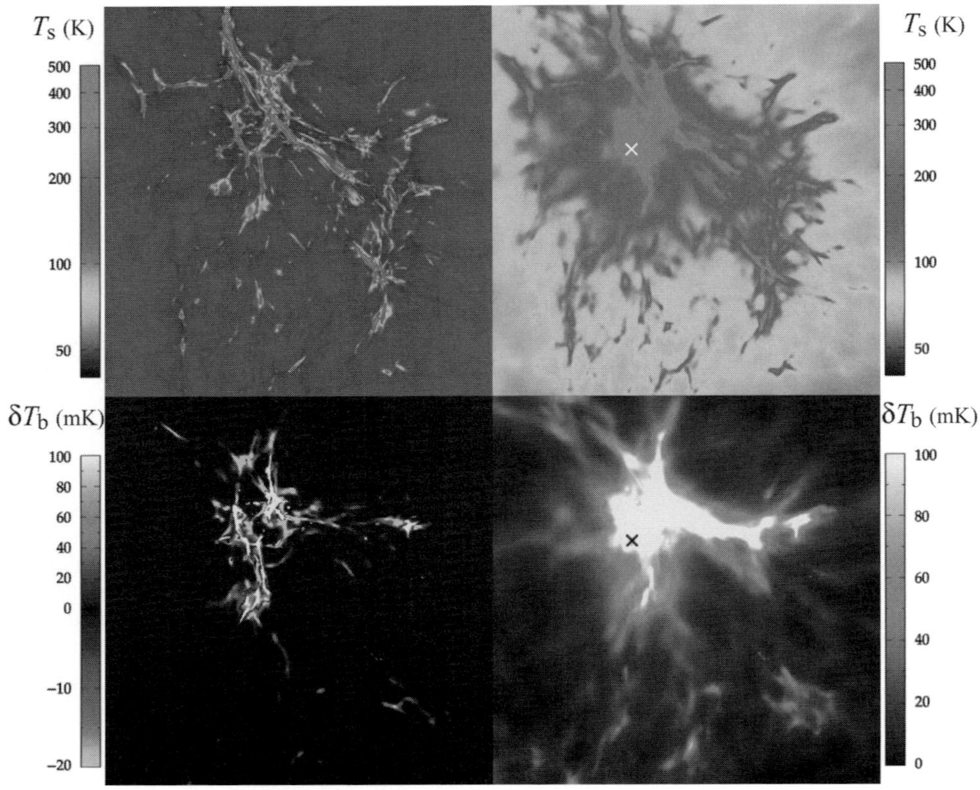

FIGURE 3.21. Projected (mass-weighted) spin temperature (upper panels, logarithmic scale) and 21-cm differential brightness temperature (lower panels, composite linear scale) in a 0.5-Mpc simulation box for runs "NS" (left) and "MQ" (right) at $z = 17.5$. The location of the miniquasar is indicated by crosses in the right panels.

The effect of the miniquasar on the spin temperature is clearly seen in the right panel of Figure 3.20. X-ray radiation drives the volume-averaged temperature and electron fraction (x_e) within the simulation box from $(8\,\mathrm{K}, 1.4 \times 10^{-4})$ to $(2800\,\mathrm{K}, 0.03)$, therefore producing a warm, weakly ionized medium (Kuhlen & Madau 2005). The H–H-collision term for spin exchange in the low-density IGM increases on average by a factor of $350^{0.36} \sim 8$, while the e–H-collision term grows to $C_e \sim 0.5 C_H$. Gas with $(\delta_b, T_e, x_e) = (1, 2800\,\mathrm{K}, 0.03)$ has coupling efficiency $y = 0.008$ and spin temperature $T_S = 73\,\mathrm{K} > T$, and can now be detected in emission against the CMB. Within 150 comoving kiloparsecs from the source, the volume-averaged electron fraction rises above 10%, and e–H collisions dominate the coupling.

A beam of 21-cm radiation passing through a neutral-hydrogen patch having optical depth τ and spin temperature T_S causes absorption and induces emission. In the comoving frame of the patch, the radiative-transfer equation yields for the brightness temperature through the region $T_b = T e^{-\tau} + T_S (1 - e^{-\tau})$. We have used our numerical simulations to perform 21-cm-radiation transport calculations, including the effect of peculiar velocities and local changes in spin temperature, gas density, and neutral-hydrogen fraction. The resulting 21-cm radio signal is shown in Figure 3.21 (lower panels), together with an image of the *projected* hydrogen spin temperature (upper panels): the latter highlights the abundance of structure within our simulation box on scales up to hundreds of kiloparsecs. Owing to Hubble and peculiar velocity shifts not all of this structure contributes to the δT_b map. In the NS simulation, coherent features in the IGM can be discerned in emission ($T_S > T$, $\delta T_b > 0$): this filamentary shock-heated structure is typically surrounded by mildly over-dense gas that is still colder than the CMB and appears in absorption ($T_S < T$, $\delta T_b < 0$). The covering factor of material with $\delta T_b \leq -10$ mK is 1.7%, which is comparable to that of material with $\delta T_b \geq +10$ mK: only about 1% of the pixels are brighter than +40 mK. While low-density gas (black color in the left-lower panel) remains invisible against the CMB in the NS run, *the entire box glows in 21-cm emission after being irradiated by X-rays.* The fraction of sky emitting with $\delta T_b > (+10, +20, +30, +50, +100)$ mK is now (0.57, 0.31, 0.19, 0.1, 0.035).

The above calculations show that, even in the absence of external heating sources, spin exchange by H–H collisions can make filamentary structures in the IGM (heated by adiabatic compression or shock-heating) observable in 21-cm emission at redshifts $z \leq 20$. Some cold gas with over-densities in the range 5–100 is still detectable in absorption at a level of $\delta T_b \leq -10$ mK, with a signal that grows as T_S^{-1} and covers a few percent of the sky. X-ray radiation from miniquasars preheats the IGM to a few thousand kelvins and increases the electron fraction: this boosts both the H–H and the e–H collisional coupling between T_S and T_e, making even low-density gas visible in 21-cm emission well before the Universe is significantly reionized. Any absorption signal has disappeared, and as much as 30% of the sky is now shining with $\delta T_b \geq +20$ mK. As pointed out by Nusser (2005), the enhanced e–H coupling makes the spin temperature very sensitive to the free-electron fraction: the latter is also a tracer of the H_2 molecular fraction in the IGM.

3.4.4 *Concluding remarks*

Since hierarchical clustering theories provide a well-defined framework in which the history of baryonic material can be tracked through cosmic time, probing the reionization epoch may then help constrain competing models for the formation of cosmic structures. Quite apart from uncertainties in the primordial power spectrum on small scales, however, it is the astrophysics of baryons that makes us unable to predict when reionization actually occurred. Consider the following illustrative example.

Photoionization of hydrogen requires more than one photon above 13.6 eV per hydrogen atom: of order $t/t_{rec} \sim 10$ (where t_{rec} is the volume-averaged hydrogen-recombination timescale) extra photons appear to be needed to keep the gas in over-dense regions and filaments ionized against radiative recombinations (Gnedin 2000; Madau *et al.* 1999). A "typical" stellar population produces during its lifetime about 4000 Lyman-continuum (ionizing) photons per stellar proton. A fraction $f \sim 0.25\%$ of cosmic baryons must then condense into stars to supply the requisite UV flux. This estimate assumes a standard (Salpeter) IMF, which determines the relative abundance of hot, high-mass stars versus cold, low-mass ones. The very first generation of stars ("Population III") must have formed, however, out of unmagnetized metal-free gas: characteristics, these, which may have led to a "top-heavy" IMF biased toward very massive stars (i.e. stars a few hundred times more massive than the Sun), quite different from the present-day Galactic case. Population III stars emit about 10^5 Lyman-continuum photons per stellar baryon (Bromm *et al.* 2001), approximately 25 times more than a standard stellar population. A correspondingly smaller fraction of cosmic baryons would have to collapse then into Population III stars to reionize the Universe, $f \sim 10^{-4}$. There are of course further complications. Since, at zero metallicity, mass loss through radiatively driven stellar winds is expected to be negligible, Population III stars may actually die losing only a small fraction of their mass. If they retain their large mass until death, stars with masses

$140M_\odot \leq m_* \leq 260M_\odot$ will encounter the electron–positron-pair instability and disappear in a giant nuclear-powered explosion (Fryer *et al.* 2001), leaving no compact remnants and polluting the Universe with the first heavy elements. In still-heavier stars, however, oxygen and silicon burning is unable to drive an explosion, and complete collapse to a black hole will occur instead (Bond *et al.* 1984). Thin-disk accretion onto a Schwarzschild black hole releases about 50 MeV per baryon. The conversion of a trace amount of the total baryonic mass into early black holes, $f \sim 3 \times 10^{-6}$, would then suffice to at least partially ionize and reheat the Universe.

The above discussion should make it clear that, despite much recent progress in our understanding of the formation of early cosmic structure and the high-redshift Universe, the astrophysics of first light remains one of the missing links in galaxy formation and evolution studies. We are left very uncertain about the whole era from 10^8 to 10^9 yr – the epoch of the first galaxies, stars, supernovae, and massive black holes. Some of the issues reviewed here are likely to remain a topic of lively controversy until the launch of the James Webb Space Telescope (JWST), which will be ideally suited to image the earliest generation of stars in the Universe. If the first massive black holes form in pregalactic systems at very high redshifts, they will be incorporated through a series of mergers into larger and larger halos, sink to the center owing to dynamical friction, accrete a fraction of the gas in the merger remnant to become supermassive, and form binary systems (Volonteri *et al.* 2003). Their coalescence would be signaled by the emission of low-frequency gravitational waves detectable by the planned Laser Interferometer Space Antenna (LISA). An alternative way to probe the end of the dark ages and discriminate between different possible reionization histories is through 21-cm tomography. Prior to the epoch of full reionization, 21-cm spectral features will display angular structure as well as structure in redshift space due to inhomogeneities in the gas-density field, ionized fraction of hydrogen, and spin temperature. Radio maps will show a patchwork (both in angle and in frequency) of emission signals from H I zones modulated by H II regions where no signal is detectable against the CMB (Ciardi & Madau 2003). The search at 21 cm for the epoch of first light has become one of the main science drivers for the next generation of radio arrays.

While many of the cosmological puzzles we have discussed can be tackled directly by studying distant objects, it has also become clear that many of today's "observables" within the Milky Way and nearby galaxies relate to events occurring at very high redshifts, during and soon after the epoch of reionization. In this sense, studies of galaxies in the Local Group ("near-field cosmology") can provide a crucial diagnostic link to the physical processes that govern structure formation and evolution in the early Universe ("far-field cosmology"). It is now well established, for example, that the hierarchical mergers that form the halos surrounding galaxies are rather inefficient, leaving substantial amounts of stripped-halo cores or "subhalos" orbiting within these systems (see Figure 3.22). Small halos collapse at high redshift when the Universe is very dense, so their central densities are correspondingly high. When these merge into larger hosts, their high densities allow them to resist the strong tidal forces that act to destroy them. Gravitational interactions appear to unbind most of the mass associated with the merged progenitors, but a significant fraction of these small halos survives as distinct substructure.

The Via Lactea simulation has recently shown that, in the standard CDM paradigm, galaxy halos should be filled with tens of thousands of subhalos that appear to have no optically luminous counterpart: this is more than an order of magnitude more than found in previous simulations. Their cumulative mass function is well fit by $N(>M_{\rm sub}) \propto M_{\rm sub}^{-1}$ down to $M_{\rm sub} = 4 \times 10^6 M_\odot$. Sub-substructure is apparent in all the larger satellites, and

FIGURE 3.22. A projected dark-matter density-squared map of the Via Lactea halo at the present epoch. The image covers an area of 800 kpc × 600 kpc, and the projection goes through a 600-kpc-deep cuboid containing a total of 110 million particles. The logarithmic color scale covers 20 decades in density squared.

a few dark-matter lumps are now resolved even in the Solar vicinity. In Via Lactea, the number of dark satellites with peak circular velocities above $5\,\mathrm{km\ s^{-1}}$ ($10\,\mathrm{km\ s^{-1}}$) exceeds 800 (120). As shown in Figure 3.23, such a finding appears to exacerbate the so-called "missing-satellite problem," the large mismatch between the 20 or so dwarf satellite galaxies observed around the Milky Way and the predicted large number of CDM

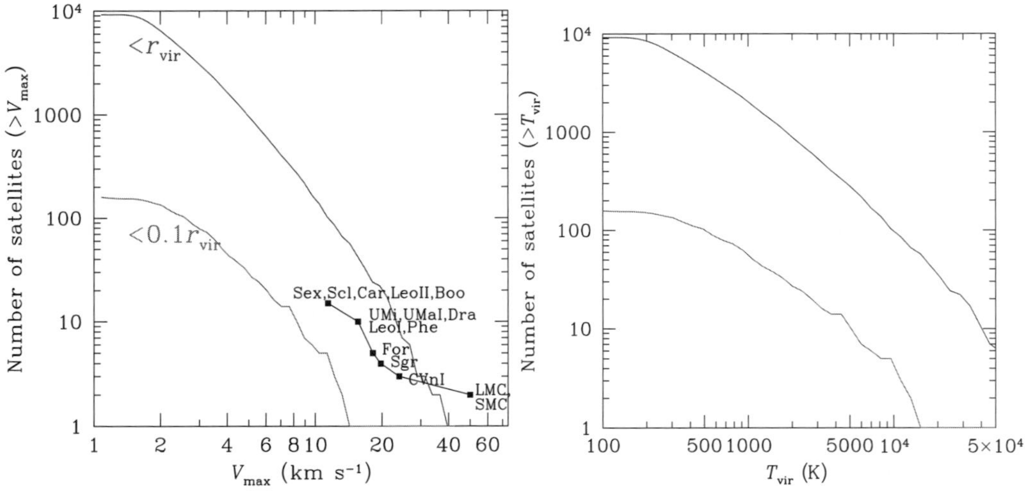

FIGURE 3.23. *Left:* Cumulative peak circular velocity function for all subhalos within Via Lactea's $r_{\rm vir}$ (*upper curve*) and for the subpopulation within the inner $0.1\,r_{\rm vir}$ (*lower curve*). *Solid line with points:* observed number of dwarf galaxy satellites around the Milky Way. *Right:* Same plotted versus virial temperature $T_{\rm vir}$.

subhalos (Moore *et al.* 1999; Klypin *et al.* 1999). Solutions involving feedback mechanisms that make halo substructure very inefficient at forming stars offer a possible way out (e.g. Bullock *et al.* 2000; Kravtsov *et al.* 2004; Moore *et al.* 2006). Even if most dark-matter satellites have no optically luminous counterparts, the substructure population may be detectable via flux-ratio anomalies in strong gravitational lenses (Metcalf & Madau 2001), through its effects on stellar streams (Ibata *et al.* 2002), or possibly via γ-rays from dark-matter annihilation in their cores (e.g. Bergstrom *et al.* 1999; Colafrancesco *et al.* 2006). We are entering a new era of studies of galaxy formation and evolution, in which fossil signatures accessible today within nearby galaxy halos will allow us to probe back to early epochs, and in which the basic building blocks of galaxies will become recognizable in the near field. Understanding galaxy formation is largely about understanding the survival of substructure and baryon dissipation within the CDM hierarchy.

Acknowledgments

I would like to thank my collaborators, Jürg Diemand and Michael Kuhlen, for innumerable discussions and for their years of effort on the science presented here. It is my pleasure to thank the organizers of "The Emission-Line Universe" for their hospitality and patience in waiting for this text, and all the students for making this a very enjoyable winter school. Support for this work was provided by NASA grant NNG04GK85G.

REFERENCES

Abel, T., Bryan, G., Norman, M. L. 2000, ApJ, 540, 39

Abel, T., Bryan, G., Norman, M. L. 2002, Science, 295, 93

Bardeen, J. M., Bond, J. R., Kaiser, N., Szalay, A. S. 1986, ApJ, 304, 15

Barkana, R., Haiman, Z., Ostriker, J. P. 2001, ApJ, 558, 482

Barkana, R., Loeb, A. 2001, PhR, 349, 125

Barnes, J. E. 1988, ApJ, 331, 699

Bergstrom, L., Edsjo, J., Gondolo, P., Ullio, P. 1999, Phys. Rev. D, 59, 043506

Bond, J. R., Arnett, W. D., Carr, B. J. 1984, ApJ, 280, 825

Bromm, V., Coppi, P. S., Larson, R. B. 2002, ApJ, 564, 23

Bromm, V., Kudritzki, R. P., Loeb, A. 2001, ApJ, 552, 464

Bromm, V., Larson, R. B., 2004, ARAA, 42, 79

Bryan, G. L., Norman, M. L. 1998, ApJ, 495, 80

Bullock, J. S., Kravtsov, A. V., Weinberg, D. H. 2000, ApJ, 539, 517

Bullock, J. S., Kravtsov, A. V., Weinberg, D. H. 2001, ApJ, 548, 33

Cen, R. 2003, ApJ, 591, 12

Ciardi, B., Ferrara, A. 2005, SSRv, 116, 625

Ciardi, B., Ferrara, A., White, S. D. M. 2003, MNRAS, 344, L7

Ciardi, B., Madau, P. 2003, ApJ, 596, 1

Colafrancesco, S., Profumo, S., Ullio, P. 2006, AA, 455, 21

Couchman, H. M. P., Rees, M. J. 1986, MNRAS, 221, 53

Diemand, J., Moore B., Stadel J. 2005, Nature, 433, 389

Diemand, J., Madau, P., Moore, B. 2005b, MNRAS, 364, 367

Diemand, J., Kuhlen, M., Madau, P. 2006, ApJ, 649, 1

Diemand, J., Kuhlen, M., Madau, P. 2007a, ApJ, 657, 262

Diemand, J., Kuhlen, M., Madau, P. 2007b, ApJ, submitted (astro-ph/0703337)

Fan, X. 2006, NewAR, 60, 665

Field, G. B. 1958, Proc. IRE, 46, 240

Field, G. B. 1959, ApJ, 129, 525

Fixsen, D. J., Cheng, E. S., Gales, J. M., Mather, J. C., Shafer, R. A., Wright, E. L. 1996, ApJ, 473, 576

Fryer, C. L., Woosley, S. E., Heger, A. 2001, ApJ, 550, 372

Fuller, T. M., Couchman, H. M. P. 2000, ApJ, 544, 6

Furlanetto, S. R., Oh, S. P., Briggs, F. H. 2006, PhR, 433, 181

Galli, D., Palla, F. 1998, AA, 335, 403

Gao, L., White, S. D. M., Jenkins, A., Frenk, C. S., Springel, V. 2005, MNRAS, 363, 379

Gnedin, N. Y. 2000, ApJ, 542, 535

Gnedin, N. Y., Hui, L. 1998, MNRAS, 296, 44

Green, A. M., Hofmann, S., Schwarz, D. J. 2004, MNRAS, 353, L23

Haiman, Z., Rees, M. J., Loeb, A. 1996, ApJ, 467, 52 5222

Haislip, J. *et al.* 2006, Nature, 440, 181

Helly, J. C., Cole, S., Frenk, C. S., Baugh, C. M., Benson, A., Lacey, C. Pearce, F. R. 2003, MNRAS, 338, 903

Hernquist, L. 1992, ApJ, 400, 460

Hirata, C. M. 2006, MNRAS, 367, 259

Hogan, C. J., Rees, M. J. 1979, MNRAS, 188, 791

Ibata, R. A., Lewis, G. F., Irwin, M. J., Quinn, T. 2002, MNRAS, 332, 915

Iliev, I. T., Shapiro, P. R., Ferrara, A., Martel, H. 2002, ApJ, 572, L123

Iye, M. *et al.* 2006, Nature, 443, 186

Klypin, A., Kravtsov, A. V., Valenzuela, O., Prada, F. 1999, ApJ, 522, 82

Kravtsov, A. V., Gnedin, O. Y., Klypin, A. A. 2004, ApJ, 609, 482

Kuhlen, M., Madau, P. 2005, MNRAS, 363, 1069

Kuhlen, M., Madau, P., Montgomery, R. 2006, ApJ, 637, L1

Lahav, O., Lilje, P. B., Primack, J. R., Ress, M. J. 1991, MNRAS, 251, 128

Lepp, S., Shull, J. M. 1984, ApJ, 280, 465

Loeb, A., Zaldarriaga, M. 2004, PhRvL, 92, 211301

Machacek, M. M., Bryan, G. L., Abel, T. 2001, MNRAS, 548, 509

Machacek, M. M., Bryan, G. L., Abel, T. 2003, MNRAS, 338, 273

Madau, P., Ferrara, A., Rees, M. J. 2001, ApJ, 555, 92

Madau, P., Haardt, F., Rees, M. J. 1999, ApJ, 514, 648

Madau, P., Meiksin, A., Rees, M. J. 1997, ApJ, 475, 492

Madau, P., Rees, M. J., Volonteri, M., Haardt, F., Oh, S. P. 2004, ApJ, 604, 484

Metcalf, R. B., Madau, P. 2001, ApJ, 563, 9

Moore, B., Diemand, J., Madau, P., Zemp, M., Stadel, J. 2006, MNRAS, 368, 563

Moore, B., Ghigna, S., Governato, F., Lake, G., Quinn, T., Stadel, J., Tozzi, P. 1999, ApJ, 524, L19

Navarro, J. F., Frenk, C. S., White, S. D. M. 1997, ApJ, 490, 493

Nusser, A. 2005, MNRAS, 359, 183

Peebles, P. J. E. 1968, ApJ, 153, 1

Peebles, P. J. E. 1993, *Principles of Physical Cosmology* (Princeton, MA: Princeton University Press)

Peebles, P. J. E., Dicke, R. H. 1968, ApJ, 154, 891

Pettini, M., Madau, P., Bolte, M., Prochaska, J. X., Ellison, S. L., Fan, X. 2003, ApJ, 594, 695

Rauch, M. 1998, ARAA, 36, 267

Reed, D. S., Bower, R., Frenk, C. S., Gao, L., Jenkins, A., Theuns, T., White, S. D. M. 2005, MNRAS, 363, 393

Reiprich, T. H., Böhringer, H. 2002, ApJ, 567, 716

Ricotti, M., Ostriker, J. P., Gnedin, N. Y. 2005, MNRAS, 357, 207

Ryan-Weber, E. V., Pettini, M., Madau, P. 2006, MNRAS, 371, L78

Ryden, B. 2003, *Introduction to Cosmology* (San Fransciso, CA: Addison Wesley)

Saslaw, W. C., Zipoy, D. 1967, Nature, 216, 967

Scott, D., Rees, M. J. 1990, MNRAS, 247, 510

Seager, S., Sasselov, D. D., Scott, D. 1999, ApJ, 523, L1

Somerville, R. S., Bullock, J. S., Livio, M. 2003, ApJ, 593, 616

Songaila, A. 2001, ApJ, 561, 153

Spergel, D. N. *et al.* 2003, ApJS, 148, 175

Spergel, D. N. *et al.* 2006, ApJ, in press (astro-ph/0603449)

Stadel, J. 2001, PhD thesis, University of Washington

Stoehr, F., White, S. D. M., Springel, V., Tormen, G., Yoshida, N. 2003, MNRAS, 345, 1313

Sugiyama, N. 1995, ApJS, 100, 281

Sunyaev, R. A., Zel'dovich, Ya. B. 1975, MNRAS, 171, 375

Tegmark, M., Silk, J., Rees, M. J., Blanchard, A., Abel, T., Palla, F. 1997, ApJ, 474, 1

Tozzi, P., Madau, P., Meiksin, A., Rees, M. J. 2000, ApJ, 528, 597

Volonteri, M., Haardt, F., Madau, P. 2003, ApJ, 582, 559

Wouthuysen, S. A. 1952, AJ, 57, 31

Wyithe, J. S. B., Loeb, A. 2003, ApJ, 588, L69

Yoshida, N., Abel, T., Hernquist, L., Sugiyama, N. 2003, ApJ, 592, 645

Zel'dovich, Y. B., Kurt, V. G., Sunyaev, R. A. 1969, JETP, 28, 146

Zentner, A. R., Bullock, J. S. 2002, PhRD, 66, 043003

Zhao, H. S., Hooper, D., Angus, G. W., Taylor, J. E., Silk, J. 2007, ApJ, 654, 697

4. Primeval galaxies

DANIEL SCHAERER

4.1. Introduction

What do we mean by primeval? According to the Webster dictionary "Primeval: adj. [primaevus, from: primus first + aevum age] of or relating to the earliest ages (as of the world or human history)". We will follow this definition and mostly discuss topics related to galaxies in the "early" Universe, whose limit we somewhat arbitrarily define at redshifts $z \gtrsim 6$, corresponding approximately to the first billion years (Gyr) after the Big Bang. In contrast the frequently employed adjective "primordial", defined as "Primordial: adj. [primordialis, from primordium origin, from primus first + ordiri to begin] a) first created or developed b) existing in or persisting from the beginning (as of a solar system or universe) c) earliest formed in the growth of an individual or organ", should not be used synonymously, for obvious reasons. Luckily "primeval" encompasses more than "primordial", otherwise there would not be much in the way of observational aspects to discuss (now in 2006–2007) in these lectures!

If we follow the history of discoveries of quasars and galaxies over the last few decades it is indeed impressive to see how progress has been made in detecting ever-more-distant objects, increasing samples at a given redshift and their analysis and interpretation. During the last decade, approximately since the pioneering observations of the Hubble Deep Field in 1996 (Williams *et al.* 1996) and the spectroscopic studies of a large sample of star-forming galaxies at redshift 3 by Steidel and collaborators (Steidel *et al.* 1996), the observational limits have continuously been pushed further, reaching now record redshifts of $z \sim 7$ (secure) (Iye *et al.* 2006) but maybe up to $\sim$10 (Pelló *et al.* 2004; Richard *et al.* 2006; Stark *et al.* 2007).

Most of this progress has been possible only thanks to the Hubble Space Telescope, the availability of 10-m class telescopes (Keck, VLT, SUBARU) and continuous improvements in detector technologies, especially in the optical and near-IR domain. Recently the IR Spitzer Space Telescope, with its 60-cm mirror, has begun to play an important role in characterizing the properties of the highest-redshift galaxies.

Not only have observations progressed tremendously, but also theory and numerical simulations now provide very powerful tools and great insight into the physics in the early Universe, first stars and galaxies. Within the model of hierarchical structure formation we have the following simplified global picture of primeval galaxies, their formation and interactions with the surrounding medium. Schematically, following the growth of quantum fluctuations after the Big Bang, one has, in parallel, structure formation (hierarchical), star formation in sufficiently massive haloes, "local" and "global" chemical evolution (including dust formation) and "local" and "global" reionization.† These different processes are coupled via several feedback mechanisms (radiation, hydrodynamics). In this way the first stars and galaxies are thought to form, begin their evolution, contribute to

† By *local* we here mean within a dark-matter (DM) halo, proto-cluster, or galaxy, i.e. at scales corresponding to the interstellar medium (ISM), intra-cluster medium (ICM), or the "nearby" intergalactic medium (IGM). The *global* scale refers here to cosmic scales, i.e. scales of the IGM.

The Emission-Line Universe, ed. J. Cepa. Published by Cambridge University Press.

the chemical enrichment and dust production and gradually reionize the Universe from shortly after the Big Bang to approximately 1 Gyr after that.

This global scenario and its various physical ingredients have been presented in depth in several excellent reviews, to which the reader is referred (Barkana & Loeb 2001; Bromm & Larson 2004; Ciardi & Ferrara 2005; Ferrara 2006). In these lectures I shall only briefly outline the most important theoretical aspects concerning the first stars and galaxies and their expected properties (Section 4.2). In Section 4.3 I will introduce and discuss Lyα, one of the strongest emission lines, if not the very strongest, in distant star-forming galaxies, and review numerous results concerning this line and its use as a diagnostic tool. Finally I will present an overview of our current observational knowledge about distant galaxies, mostly Lyα emitters and Lyman-break galaxies (Section 4.4). Questions remaining to be answered and some perspectives for the future are discussed in Section 4.4.4. It is the hope that these considerations may be helpful for students and other researchers who wish to acquire an overview of this very active and rapidly changing field and basics for its understanding, and maybe will also provide some stimulation for persons working on related topics to explore the rich connections between different fields intertwined in the early Universe and contributing to the richness of astrophysics.

4.2. Population III stars and galaxies: a "top-down" theoretical approach

We shall now briefly summarize the expected theoretical properties governing the first generations of stars and galaxies, i.e. objects that are of primordial composition or very metal-poor.

4.2.1 *Primordial star formation*

In present-day gas, with a heavy-element mass fraction (metallicity) up to $\sim$2%, C^+, O, CO and dust grains are excellent radiators (coolants) and the thermal equilibrium timescale is much shorter than the dynamical timescale. Hence large gas reservoirs can cool and collapse rapidly, leading to clouds with typical temperatures of $\sim$10 K. In contrast, primordial gas clouds would evolve almost adiabatically, since heavy elements are absent and H and He are poor radiators for $T < 10^4$ K. However, molecules such as H_2 and HD can form and cool the gas under these conditions. Approximately, it is found that at metallicities $Z \lesssim Z_{\rm crit} = 10^{-5\pm1} Z_\odot$ these molecules dominate the cooling (e.g. Schneider *et al.* 2002, 2004).

Starting from the largest scale relevant for star formation (SF) in galaxies, i.e. the scale of the DM halo, one can consider the conditions necessary for star formation (e.g. Barkana & Loeb 2001; Ferrara 2007). Such estimates usually rely on timescale arguments. Most importantly, the necessary condition for fragmentation that the cooling timescale is shorter than the free-fall timescale, $t_{\rm cool} \ll t_{\rm ff}$, translates to a minimum mass $M_{\rm crit}$ of the DM halo for SF to occur as a function of redshift. A classical derivation of $M_{\rm crit}$ is found in Tegmark *et al.* (1997); typical values of $M_{\rm crit}$† are about $10^7 M_\odot$ to $10^9 M_\odot$ from $z \sim 20$ to 5. However, the value of $M_{\rm crit}$ is subject to uncertainties related to the precise cooling function and to the inclusion of other physical mechanisms (e.g. ultra-high-energy cosmic rays), as discussed e.g. in the review of Ciardi & Ferrara (2005).

After SF has started within a DM halo, the "final products" may be quite diverse, depending in particular strongly on a variety of radiative and mechanical feedback processes. Schematically, taking fragmentation and feedback into account, one may foresee

† Remember that this denotes the total DM mass, not the baryonic mass.

the following classes of objects according to Ciardi *et al.* (2000): "normal" gaseous galaxies, naked star clusters (i.e. "proto-galaxies" that have blown away all their gas) and dark objects (where no stars formed, or where SF was rapidly turned off due to negative radiative feedback). At very high redshift ($z > 10$) naked star clusters may be more numerous than gaseous galaxies.

How does SF proceed within such a small "proto-galaxy" and what stars will be formed? Fragmentation may continue down to smaller scales. In general mass of the resulting stars will depend on the fragment mass, accretion rate, radiation pressure and other effects such as rotation, outflows, competitive accretion etc., forming a rich physics that cannot be described here; see e.g. reviews by Bromm & Larson (2004) and Ciardi & Ferrara (2005) and references therein. Most recent numerical simulations following early star formation at very low metallicities agree that at $Z \lesssim Z_{\rm crit}$ the smallest fragments are quite massive and that they undergo a runaway collapse accompanied by a high accretion rate resulting in (very) massive stars, (10–100)$M_\odot$ or larger, compared with a typical mass scale of $\sim M_\odot$ at "normal" (higher) metallicities (cf. Bromm & Larson 2004). This suggests that the stellar initial mass function (IMF) may differ significantly from the present-day distribution at $Z \lesssim Z_{\rm crit} = 10^{-5\pm1} Z_\odot$. The value of the critical metallicity is found to be determined mostly by fragmentation physics; in the transition regime around $Z_{\rm crit}$ the latter may in particular also depend on dust properties (cf. Schneider *et al.* 2002, 2004).

Determining the IMF at $Z < Z_{\rm crit}$ observationally is difficult and relies mostly on indirect constraints (e.g. Schneider *et al.* 2006). The most-direct approaches use the most-metal-poor Galactic halo stars found. From counts (metallicity distributions) of these stars, Hernandez & Ferrara (2001) find indications for an increase of the characteristic stellar mass at very low Z. Similar results have been obtained by Tumlinson (2006), using also stellar abundance patterns. However, no signs of very massive ($>130 M_\odot$) stars giving rise to pair-instability supernovae (see Section 4.2.4) have been found yet (cf. Tumlinson 2006). In Section 4.4 we will discuss attempts to detect Population III stars and to constrain their IMF *in situ* in high-redshift galaxies.

4.2.2 *Primordial stars: properties*

Now that we have formed individual (massive) stars at low metallicity, what are their internal and evolutionary properties? Basically these stars differ on two main points from their normal-metallicity equivalents: the initial source of nuclear burning and the opacity in their outer parts. Indeed, since Population III stars (or more precisely stars with metallicities $Z \lesssim 10^{-9} = 10^{-7.3} Z_\odot$) cannot burn on the CNO cycle like normal massive stars, their energy production has to rely initially on the less-efficient p–p chain. Therefore these stars have higher central temperatures. Under these conditions ($T \gtrsim 10^{8.1}$ K) and after the build-up of some amount of He, the 3α reaction becomes possible, leading to the production of some amounts of C. In this way the star can then "switch" to the more-efficient CNO cycle for the rest of H-burning, and its structure (convective interior, radiative envelope) is then similar to that of "normal" massive stars. Given the high central temperature and the low opacity (dominated by electron scattering throughout the entire star due to the lack of metals), these stars are more compact than corresponding Population II and I stars. Their effective temperatures are therefore considerably higher, reaching up to $\sim 10^5$ K for $M \gtrsim 100 M_\odot$ (cf. Schaerer 2002). The lifetimes of Population III stars are "normal" (i.e. $\sim$3 Myr at a minimum), since $L \sim M$, i.e. since the luminosity increases approximately linearly with the increase of the fuel reservoir. Other properties of "canonical" Population III stellar-evolution models are discussed in detail in Marigo *et al.* (2001), Schaerer (2002) and references therein.

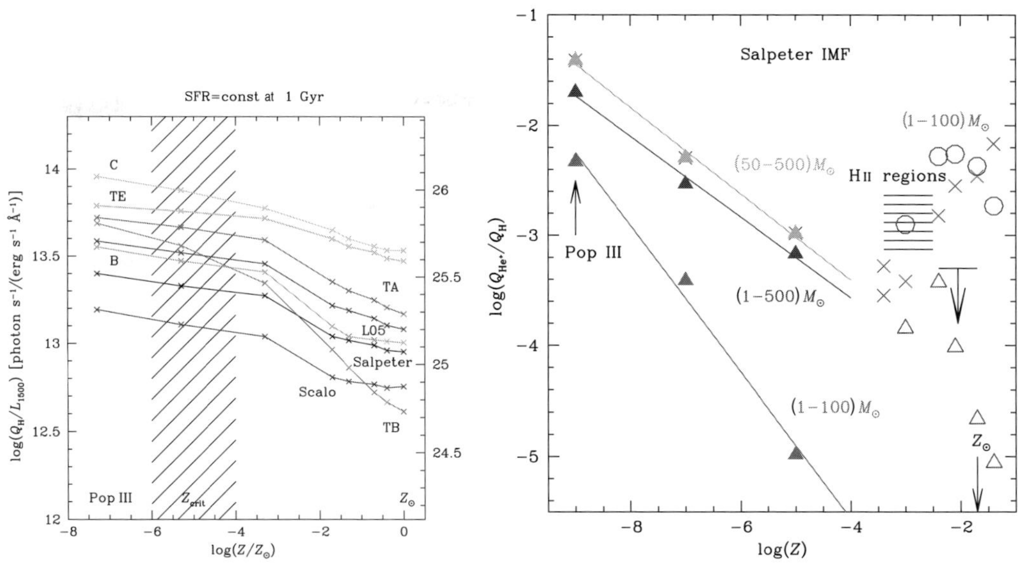

FIGURE 4.1. Left: relative output of hydrogen-ionizing photons to UV light, measured in the 1500-Å restframe, $Q_{\rm H}/L_{1500}$, as a function of metallicity for constant star formation over 1 Gyr. Results for different IMFs, including a Salpeter, Scalo and top-heavier cases, are shown. The shaded area indicates the critical metallicity range where the IMF is expected to change from a "normal" Salpeter-like regime to a more-massive IMF. Right: hardness $Q({\rm He}^+)/Q({\rm H})$ of the $\rm He^+$-ionizing flux for constant star formation as a function of metallicity (in mass fraction) and for different IMFs. At metallicities above $Z \geq 4 \times 10^{-4}$ the predictions from our models (crosses), as well as those of Leitherer *et al.* (1999, open circles), and Smith *et al.* (2002, open triangles) are plotted. The shaded area and the upper limit (at higher Z) indicate the range of the empirical hardness estimated from H II-region observations. From Schaerer (2003).

More-sophisticated stellar evolution models including many physical processes related to stellar rotation are now being constructed (cf. Meynet *et al.* 2006; Ekström *et al.* 2006). Whereas before it was thought that mass loss would be negligible for Population III and very-metal-poor stars, since radiation pressure is very low and pulsational instabilities may occur only during a very short phase, cf. Kudritzki (2002) and Baraffe *et al.* (2001), fast rotation – due to fast initial rotation and inefficient transport of angular momentum – may lead to mechanical mass loss, when these stars reach critical (break-up) velocity. Rotation also alters the detailed chemical yields, may lead to an evolution at hotter $T_{\rm eff}$, even to Wolf–Rayet (WR) stars, and may alter the final fate of Population III/very-metal-poor stars, which may in this way even avoid the outcome of becoming a "classical" pair-instability supernova (PISN, cf. below). Many details and the implications of these models on observable properties of metal-free/very-metal-poor populations still remain to be worked out.

4.2.3 *Primordial stars and galaxies: observable properties*

The observable properties of individual Population III and metal-poor stars and of an integrated population of such stars can be predicted using stellar-evolution models, appropriate non-local-thermodynamic-equilibrium (NLTE) stellar atmospheres and evolutionary synthesis techniques; see e.g. Tumlinson *et al.* (2001), Bromm *et al.* (2001) and detailed discussions in Schaerer (2002, 2003).

Given the exceptionally high effective temperatures of Population III stars on the zero-age main sequence, such objects emit a larger fraction of the luminosity in the Lyman

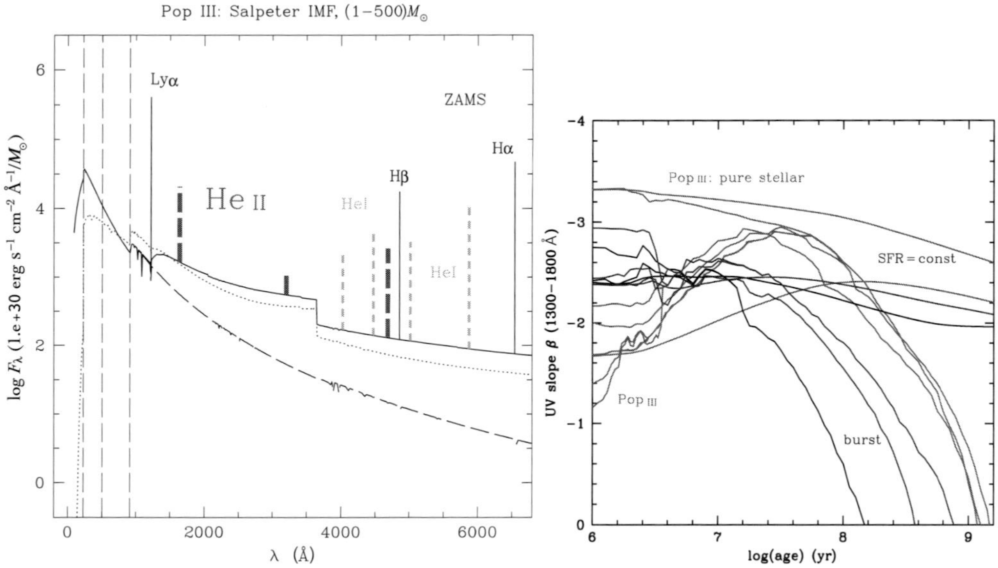

FIGURE 4.2. Left: the spectral energy distribution of a very young Population III galaxy including H and He recombination lines. The pure stellar continuum (neglecting nebular emission) is shown by the dashed line. For comparison the SED of the $Z = (1/50)Z_\odot$ population (model ZL: Salpeter IMF from $M_\odot$ to $150M_\odot$) is shown by the dotted line. The vertical dashed lines indicate the ionization potentials of H, He^0 and He^+. Note the presence of the unique He II features (shown as thick dashed lines) and the importance of nebular continuous emission. From Schaerer (2002). Right: Temporal evolution of the UV slope β measured between 1300 and 1800 Å from synthesis models of various metallicities and for instantaneous bursts (solid lines) and constant SF (long dashed lines). Full lines show Solar-metallicity models, metallicities between $Z = 10^{-5}$ and zero (Population III) and intermediate cases of $Z = 0.004$ and 0.0004. The dotted lines show β if nebular continuous emission is neglected, i.e. assuming pure stellar emission. Note especially the strong degeneracies of β in age and metallicity for bursts, the insensitivity of β to Z for constant SF and the rather red slope for young very-metal-poor bursts. From Schaerer & Pelló (2005).

continuum and have a much-harder ionizing spectrum than do higher-metallicity stars. For example, a Population III star of $5M_\odot$ is still an ionizing source! In other words, stellar populations at low metallicity are characterized by a high ionization efficiency (per unit stellar mass formed) and by a hard spectrum, as illustrated in Figure 4.1. For an unchanged IMF, e.g. Salpeter, the ionizing output normalized with respect to the UV flux density increases by a factor of ∼2 or more on going from Solar metallicity to Population III. However, this increase may be much more substantial if the IMF favours massive stars at low Z, as argued before.

The predicted integrated spectrum of a very-young zero-age main-sequence (ZAMS) ensemble of Population III stars is shown in Figure 4.2. Its main characteristics are the presence of strong H emission lines (in particular strong Lyα, cf. below) due to the strong ionizing flux, He^+ recombination lines (especially He II λ1640) due to spectral hardness and strong/dominating nebular continuum emission (cf. Schaerer 2002). The strength of Lyα can be used to identify interesting Population III or very-metal-poor galaxy candidates (cf. Section 4.3). The detection of nebular He II λ1640, if shown to be due to stellar photoionization, i.e. non-active-galactic-nucleus (AGN) origin, would be a very interesting signature of primordial (or very close to primordial) stars. Indeed, as shown on the right of Figure 4.1, very hard spectra are predicted only at $Z \lesssim 10^{-5 \cdots -6} Z_\odot$.

It is often heard that Population III, primeval or similar galaxies should be distinguished by bluer colours, e.g. measured in the rest-frame UV, as one would naively expect. Although the colours of stars do indeed get bluer on average with decreasing metallicity, this is no longer the case for the integrated spectrum of such a population, since nebular continuum emission (originating from the H II regions surrounding the massive stars) may dominate the spectrum, even in the UV. This leads to a much-redder spectrum, as shown in Figure 4.2 (left). Taking this effect into account leads in fact to a non-monotonic behaviour of the slope (colour) of the UV spectrum with metallicity, as illustrated in Figure 4.2 (right). This fact, and the dependence of the UV slope on the star-formation history on timescales shorter than 10^8–10^9 yr, corresponding to 10%–100% of the Hubble time at $z \gtrsim 6$, show that the interpretation of the UV slope (or colour) of primeval galaxies must be performed with great caution.

4.2.4 *Final fate*

The end stages of very-metal-poor and Population III stars may also differ from those at higher metallicity, with several interesting consequences also for the observational properties of primeval galaxies. In particular such massive stars may at the end of their evolution exhibit such conditions in terms of central temperature and density that the creation of electron–positron pairs occurs, leading to an instability that will completely disrupt the star. This phenomenon is known as a pair-instability supernova (PISN),† and there exists a rich literature about the phenomenon and its many implications. Here we shall only summarize the main salient points and recent findings.

A recent overview of the various "final events" and remnants is found in Heger *et al.* (2003). PISNs are thought to occur for stars with initial masses of $M \sim (140–260)M_\odot$ at very low Z. Owing to their high energy and to non-negligible time dilation, which increases the duration of their "visibility", PISNs are potentially detectable out to very high redshift (e.g. Weinmann & Lilly 2005; Scannapieco *et al.* 2005; Wise & Abel 2005). Large amounts of gas are ejected, as the event disrupts the star completely. Furthermore, the processed matter contains peculiar nucleosynthesic signatures, which may in principle be distinguished from those of normal SNs (cf. below). Finally PISNs are also thought to be the first dust-production factories in the Universe (cf. Schneider *et al.* 2004). Thus PISNs may be observable directly and indirectly, which would be very important to confirm or disprove the existence of such massive stars, i.e. to constrain the IMF of the first stellar generations. Currently, however, there is no such confirmation, as we will go on to discuss.

4.2.5 *Nucleosynthesis and abundance pattern*

Among the particularities of PISNs are the production of large quantities of O and Si, which translate e.g. into large O/C and Si/C abundance ratios potentially measurable in the IGM. More generally one expects roughly solar abundance of nuclei with even nuclear charge (Si, S, Ar, ...) and deficiencies in odd nuclei (Na, Al, P, V, ...) i.e. a strong so-called odd/even effect, and no elements heavier than Zn, due to the lack of s- and r-processes; see Heger & Woosley (2002) for recent predictions.

Results from abundance studies of the most-metal-poor halo stars in the Galaxy do not show the odd/even effect predicted for PISNs. In the face of our current knowledge, in particular on nucleosynthesis, quantitative analysis of the observed abundance pattern thus disfavours IMFs with a large fraction of stars with masses $M \sim (140–260)M_\odot$ (Tumlinson 2006). However, the abundance pattern and other constraints are compatible

† Sometimes the terms pair-creation SN and pair-production SN are also used.

with a qualitative change of the IMF at $Z \lesssim 10^{-4} Z_{\odot}$ as suggested by simulations (cf. above).

4.2.6 *Dust at high* z

Dust is known to be present out to the highest redshifts from damped-Lyα absorbers (DLA), from sub-millimetre emission in $z \sim 6$ quasars (e.g. Walter *et al.* 2003), from a gamma-ray-burst (GRB) host galaxy at $z = 6.3$ (Stratta *et al.* 2007) and possibly also from the spectral energy distribution (SED) of some normal galaxies at $z \sim 6$ (Schaerer & Pelló 2005). We also know that dust exists in the most-metal-poor galaxies, as testified e.g. by the nearby galaxy SBS 0335-052 with a metallicity of $\sim(1/50)Z_{\odot}$.

Since the age of the Universe at $z > 6$ is ~1 Gyr at most, longer-lived stars cannot be invoked to explain the dust production in primeval galaxies. Among the possible "short-lived" dust producers are SN II, PISN and maybe also WR stars or massive AGB stars. SN II are known dust producers (e.g. SN1987A), albeit maybe not producing enough dust. Efficient dust production is found in explosions of SN II and PISNs (e.g. Todini & Ferrara 2001; Schneider *et al.* 2004). At zero metallicity PISNs may provide a very efficient mechanism, converting up to 7%–20% of PISN mass into dust.

Evidence for dust produced by SN has been found from the peculiar extinction curve in the BAL QSO SDSS1048+46 at $z = 6.2$, which is in good agreement with SN dust models (Maiolino *et al.* 2004). Similar indications have been obtained recently from a GRB host galaxy at $z = 6.3$ (Stratta *et al.* 2007). Whether this is a general feature remains, however, to be established. Furthermore, the most important questions, including how common dust is in high-z galaxies, in what quantities and up to what redshift, have not yet been touched upon. Forthcoming IR to sub-millimetre facilities such as Herschel and especially ALMA will allow us to address these important issues.

4.3. Lyα physics and astrophysics

Since Lyα, one of the strongest emission lines in the UV, plays an important role in searches for and studies of distant and primeval galaxies, we wish to discuss this line, the basic principles governing it, its diagnostics and possible difficulties, empirical findings etc. To the best of my knowledge few if any reviews or lectures summarizing these topics in a single text exist.

4.3.1 *ISM emission and "escape"*

All ionized regions, i.e. H II regions, the diffuse ISM and like regions in galaxies, emit numerous emission lines, including recombination lines from H, He and other atoms, and forbidden semi-forbidden and fine-structure metal lines resulting from de-excitations of these atoms; see the textbooks of Osterbrock & Ferland (2006) and Dopita & Sutherland (2003), and Chapter 1 in this volume. All galaxies with ongoing massive-star formation (somewhat loosely called "starbursts" hereafter) emitting intense UV radiation and an ionizing flux (i.e. energy at >13.6 eV) will thus "intrinsically", viz. at least in their H II regions, exhibit Lyα emission.

From quite simple considerations one can find that the luminosity in a given H recombination line is proportional to the number of ionizing photons (i.e. Lyman-continuum photons), $L(\mathrm{Ly}\alpha, \mathrm{H}\alpha, \ldots) = c_l Q_{\mathrm{H}}$, where Q_{H} is the Lyman-continuum flux in photons s^{-1} and c_l a "constant" depending somewhat on the nebular temperature T_e and the electron density n_e. For hydrogen, about two thirds of the recombinations lead to the emission of a Lyα photon, corresponding to the transition from level 2 to the ground state (Spitzer 1978; Osterbrock & Ferland 2006).

Furthermore the relative intensities of two different H recombination lines are known and relatively slowly varying functions of temperature and density, e.g. $I(\mathrm{Ly}\alpha)/I(\mathrm{H}n) = c(T, n_e)$.

Already in the sixties it was recognized that Lyα could be important in searches for primeval galaxies (e.g. Partidge & Peebles 1967). Indeed, at (very) low metallicities the Lyα line is expected to be strong, if not dominant, for several reasons: there is an increasing ionizing flux from stellar populations; Lyα can become the dominant cooling line when few metals are present; and one expects greater emissivity due to collisional excitation in a nebula with higher temperature. As a result up to ~10% of the bolometric luminosity may be emitted in Lyα, rendering the line potentially detectable out to the highest redshifts!

This prospect triggered various searches for distant Lyα emitters, which remained, however, basically unsuccessful until the 1990s (see Section 4.4), for the reasons discussed below. In any case, it is interesting to note that most of the observational features predicted nowadays for Population III galaxies (Section 4.2.3) were predicted by early calculations, such as Partridge & Peebles' (1967), including of course the now-famous Lyman break (Section 4.4).

To anticipate it is useful to mention here the basics of the Lyα escape problem. In short, even for very low column densities of $N_{\mathrm{H\,I}} \gtrsim 10^{13}\,\mathrm{cm}^{-2}$ the Lyα line is optically thick. Therefore radiation transfer within the galaxy determines the emergent line profile and the Lyα "transmission"! Furthermore, dust may destroy Lyα photons. Overall, the fate of Lyα photons emitted in a galaxy can be one of the following: (1) scattering until escape, forming thus an extended Lyα "halo"; (2) destruction by dust; or (3) destruction through two-photon emission. However, process (3) is possible only in the ionized region.

4.3.2 *Lyα: the observational problem*

As already mentioned, there were several unsuccessful searches for Lyα emission from $z \sim$ 2–3 "primordial" galaxies in the 1980s and 1990s (Pritchet 1994). Why these difficulties occurred could be understood from observations of nearby starbursts, which revealed one or two puzzles, namely a small number of Lyα-emitting galaxies and/or lower-than-expected Lyα emission. The second puzzle could, of course, in principle explain the first one. In particular UV spectra of nearby starbursts (Lyα) taken with the IUE satellite and optical spectra (Hα, Hβ) showed that (i) after extinction correction, the relative line intensity of e.g. $I(\mathrm{Ly}\alpha)/I(\mathrm{H}\beta)$ was much smaller than the expected case-B value and the Lyα equivalent width $W(\mathrm{Ly}\alpha)$ smaller than expected from evolutionary synthesis models, and (ii) these findings do not depend on metallicity (e.g. Meier & Terlevich 1981; Hartmann *et al.* 1984; Deharveng *et al.* 1986).

Among the possible explanations put forward were (a) dust, which would destroy the Lyα photons (Charlot & Fall 1993); (b) an inhomogeneous ISM geometry, not dust, as primary determining factor (Giavalisco *et al.* 1996); and (c) a short "duty cycle" of SF to explain the small number of Lyα emitters. Also (d), Valls-Gabaud (1993) argued that with an appropriate, i.e. metallicity-dependent, extinction law (i) was no problem. Also, he stressed the importance of underlying stellar Lyα absorption.

Dust as a sole explanation was rapidly ruled out by the observations of I Zw 18 and SBS 0335-052, the most-metal-poor starbursts known, which exhibit no Lyα emission, actually having even a damped-Lyα-absorption profile (Kunth *et al.* 1994; Thuan & Izotov 1997). However, we now know (from ISO and Spitzer observations) that these objects contain also non-negligible amounts of dust (Thuan *et al.* 1999; Wu *et al.* 2007), although it is not clear whether and how it is related to the line-emitting regions, in particular spatially. From the absence of correlations between different measurements of

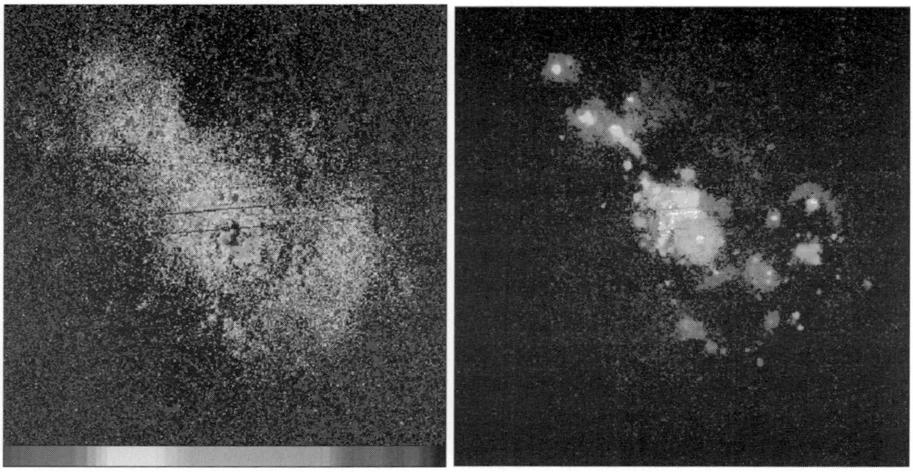

FIGURE 4.3. Observations of the nearby Blue Compact Galaxy ESO 338-IG04 from Hayes *et al.* (2005). Left: Lyα equivalent-width map. Regions of high equivalent width show up in dark colours. Particularly visible are the diffuse emission regions outside the starburst region. Much local structure can be seen, particularly around knot A (the main UV know) and the other bright continuum sources. Right: a false colour image showing [O III] in red, the UV continuum in green and the continuum-subtracted Lyα image in blue.

extinction, Giavalisco *et al.* (1996) suggest that an inhomogeneous ISM geometry must be the primary determining factor, not dust. However, no quantification of this effect was presented or proposed. More detailed observations of local starbursts have since then provided important new pieces of information that we will now briefly summarize.

4.3.3 *Lessons from local starbursts*

High-dispersion spectroscopy with the HST has shown the presence of neutral gas outflows in four starbursts with Lyα in emission (P-Cygni profiles), whereas other starbursts with broad damped-Lyα absorption do not exhibit velocity shifts between the ionized emitting gas and the neutral ISM traced by O I or Si II (Kunth *et al.* 1998). The metallicities of these objects range from $12 + \log(\mathrm{O/H}) \sim 8.0$ to Solar; their extinction is $E_{B-V} \sim 0.1$–0.55. From these observations Kunth *et al.* (1998) suggest that outflows and superwinds are the main determining factor for Lyα escape.

Two- and three-dimensional studies of Lyα emission and related properties in nearby starbursts have been carried out with the HST (UV) and integral-field spectroscopy (optical) to analyse at *high spatial resolution* the distribution and properties of the relevant components determining Lyα, i.e. the young stellar populations, their UV slope (a measurement of the extinction), the ionized gas and the resulting Lyα emission, absorption and local line profile (e.g. Mas-Hesse *et al.* 2003; Kunth *et al.* 2003; Hayes *et al.* 2005). In ESO 338-IG04 (Tol 1914-416), for example, diffuse Lyα corresponding to about two thirds of the total flux observed in large apertures (e.g. the IUE) is observed, thus confirming the existence of a Lyα resonant-scattering halo; see Figure 4.3 (Hayes *et al.* 2005). No clear spatial correlation between stellar ages and Lyα is found. However, correlations between the Lyα line kinematics and other kinematic tracers (NaID and Hα) are found.

Another interesting case is ESO 350-IG038, where Kunth *et al.* (2003) find two young star-forming knots (B and C) with similar, high extinction, one exhibiting Lyα emission, the other not. Hence dust absorption cannot be the dominant mechanism here. From the observed Hα velocity field, Kunth *et al.* suggest that kinematics is primarily responsible for the observed differences between the two regions.

A "unifying" scenario to explain the observed diversity of Lyα profiles in terms of an evolutionary sequence of starburst-driven supershells/superwind has been presented by Tenorio-Tagle *et al.* (1999) and confronted with local observations in the same paper and, in more depth, by Mas-Hesse *et al.* (2003).

In short we retain the following empirical results from nearby starbursts on Lyα: W(Lyα) and Lyα/Hβ are often smaller than the case-B prediction. No clear correlation of Lyα with metallicity, dust and other parameters is found. Strong variations of Lyα are observed within a galaxy. A Lyα-scattering "halo" is observed. Starbursts have complex structure (super star clusters plus diffuse ISM), and outflows are ubiquitous. From the various observations it is clear that the formation of Lyα is affected by (1) ISM kinematics, (2) ISM (H I) geometry and (3) dust. However, the precise order of importance remains unknown and may well vary among objects.

New, more-complete high-spatial-resolution observations are needed. In parallel quantitative modelling including known constraints (stars, emitting gas, H I, dust plus kinematics) with a three-dimensional radiation-transfer model remains to be done.

4.3.4 *Lyα radiation transfer*

4.3.4.1 Basic line-formation processes and examples

To gain insight into the physical processes affecting Lyα, to understand the variety of observed line profiles and their nature, and hence to develop quantitative diagnostics using Lyα, it is important to understand the basics of Lyα radiation transfer. To do so we rely on the recent paper by Verhamme *et al.* (2006), where more details and numerous references to earlier papers can be found. Among recent papers shedding new light on Lyα radiation transfer we mention here the work of Hansen & Oh (2006) and Dijkstra *et al.* (2006ab).

The Lyα line's optical depth can be written as

$$\tau_x(s) = 1.041 \times 10^{-13} T_4^{-1/2} N_{\rm H} \frac{H(x,a)}{\sqrt{\pi}}, \tag{4.1}$$

where T_4 is the temperature in units of 10^4 K, $N_{\rm H}$ the neutral-hydrogen column density and $H(x,a)$ the Hjerting function describing the Voigt absorption profile. Here x describes the frequency shift in Doppler units, $x = (\nu - \nu_0)/\Delta\nu_{\rm D} = -V/b$, where the second equation gives the relation between x and a macroscopic velocity component V measured along the photon propagation (i.e. parallel to the light path and in the same direction). b is the usual Doppler parameter, $b = \sqrt{V_{\rm th}^2 + V_{\rm turb}^2}$. Equation (4.1) shows that Lyα is very rapidly optically thick at line centre, i.e. even for modest column densities ($N_{\rm H} > 3 \times 10^{13}\,{\rm cm}^{-2}$). For $N_{\rm H} = 10^{20}$ a very large number of scatterings ($\sim 10^7$) is required for a photon to escape. However, velocity fields or an inhomogeneous medium can ease the escape (cf. below).

As is true for other lines, the scattering of photons in the Lyα line is not a random walk: it corresponds to a walk in coupled spatial and frequency space, where transport is dominated by excursions to the line wings. In other words, photons propagate only over large distances allowing (long mean free path) them to escape when they are in the wings, where the opacity is lower. This already suffices to allow us to understand the formation of double-peak Lyα line profiles in the case of Lyα emission surrounded (or covered) by a static medium, as shown in Figure 4.4 (left): all photons initially emitted

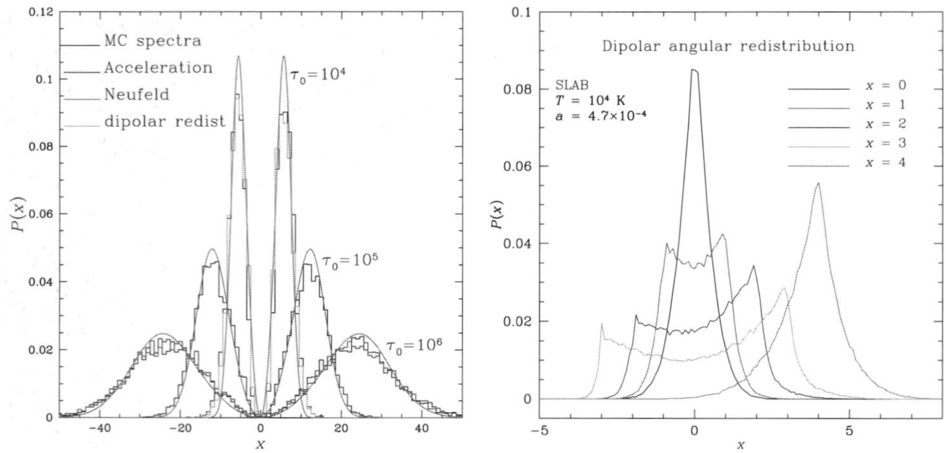

FIGURE 4.4. Left: the predicted Lyα line profile for a monochromatic source embedded in a static medium with various $N_{\rm H}$ column densities. Note the characteristic symmetrical double-peak profile. The separation between the two peaks depends in particular on the total optical depth, i.e. on $N_{\rm H}$. Right: the angle-averaged frequency-redistribution function for specific conditions (T and Voigt parameter a), shown as the probability distribution function for input frequencies $x = 0$ (line centre) to 4 ("wing"). Figures adapted from Verhamme *et al.* (2006).

at line centre (for illustration) are absorbed and "redistributed" to the wings, where they can escape. The higher the total optical depth, the larger the separation of the two peaks becomes. Asymmetries between the two peaks are of course introduced with shifts of the intrinsic emission frequency, or – equivalently – with an approaching/receding medium. These cases and many variations thereof are discussed in detail by Neufeld (1990).

In contrast to other scattering processes, Lyα scattering is neither coherent nor isotropic. The frequency redistribution, e.g. described by the angle-averaged frequency-redistribution functions R_{II} of Hummer (1962), is illustrated in Figure 4.4 (right). Schematically, for input frequencies $x_{\rm in}$ close to the core the emergent photon has its frequency redistributed over the interval $\sim[-x_{\rm in}, +x_{\rm in}]$. Once photons are sufficiently far into the wing they are re-emitted at close to their input frequency, i.e. scattering is close to coherent in the comoving frame. This behaviour is fundamental to understanding e.g. the formation of the emergent line profile for expanding shells, which is illustrated in Figure 4.5. There detailed radiation-transfer calculations show that the peak of the asymmetric Lyα profile is located approximately at the frequency Doppler-shifted by twice the expansion velocity resulting from photons from the back side of the shell (Verhamme *et al.* 2006). This mostly results from two facts. The first is the re-emission of the photons after their first scattering in the shell peaks at a Doppler shift of $\sim v_{\rm exp}$ in the comoving reference frame of the shell, since the original Lyα photon emitted at line centre ($x = 0$) is seen in the wing by the material in the shell (re-emission close to coherence). Secondly, in the external frame these photons have then frequencies between $x \sim 0$ and $-2x(v_{\rm exp})$. Now, the escape of the photons with the largest redshift being favoured, this will preferentially select photons from the back of the shell, thus creating a peak at $-2x(v_{\rm exp})$. The interplay between these different probabilities imprints the detailed line shape, as discussed in more detail in Verhamme *et al.* (2006). For a given geometry, e.g. an expanding shell appropriate to model outflows in starbursts, a wide variety of Lyα profiles can be obtained, depending on the shell velocity and its temperature, the column density, the relative strength of the initial Lyα emission with respect to the continuum

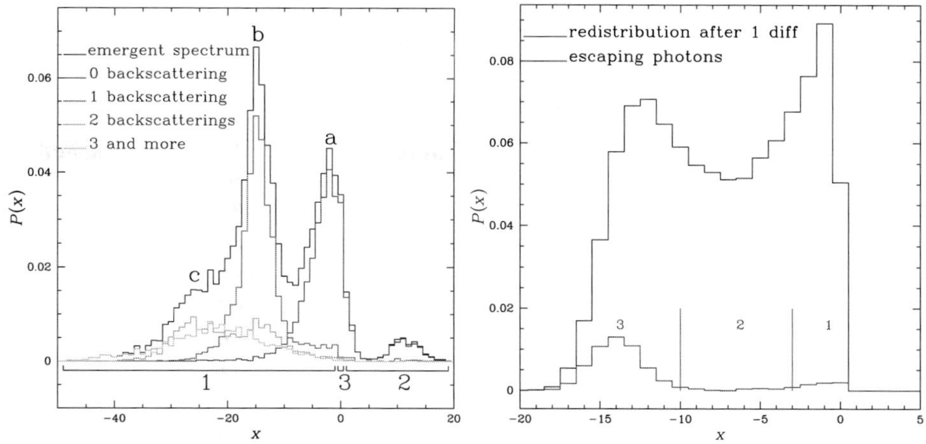

FIGURE 4.5. Left: an emergent Lyα profile from an expanding shell with central monochromatic source. The different shapes can be described with the number of backscatterings that photons undergo: bumps 1a and 2 are built up with photons that did not undergo any backscattering, the highest peak located at $x = -2v_{\rm exp}/b$ (feature 1b) is composed of photons that undergo exactly one backscattering, and the red tail 1c is made of photons that undergo two or more backscatterings. See Verhamme *et al.* (2006) for more details. Right: the frequency distribution of the photons in the expanding shell after the first scattering. The solid curve contains all photons, whereas the dotted one represents the histogram of those photons which escaped after only one scattering. They form a bump around $x \sim -2x(v_{\rm exp})$, which explains the appearance of feature 1b. See the description in the text. $V_{\rm exp} = 300\,{\rm km\,s^{-1}}$, $b = 40\,{\rm km\,s^{-1}}$ and $N_{\rm H\,I} = 2 \times 10^{20}\,{\rm cm^{-2}}$. Adapted from Verhamme *et al.* (2006).

and the presence of dust; see Verhamme *et al.* (2006) for an overview. Let us now briefly discuss how dust affects the Lyα radiation transfer.

4.3.4.2 Lyα transfer with dust

A simple consideration of the probability of Lyα photons interacting with dust, $P_{\rm d} = n_{\rm d}\sigma_{\rm d}/(n_{\rm H}\sigma_{\rm H}(x) + n_{\rm d}\sigma_{\rm d}$, shows that this event is quite unlikely, especially in the line core, where the Lyα cross section exceeds that of dust, $\sigma_{\rm d}$, by several orders of magnitude. Despite this, interactions with dust particles occur, especially in the wings, but also closer to line centre since the overall probability of a photon interacting with dust is increased by the large number of line scatterings occurring there. For this reason it is immediately clear that the destruction of Lyα photons by dust depends also on the kinematics of the H I gas, where supposedly the dust is mixed in, although *per se* the interaction of UV photons with dust is independent of the gas kinematics.

The net result is a fairly efficient destruction of Lyα photons by dust, as e.g. illustrated for static cases by Neufeld (1990) and for expanding shells by Verhamme *et al.* (2006). In the latter case the escape of Lyα photons is typically reduced by a factor of ~2–4 with respect to a simple reduction by $\exp(-\tau_{\rm a})$, where $\tau_{\rm a}$ is the dust-absorption optical depth. Finally it is also interesting to note that dust not only reduces the Lyα emission (or the line's equivalent width), but also alters the line profile somewhat in a non-grey manner (Ahn 2004; Hansen & Oh 2006), since its effect depends on Lyα scattering. See Verhamme *et al.* (2006) for illustrations.

4.3.4.3 Lyα transfer: geometrical effects

Given the scattering nature of Lyα, it is quite clear that the observed Lyα properties of galaxies depend also in particular on geometry. By this we mean the intrinsic geometry of the object, i.e. the spatial location of the "initial" Lyα emission in the H II gas, the distribution and kinematics of the scattering medium (namely the H I), but also the spatial region of this object which is ultimately observed. In other words the observed Lyα-line properties (equivalent width and line profile) will in principle also vary if the observations provide an integrated spectrum of the entire galaxy or only a region thereof.

In an inhomogeneous ISM, UV-continuum and Lyα-line photons will also propagate in different ways, since their transmission/reflection properties differ. Such cases were discussed e.g. by Neufeld (1991) and Hansen & Oh (2006), who show that this can lead to higher Lyα equivalent widths.

In non-spherical cases, including for example galaxies with strong outflows and galactic winds with complex geometries and velocity structures, one may of course also expect significant orientation effects on the observed Lyα line. Such cases remain largely to be investigated in realistic three-dimensional radiation-transfer simulations.

4.3.5 *Lessons from Lyman-break galaxies*

Having already discussed relatively nearby starburst galaxies, where spatial information is available, it is of interest to examine the empirical findings related to Lyα of more-distant spatially unresolved objects, the so-called Lyman-break galaxies (LBGs) discussed also in more detail in Section 4.4 and in Chapter 2 of this volume. These different categories of objects may help us understand in particular Lyα emission and stellar populations in distant and primeval galaxies.

LBGs are galaxies with intense ongoing star formation, selected from their UV (rest-frame) emission. In 2003 approximately 1000 LBGs with spectroscopic redshifts were known, mostly ones studied by the group of Steidel (Shapley *et al.* 2003). Since then the number has grown, but this study remains the most comprehensive one on $z \sim 3$ LBGs. The rest-frame UV spectra of LBGs include stellar, interstellar and nebular lines testifying to the presence of massive stars. A diversity of Lyα line profiles, ranging from emission, over P-Cygni to broad absorption-line profiles, and strengths is observed. Interstellar (IS) lines are found blueshifted with respect to the stellar lines (defining the object's redshift, when detected) by $\Delta v(\mathrm{abs} - \star) = -150 \pm 60\,\mathrm{km\,s^{-1}}$. A shift of $\Delta v(\mathrm{em} - \mathrm{abs}) \sim$ 450–650 $\mathrm{km\,s^{-1}}$ is also observed between the IS absorption lines and Lyα. Finally Shapley *et al.* (2003) find several correlations relating the extinction, $W(\mathrm{Ly}\alpha)$, $W(\mathrm{IS})$ and the star-formation rate (SFR), which have been understood poorly or not at all, at least until very recently; see Ferrara & Ricotti (2007) for a possible explanation.

From Lyα radiation-transfer modelling discussed before, the observed shifts of stellar, IS and Lyα lines are naturally understood if the geometry is that of a "global" expanding shell (Verhamme *et al.* 2006). The IS lines are then formed by absorption of the UV-continuum light from a central starburst in the shell along the line of sight towards the observer. Their blueshift with respect to the stars measures thus the expansion velocity v_{exp}. One then obtains naturally $\Delta v(\mathrm{em} - \mathrm{abs}) \sim 3|\Delta v(\mathrm{abs} - \star)| = 3v_{\mathrm{exp}}$, since Ly$\alpha$ originates from the back of the shell redshifted by $2v_{\mathrm{exp}}$. This result indicates that large-scale, fairly symmetrical shell structures must be a good description of the outflows in LBGs.

What causes the variety of observed Lyα line profiles and what does this tell us about these galaxies? Using the radiation-transfer code described in Verhamme *et al.* (2006) we have recently undertaken the first detailed modelling of typical LBGs at $z \sim 3$, in particular objects from the FORS Deep Field observed by Tapken *et al.* (2007) at a spectral resolution $R \sim 2000$, which is sufficient for one to do detailed line-profile fitting. Assuming the spherically expanding shell model motivated in particular by the correct velocity shifts just mentioned, the full variety of profiles can be reproduced for the observed values of v_{exp} and extinction, and by varying N_{H} and intrinsic Lyα-line parameters (W and the FWHM).

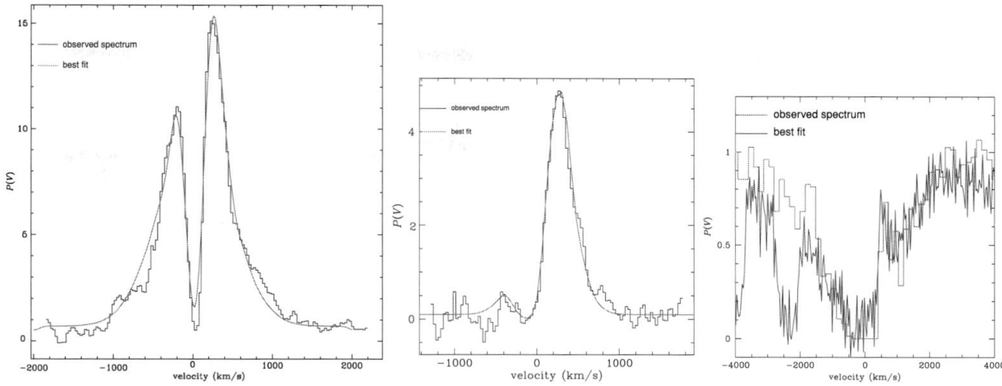

FIGURE 4.6. Comparison of observed and modelled Lyα line profiles of $z \sim 3$ LBGs with a variety of line-profile morphologies, from double-peaked, over P-Cygni, to broad absorption. See the discussion in the text. Adapted from Verhamme *et al.* (2007).

Three such examples are illustrated in Figure 4.6. Fitting the double-peak profile of FDF 4691 (left) is possible only with low velocities, i.e. conditions close to a static medium (cf. Figure 4.4). Such Lyα profiles are relatively rare; other cases with such double-peak profiles include the Lyα blob observed by Wilman *et al.* (2005) and interpreted by them as a "stalled" expanding shell, or even as a collapsing proto-galaxy (Dijkstra *et al.* 2006b). The profile of FDF 4454 (middle), which is quite typical of LBGs, indicates a typical expansion velocity of $v_{\rm exp} \sim 220\,{\rm km\,s^{-1}}$ and a low extinction, compatible with its very blue UV slope. Finally, the profile of the lensed galaxy cB58 (right) from Pettini *et al.* (2000) is well reproduced with the observed expansion velocity and extinction ($v_{\rm exp} \sim 255\,{\rm km\,s^{-1}}$, $E_{B-V} = 0.3$). The fits yield in particular constraints on the column density $N_{\rm H}$ and the intrinsic Lyα-line parameters (W and the FWHM). This allows us to examine the use of Lyα as a SFR indicator, to provide constraints on the SF history and age of these galaxies, and to shed new light on the observed correlations between Lyα and other properties of LBGs (Verhamme *et al.* 2007). Understanding Lyα in galaxies for which sufficient observations are available and that are located at various redshifts is of great interest also in order to learn how to exploit the more-limited information available for objects at higher z, including primeval galaxies (see Section 4.4).

4.3.6 *Lyα through the intergalactic medium*

Having discussed the properties of Lyα-line formation and radiation-transfer effects in galaxies, we will now examine how the Lyα profile is transformed/transmitted on its way to the observer, i.e. through the intergalactic medium (IGM).

In this situation we consider radiation from a distant background source passing through one or several "H I clouds". This geometry leads to a very simple case in which Lyα photons are absorbed and then either scattered out of the line of sight or absorbed internally by dust. In other words *no true radiation transfer needs to be computed*, and the resulting Lyα profile of the radiation emerging from the cloud is simply the input flux attenuated by a Voigt absorption profile characteristic of the cloud properties. For a given density and (radial) velocity – or equivalently redshift – distribution along the line of sight, the computation of the total attenuation and hence of the observed spectrum is thus straightforward.

The observational consequences for a distant source will thus be (1) the imprint of a number of (discrete) absorption components on top of the background source spectrum due to intervening H I clouds or filaments and (2) an alteration of the emergent

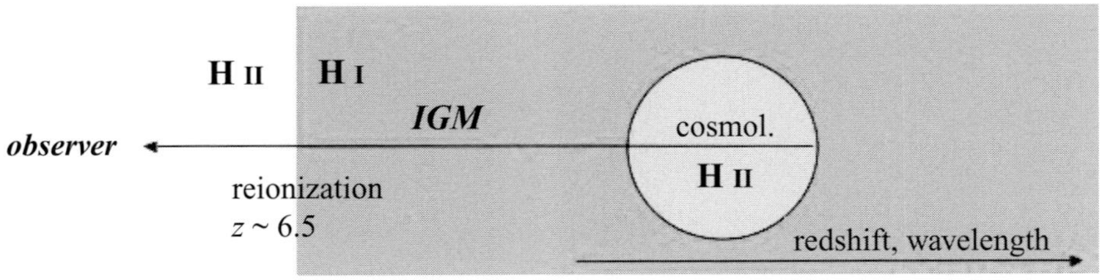

FIGURE 4.7. A schematic representation of a star-forming galaxy situated beyond the reionization redshift (here indicated at $z_r \sim 6.5$), its surrounding cosmological H II region, the neutral IGM down to z_r and the transparent (ionized) IGM towards observer. Redshift and observed Lyα wavelength increase to the right.

galactic Lyα profile plus a reduction of the Lyα flux if neutral H is present close in velocity/redshift space to the source. The first is well known observationally as the Lyα forest, leading even to complete absorption (the so-called Gunn–Peterson trough) in distant ($z \sim 6$) quasars; see the review by Fan *et al.* (2006). The appearance of a complete Gunn–Peterson trough in high-z quasars implies a quantitative change of the ionization of the IGM, possibly tracing the end of the epoch of cosmic reionization (Fan *et al.* 2006). The second effect leads e.g. to the alteration of the Lyα profile and to a strong reduction of the Lyα flux in high-z quasars, due to absorption by the red damping wing of Lyα by nearby H I (Miralda-Escudé 1998; Fan *et al.* 2003).

The two effects just discussed have the following immediate implications.

- The SED of high-z galaxies is altered by Lyman-forest attenuation at wavelengths shorter than Lyα (<1216 Å). A statistical description of this attenuation is given by Madau (1995).
- For $z \gtrsim 4$–5 the Lyman forest attenuation is so strong that it effectively leads to a spectral break at Lyα, replacing therefore the "classical" Lyman break (at 912 Å) due to photoelectric absorption by H I. The Lyα break becomes then the determining feature for photometric redshift estimates.
- The reduction of the Lyα flux implies that (a) determinations of the SFR from this line will underestimate the true SFR, (b) the observed Lyα-luminosity function (LF) does not correspond to the true (intrinsic) one, and (c) the detectability of high-z Lyα emitters (hereafter LAEs) is reduced.
- The Lyα profile, Lyα transmission and Lyα LF contain information on the ionization fraction of hydrogen and can hence in principle constrain cosmic reionization.

We will now discuss how/whether it is still possible to observe LAEs beyond the reionization redshift.

4.3.7 *Lyα from sources prior to reionization*

How is it possible to observe Lyα emission from sources "beyond the end of reionization", i.e. at very high redshift where the IGM contains a significant fraction of neutral hydrogen, which absorbs the Lyα emission? The way to achieve this is in principle quite simple and sketched in Figure 4.7. It suffices to create around the Lyα source a "cosmological" H II region big enough that no or very little H I is present at velocities – i.e. redshifts – close to the source. In this way attenuation close to the Lyα emission is avoided and the line flux from this distant source can propagate freely to the observer, since it comes from the most-redshifted part along the line of sight.

So, how are these cosmological H II regions created? Obviously this requires one or several sources (galaxies or quasars) producing ionizing photons that are able to escape

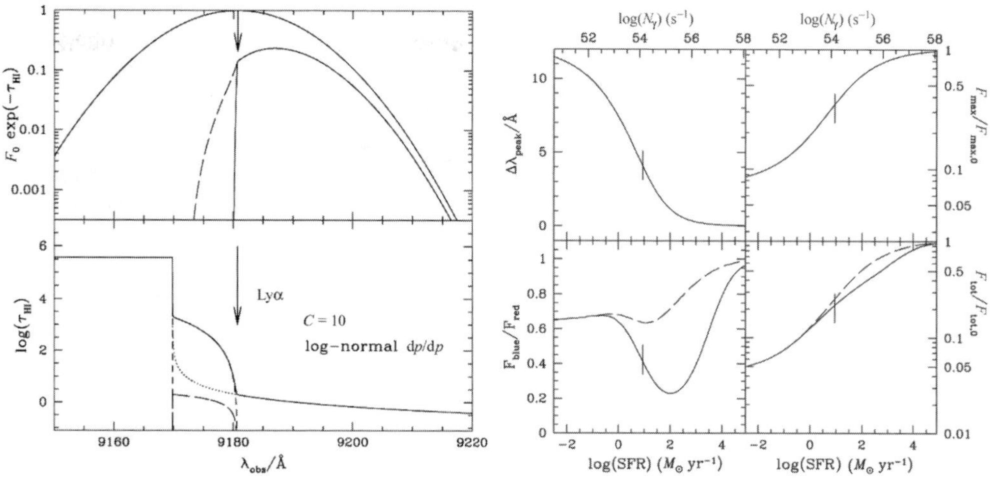

FIGURE 4.8. Predicted Lyα line profile, Lyα transmission and other properties from the model of a $z = 6.56$ lensed galaxy taking IGM absorption into account. Adapted from Haiman (2002). Left: intrinsic and resulting line profiles (top), opacities leading to the Lyα attenuation. Right: parameters, such as asymmetry, peak position and total transmission (bottom right) of the predicted Lyα line as a function of the SFR.

the galaxy and can then progressively ionize the surrounding IGM. This is referred to as the "proximity effect". The properties and the evolution of cosmological H II regions have been studied and described analytically in several papers; see e.g. Shapiro & Giroux (1987), Cen & Haiman (2000), and the review by Barkana & Loeb (2001). For example, neglecting recombinations in the IGM (since for the low IGM densities the recombination timescale is much longer than the Hubble time) and assuming that the ionizing source is "turned on" and constant during the time $t_{\rm Q}$ the Strömgren radius (size) of the H II region becomes

$$R_{t_{\rm Q}} = \left(\frac{3\dot{N}_{\rm ph} t_{\rm Q}}{4\pi \langle n_{\rm H} \rangle} \right)^{1/3}, \tag{4.2}$$

where $\dot{N}_{\rm ph} = f_{\rm esc} Q_{\rm H}$ is the escaping ionizing flux and $\langle n_{\rm H} \rangle$ the mean IGM density, taking possibly a non-uniform density distribution into account. The residual H I fraction inside the H II region is given by photoionization equilibrium and can also be computed. Then the resulting attenuation $e^{-\tau}$ can be computed by integrating the optical depth along the line of sight:

$$\tau(\lambda_{\rm obs}, z_{\rm s}) = \int_{z_{\rm r}}^{z_{\rm s}} {\rm d}z \, c \frac{{\rm d}t}{{\rm d}z} n_{\rm H}(z) \sigma_\alpha(\lambda_{\rm obs}/(1+{\rm z})). \tag{4.3}$$

Here $z_{\rm s}$ is the source redshift, $z_{\rm r}$ a limiting redshift (the redshift of reionization in Figure 4.7) below which the IGM is supposed to be transparent and σ_α the Lyα absorption cross section.

For example, the observability of Lyα from a $z = 6.56$ galaxy observed by Hu *et al.* (2002) has been examined with such a model by Haiman (2002). The results are illustrated in Figure 4.8. For a source with SFR 9M$_\odot$ yr^{-1}, an age of $\sim$100 Myr and an escape fraction $f_{\rm esc} = 25\%$ the proper (comoving) radius of the H II region is approximately 0.45(3) Mpc. Assuming an intrinsic Lyα profile of FWHM width 300 km s^{-1} Haiman obtains a transmission of $\sim$16% of the Lyα flux and an asymmetrical line profile, as observed. A wider range of transmission encompassing also this value is found from an independent estimate based on stellar-population modelling (Schaerer & Pelló 2005).

In the picture described above, the Lyα transmission is expected to increase with increasing SFR, escape fraction, source lifetime and intrinsic line width, as also shown in Figure 4.8 (right). The first three increase the size of the cosmological H II region; with the latter a higher fraction of the line flux is emitted far from line centre, thus reducing the absorption by the red damping wing in the H I. Other factors also affect the Lyα transmission and the resulting line profile: IGM infall, outflows (galactic winds), peculiar velocities of the emitting gas within halo, the halo mass etc.; see Haiman (2002) and Santos (2004).

In a more-realistic setting, this simple model can be subject to several "complications" (e.g. Gnedin & Prada 2004; Furlanetto *et al.* 2004; Wyithe & Loeb 2004).

- Clustering of sources helps to create a larger H II region. Since the probability of clustering increases with z and for fainter galaxies, this could play an important role in determining the detectability of high-redshift Lyα sources.
- In a non-homogeneous structure around the source the H II regions are expected to deviate from spherical symmetry, since the ionization fronts will propagate more rapidly into directions with lower IGM density.

From this it is clear that strong variations depending on the object, its surroundings and the viewing direction are expected and the simple scaling properties of the spherical models described before might not apply. A statistical approach using hydrodynamic simulations will be needed.

In short, the answer to the question "Is Lyα emission from sources prior to reionization detectable?" is affirmative from the theoretical point of view, but the transmission depends on many factors! In any case, searches for such objects are going on (Section 4.4.2) and will provide the definitive answer.

4.3.8 *The Lyα luminosity function and reionization*

As a last illustration of the use of Lyα in distant, primeval galaxies, we shall now briefly discuss the statistics of LAEs, in particular the Lyα luminosity function LF(Lyα), how it may be used to infer the ionization fraction of the IGM at various redshifts, and difficulties affecting such approaches.

Since, as discussed above, the presence of neutral hydrogen in the IGM can reduce the Lyα flux of galaxies, it is clear that the Lyα LF is sensitive to the ionization fraction $x_{\mathrm{H\,I}}$. If we knew the intrinsic LF(z) of galaxies at each redshift, a deviation of the observed LF from this intrinsic distribution could be attributed to attenuation by H I, and hence be used to infer $x_{\mathrm{H\,I}}$ (see Figure 4.9). In practice the approach is of course to proceed to a differential comparison of LF(Lyα) with redshift. Indeed, from simple Lyα-attenuation models like the ones described in the previous section, a rapid decline of the LF is expected on approaching the end of reionization.

Haiman & Spaans (1999) were among the first to advocate the use of LF(Lyα) and to make model predictions. Since then, and after the detection of numerous LAEs allowing the measurement of the Lyα LF out to redshift $z = 6.5$ (see Section 4.4.2), several groups have made new predictions of the Lyα LF and have used it to constrain cosmic reionization. Some prominent examples are Malhotra & Rhoads (2004), Le Delliou *et al.* (2005, 2006) and Furlanetto *et al.* (2006).

One of the most recent of such attempts is presented by Dijkstra *et al.* (2006c), who predict the Lyα LF using a modified Press–Schechter formalism and introducing two main free parameters, a SF duty-cycle ϵ_{DC} and another parameter depending on the SF efficiency, the escape fraction and the Lyα transmission of the IGM. They find a typical

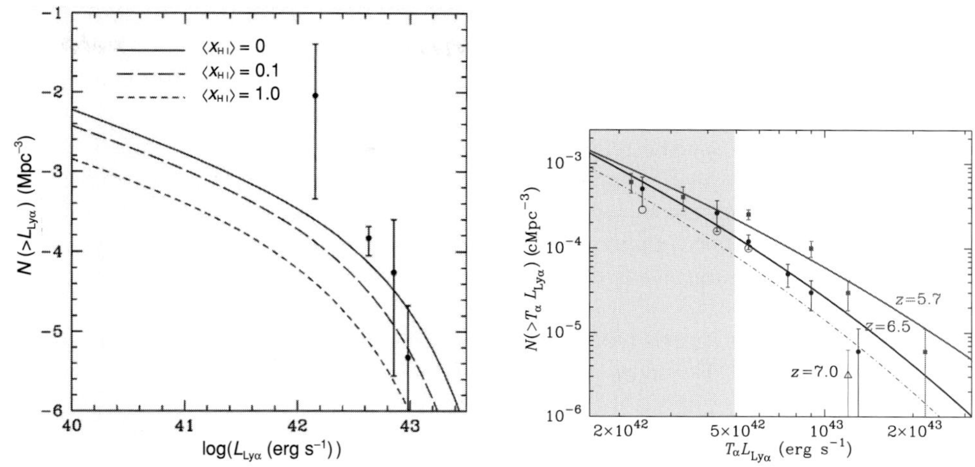

FIGURE 4.9. Left: predicted Lyα LFs for a fully ionized IGM (no attenuation case, i.e. the $z = 5.7$ LF; solid curve) and for an IGM with an increasing neutral-H fraction $x_{\rm HI}$. From Haiman & Cen (2005). Right: Predicted and observed Lyα LF at $z = 5.7$ and 6.5. The LF model is that by Dijkstra *et al.* (2006c). According to these authors the observed decline of the Lyα LF is attributed to the evolution of the halo mass function hosting the Lyα emitters.

IGM transmission of $T_\alpha \sim 30\%$ at $z = 5.7$. By adjusting the observed LFs at $z = 5.7$ and 6.5 (where quasars already indicate a significant change of the ionization fraction $x_{\rm H\,I}$ as discussed in Section 4.3.6) Dijkstra *et al.* (2006c) find good fits without the need for a strong change of the ionization state as advocated in other studies (see Figure 4.9). The observed decline of the Lyα LF between $z = 5.7$ and 6 is attributed to the evolution of the halo mass function hosting the Lyα emitters. In this case this may translate into a lower limit of ~80% for the fraction of ionized H at $z = 6.5$. This serves to illustrate the potential of LF(Lyα) analysis, but also the potential difficulties and the room for improvements.

Finally let us also note that Hu *et al.* (2005) do not find an evolution of the mean Lyα line profile between $z = 5.7$ and 6.5, in agreement with the above conclusion.

4.4. Distant/primeval galaxies: observations and main results

Before we discuss searches for distant galaxies, provide an overview of the main results and discuss briefly some remaining questions, we shall summarize the basic observational techniques used to identify high-redshift galaxies.

4.4.1 *Search methods*

The main search techniques for high-z galaxies can be classified into the two following categories.

1. The Lyman-break or dropout technique, which selects galaxies over a certain redshift interval by measuring the Lyman break, which is the drop of the galaxy flux in the Lyman continuum (at $\lambda < 912$ Å) of the Lyα break (shortwards of Lyα) for $z \gtrsim 4$–5 galaxies (cf. above). This method requires the detection of the galaxy in several (sometimes only two, but generally more) broad-band filters.
2. Emission-line searches (targeting Lyα or other emission lines). Basically three different techniques may be used: (1) narrow-band (NB) imaging (two-dimensional) e.g. of a wide field selecting a specific redshift interval with the transmission of the

NB filter; (2) long-slit spectroscopy (one-dimensional) for "blind searches" e.g. along critical lines in lensing clusters; and (3) observations with integral field units (three-dimensional) allowing one to explore all three spatial directions (two-dimensional imaging plus redshift). The first one is currently the most-used technique.

In practice, and to increase the reliability, several methods are often combined.

Surveys/searches are being carried out in blank fields or targeting deliberately gravitational-lensing clusters allowing one to benefit from gravitational magnification from the foreground galaxy cluster. For galaxies at $z \lesssim 7$ the Lyman break and Lyα are found in the optical domain. Near-IR ($\gtrsim$ 1-μm) observations are necessary to locate $z \gtrsim 7$ galaxies.

The status in 1999 of search techniques for distant galaxies has been summarized by Stern & Spinrad (1999). For more details on searches and galaxy surveys see Chapter 2 in this volume.

4.4.2 *Distant Lyα emitters*

Most of the distant known Lyα emitters (LAEs) have been found through narrow-band imaging with the SUBARU telescope, thanks to its wide-field imaging capabilities. $z \sim$ 6.5–6.6 LAE candidates are e.g. selected combining the three following criteria: an excess in a narrow-band filter (NB921) with respect to the continuum flux estimated from the broad z' filter, a 5σ detection in this NB filter, and an i-dropout criterion (e.g. $i - z' > 1.3$) making sure that these objects exhibit a Lyα break. Until recently 58 such LAE candidates had been found, with 17 of them confirmed subsequently by spectroscopy (Taniguchi *et al.* 2005; Kashikawa *et al.* 2006). The Hawaii group has found approximately 14 LAEs at $z \sim 6.5$ (Hu *et al.* 2005; Hu & Cowie 2006). The current record-holder as the most-distant galaxy with a spectroscopically confirmed redshift of $z = 6.96$ is one detected by Iye *et al.* (2006). Six candidate Lyα emitters between $z = 8.7$ and 10.2 were recently proposed by Stark *et al.* (2007) using blind long-slit observations along the critical lines in lensing clusters.

LAEs have for example been used with SUBARU to trace large-scale structure at $z = 5.7$ thanks to the large field of view (Ouchi *et al.* 2005).

Overall, quite little is known about the properties of NB-selected LAEs, their nature and their relation to other galaxy types (LBGs and others, but see Section 4.4.3), since most of them – especially the most distant ones – are detected in very few bands, i.e. their SEDs are poorly constrained. The morphology of the highest-z LAEs is generally compact, indicating ionized gas with spatial extent of ~2–4 kpc or less (e.g. Taniguchi *et al.* 2005; Pirzkal *et al.* 2006).

Although they have (SFRs) of typically $(2\text{–}50) M_\odot \text{ yr}^{-1}$, the SFR density (SFRD) of LAEs is only a fraction of that of LBGs at all redshifts. For example, at $z \sim 5\text{–}6.5$, Taniguchi *et al.* (2005) estimate the SFRD from Lyα emitters as SFRD (LAE) $\sim 0.01\times$ SFRD(LBG), or up to 10% of SFRD(LBG) at best, allowing for LF corrections. At the highest z this value could be typically three times higher if the IGM transmission of ~30% estimated by Dijkstra *et al.* (2006c) applies. Shimasaku *et al.* (2006) have found a similar space density or UV LF for LAEs and LBGs at $z \sim 6$, and argue that LAEs contribute at least 30% of the SFRD at this redshift.

The typical masses of LAEs are still unknown and being debated. For example, Lai *et al.* (2007) find stellar masses of $M_\star \sim (10^9\text{– }10^{10}) M_\odot$ for three LAEs at $z \sim 5.7$, whereas Prizkal *et al.* (2006) find much lower values of $M_\star \sim (10^6\text{–}10^8) M_\odot$ for their sample of $z \sim 5$ Lyα galaxies. Finkelstein *et al.* (2006) find masses between the two ranges for $z \sim 4.5$ LAEs. Selection criteria may explain some of these differences; e.g. the Lai *et al.* objects were selected for their detection at 3.6 and 4.5 μm with the Spitzer telescope. Mao

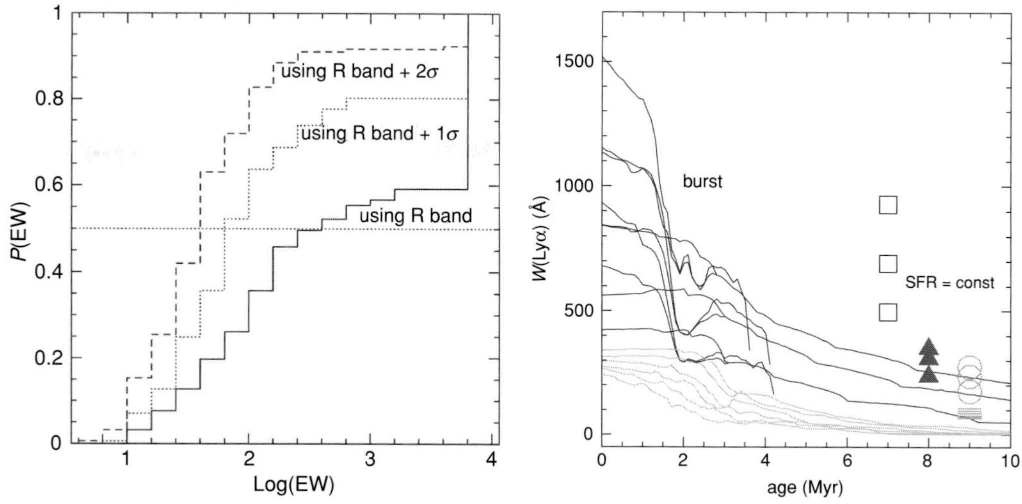

FIGURE 4.10. Left: Observed Lyα equivalent-width distribution of $z = 4.5$ sources from the LALA survey. From Malhotra & Rhoads (2002). Right: Predicted Lyα equivalent width for starbursts at various metallicities (from Solar to Population III). Normal metallicities ($Z \gtrsim (1/50) Z_{\odot}$) are shown by the pale dashed lines. The maximum value predicted in this case is $W(\mathrm{Ly}\alpha) \sim 300$ Å. From Schaerer (2003).

et al. (2006) argue that LAEs are limited to a relatively narrow mass range around $M_{\star} \sim 10^9 M_{\odot}$. Further studies will be necessary for a proper understanding of the connections between LBGs and LAEs and the evolution of the two populations with redshift.

4.4.2.1 Population III signatures in LAEs?

The Large Area Lyman-α (LALA) survey by Rhoads and collaborators, carried out on 4-m-class telescopes, has been one of the first to find a significant number of LAEs at high redshift ($z = 4.5$ and 5.7, and later also objects at $z = 6.5$). Among the most-interesting results from LALA is the finding of a large fraction of LAEs with an apparently high median Lyα equivalent width, compared with expectations from normal stellar populations (see Figure 4.10). Indeed, half of their $z = 4.5$ candidates have $W(\mathrm{Ly}\alpha)$ in excess of $\sim$ 200–300 Å (Malhotra & Rhoads 2002), a value expected only for very young starbursts, populations with extreme IMFs, or very metal-poor (or Population III) stars (cf. Schaerer 2003). Malhotra & Rhoads (2002) suggested that these could be AGNs or objects with peculiar top-heavy IMFs and/or Population III-dominated. In this context, and to explain other observations, Jimenez & Haiman (2006) also advocate a significant fraction of Population III stars, even in $z \sim$ 3–4 galaxies. Recently Hansen & Oh (2006), reviving an idea of Neufeld (1991), have suggested that the observed $W(\mathrm{Ly}\alpha)$ could be "boosted" by radiation-transfer effects in a clumpy ISM.

Follow-up observations of the LALA sources have allowed one to exclude the narrow-line AGN "option" (Wang *et al.* 2004), but have failed to provide further explanations of this puzzling behaviour. About 70% of the LALA LAEs have been confirmed spectroscopically; some high equivalent-width measurements could also be confirmed spectroscopically.† Deep spectroscopy aimed at detecting other emission lines, including the He II λ1640 line indicative of a Population III contribution (see Section 4.2.3), have been unsuccessful (Dawson *et al.* 2004), although the achieved depth (He II λ1640/Lya $<$ 13%–20%

† But aperture effects may still lead to an overestimate of $W(\mathrm{Ly}\alpha)$.

at (2–3)σ and W(He II λ1640) < 17–25 Å) might not be sufficient. The origin of these high W(Lyα) remains thus unknown.

However, there is some doubt about the reality of the LALA high equivalent widths measured from NB and broad-band imaging, or at least regarding their being so numerous even at $z = 4.5$. First of all the objects with the highest W(Lyα) have very large uncertainties since the continuum is faint or not detected. Second, the determination of W(Lyα) from a NB and a centred broad-band filter, R-band in the case of Malhotra & Rhoads (2002), may be quite uncertain, due to unknowns in the continuum shape, the presence of a strong spectral break within the broad-band filter etc.; see Hayes & Oestlin (2006) for a quantification and Shimasaku *et al.* (2006). Furthermore, other groups have not found such high-W objects (Hu *et al.* 2004; Ajiki *et al.* 2003), suggesting also that this may be related to insufficient depth of the LALA photometry.

More recently larger samples of LAEs were obtained, e.g. at $z = 5.7$; e.g. Shimasaku *et al.* (2006) has 28 spectroscopically confirmed objects. Although their *observed* rest-frame equivalent widths $W_{\rm obs}^{\rm rest}$(Lyα) (median value and W distribution) are considerably lower than those of Malhotra & Rhoads at $z = 4.5$, and only a few objects (1–3 out of 34) have $W_{\rm obs}^{\rm rest}({\rm Ly}\alpha) \gtrsim 200$ Å, it is possible that in several of these objects the maximum Lyα equivalent width of normal stellar populations is indeed exceeded. This would clearly be the case if the IGM transmission at this redshift were $T_\alpha \sim 0.3$–0.5 (cf. Shimasaku *et al.* 2006; Dijkstra *et al.* 2006c), which would imply that the true intrinsic $W^{\rm rest} = 1/T_\alpha \times W_{\rm obs}^{\rm rest}$ is ~2–3 times higher than the observed one. Shimasaku *et al.* estimate that ~30%–40% of their LAEs have $W^{\rm rest}({\rm Ly}\alpha) \geq 240$ Å and suggest that these may be young galaxies or again objects with Population III contribution. Dijkstra & Wyithe (2007), on the basis of Lyα-LF and W(Lyα) modelling, also argue for the presence of Population III stars in this $z = 5.7$ LAE sample.

Another interesting result is the increase of the fraction of large W(Lyα) LBGs with redshift, e.g. from ~2% of the objects with $W^{\rm rest}({\rm Ly}\alpha) > 100$ Å at $z \sim 3$ to ~80% at redshift 6, which is tentatively attributed to lower extinction, younger ages or an IMF change (Shimasaku *et al.* 2006; Nagao *et al.* 2007).

Despite these uncertainties it is quite clear that several very strong LAE emitters are found and that these objects are probably the most-promising candidates with which to detect direct *in situ* signatures of Population III at high redshift (see also Scannapieco *et al.* 2003). Searches are therefore going on (e.g. Nagao *et al.* 2005) and the first such discovery may be "just around the corner", or may need more-sensitive spectrographs and multi-object near-IR spectroscopy (see Section 4.4.4).

4.4.2.2 Dust properties of high-z LAEs

Although there are indications that LAEs selected through their Lyα emission are mostly young and relatively dust-free objects (e.g. Shimasaku *et al.* 2006; Pirzkal *et al.* 2006; Gawiser *et al.* 2006), it is of great interest to search for signatures of dust in distant/primeval galaxies.† Furthermore, some models predict a fairly rapid production and the presence of significant amounts of dust at high z (Mao *et al.* 2006). LAEs have the advantage of being at known redshift and of indicating the presence of massive stars. SED fits of such objects must therefore include populations of age <10 Myr, providing thus an additional constraint on modelling.

Recently the stellar populations of some high-z LAEs have been analysed with such objectives in mind. For example the $z = 6.56$ gravitationally lensed LAE discovered by

† Remember that e.g. sub-millimetre-selected galaxies – i.e. very dusty objects – or at least a subsample of them exhibit also Lyα emission (Chapman *et al.* 2003).

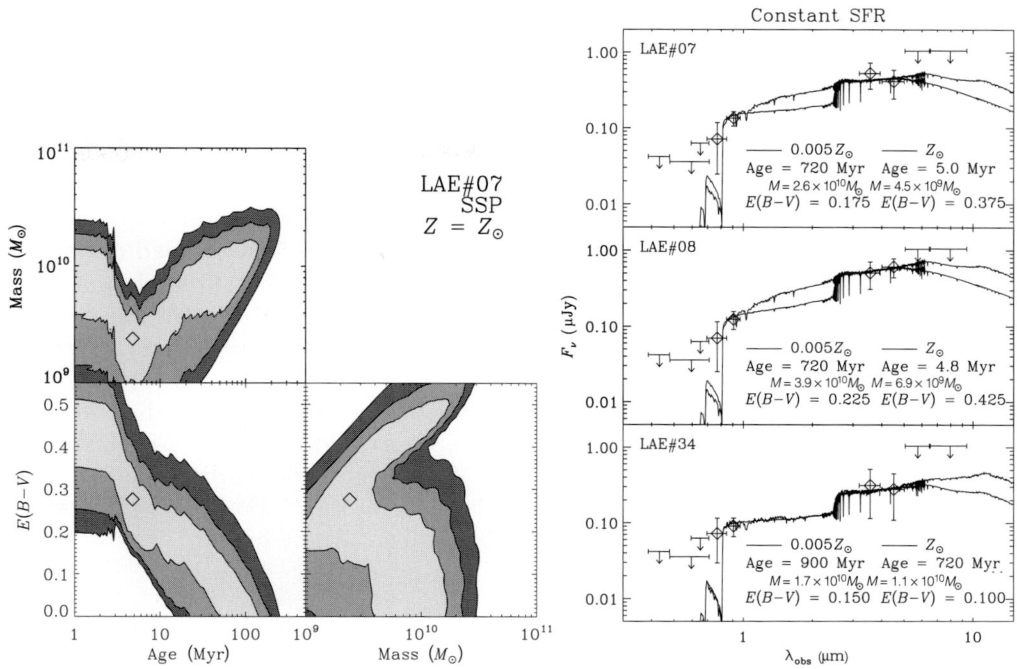

FIGURE 4.11. Illustrations of SED fits to $z = 5.7$ LAEs from Lai *et al.* (2007). Left: χ^2 contour plots showing the best solution for one object and degeneracies in the fitting parameter. Right: Comparison of best-fit SEDs with constant SFR with observations for three LAEs. These results show indications for the presence of dust in $z = 5.7$ LAEs. See the text for discussion.

Hu *et al.* (2002) has recently been analysed by Schaerer & Pelló (2005), who find that a non-negligible extinction ($A_V \sim 1.$) may be necessary to reconcile the relatively red UV-rest-frame SED and the presence of Lyα. Later this interpretation was supported by Chary *et al.* (2005), including by longer-wavelength photometry obtained with the Spitzer telescope. Three NB-selected LAEs at $z = 5.7$ detected in the optical and with the Spitzer telescope at 3.6 and 4.5 μm have recently been analysed by Lai *et al.* (2007). Overall they find SED fits degenerate in age, extinction, metallicity and SF history with stellar population ages up to 700 Myr. Most solutions require some dust extinction (see Figure 4.11). If the need for Lyα emission, i.e. for the presence of young (massive) stars, is taken into account, it seems that a constant SFR scenario is likely, together with an extinction of $E_{B-V} \sim 0.1$–0.2.

Although the evidence is not yet conclusive, these four $z \sim 5.7$–6.6 galaxies provide what are currently, to the best of my knowledge, the best indications for dust in "normal" galaxies about 1 Gyr after the Big Bang.[†] As already mentioned, these objects are probably not representative of the typical high-z LAEs, but they may be of particular interest for direct searches for high-z dust. In any case, the first attempts undertaken so far to detect dust emission from $z \sim 6.5$ galaxies in the sub-millimetre range (Webb *et al.* 2007; Boone *et al.* 2007) have provided upper limits on their dust masses of the order of $\sim(2-6) \times 10^8 M_\odot$. Future observations with more sensitive instruments and targeting gravitationally lensed objects should soon allow progress in this field.

† Dust emission has been observed in quasars out to $z \sim 6$, as discussed briefly in Section 4.2.6.

4.4.3 *Lyman-break galaxies*

In general Lyman-break galaxies (LBGs) are better known than the galaxies selected by Lyα emission (LAEs) discussed above. There is a vast literature on LBGs, summarized in an annual review paper in 2002 by Giavalisco (2002). However, progress being so fast in this area, frequent "updates" are necessary. In this last part, I shall give an overview of the current knowledge about LBGs at $z \gtrsim 6$, trying to present the main methods, results, uncertainties and controversies, and finally to summarize the main remaining questions. A more-general overview about galaxies across the Universe and out to the highest redshift is given in the lectures of Ellis (2007). Recent results from deep surveys including LBGs and LAEs can be found in the proceedings from "At the Edge of the Universe" (Afonso *et al.* 2007). Chapter 2 of this volume also covers in depth galaxy surveys.

The general principle of the LBG selection has already been mentioned above. The numbers of galaxies identified so far are approximately as follows: 4000 $z \sim 4$ galaxies (B-dropout), 1000 $z \sim 5$ galaxies (V-dropout) and 500 $z \sim 6$ galaxies (i-dropout) according to the largest dataset compiled by Bouwens and collaborators (cf. Bouwens & Illingworth 2006). The number of good candidates at $z \gtrsim 7$ is still small (see below).

4.4.3.1 i-dropout ($z \sim 6$) samples

Typically two different selections are applied to find $z \sim 6$ objects: (1) a simple $(i-z)_{\mathrm{AB}} > 1.3$–$1.5$ criterion establishing a spectral break plus optical non-detection and (2) $(i-z)_{\mathrm{AB}} > 1.3$ plus a blue UV (rest-frame) slope to select actively star-forming galaxies at these redshifts. The main samples have been found thanks to deep HST imaging (e.g. in the Hubble Ultra-Deep Field and the GOODS survey) and with SUBARU (Stanway *et al.* 2003, 2004; Bunker *et al.* 2004; Bouwens *et al.* 2003; Yan *et al.* 2006).

In general all photometric selections must avoid possible "contamination" by other sources. For the i-dropouts possible contaminants are L- or T-dwarfs, $z \sim 1$–3 extremely red objects (EROs) and spurious detections in the z band. Deep photometry in several bands (ideally as many as possible!) is required in order to minimize the contamination. The estimated contamination of i-dropout samples constructed using criterion (1) is somewhat controversial and could reach up to 25% in GOODS data, according to Bouwens *et al.* (2006) and Yan *et al.* (2006). Follow-up spectroscopy has shown quite clearly that L-dwarfs contaminate the bright end of the i-dropout samples, whereas at fainter magnitudes most objects appear to be truly at high-z (Stanway *et al.* 2004; Malhotra *et al.* 2005).

The luminosity function (LF) of $z \sim 6$ LBGs has been measured and its redshift evolution studied by several groups. Most groups find an unchanged faint-end slope of $\alpha \sim -1.7$ from $z \sim 3$ to 6. Bouwens *et al.* (2006) find a turnover at the bright end of the LF, which they interpret as being due to hierarchical build-up of galaxies. However, the results on $M_\star$ and α remain controversial. For example Sawicki & Thompson (2006) find no change of the bright end of the LF but an evolution of its faint end on going from $z \sim 4$ to $z \sim 2$, while other groups (e.g. Bunker *et al.* 2004, Yoshida *et al.* 2006, Shimasaku *et al.* 2006) find similar results to those of Bouwens *et al.* The origin of these discrepancies remains to be clarified.

The luminosity density of LBGs and the corresponding SFRD have been determined by many groups up to redshift $z \sim 6$. Most of the time this is done by integration of the LF down to a certain reference depth, e.g. $0.3L_\star(z=3)$, and at high z generally no extinction corrections are applied. Towards high z, the SFRD is found to decrease somewhat on

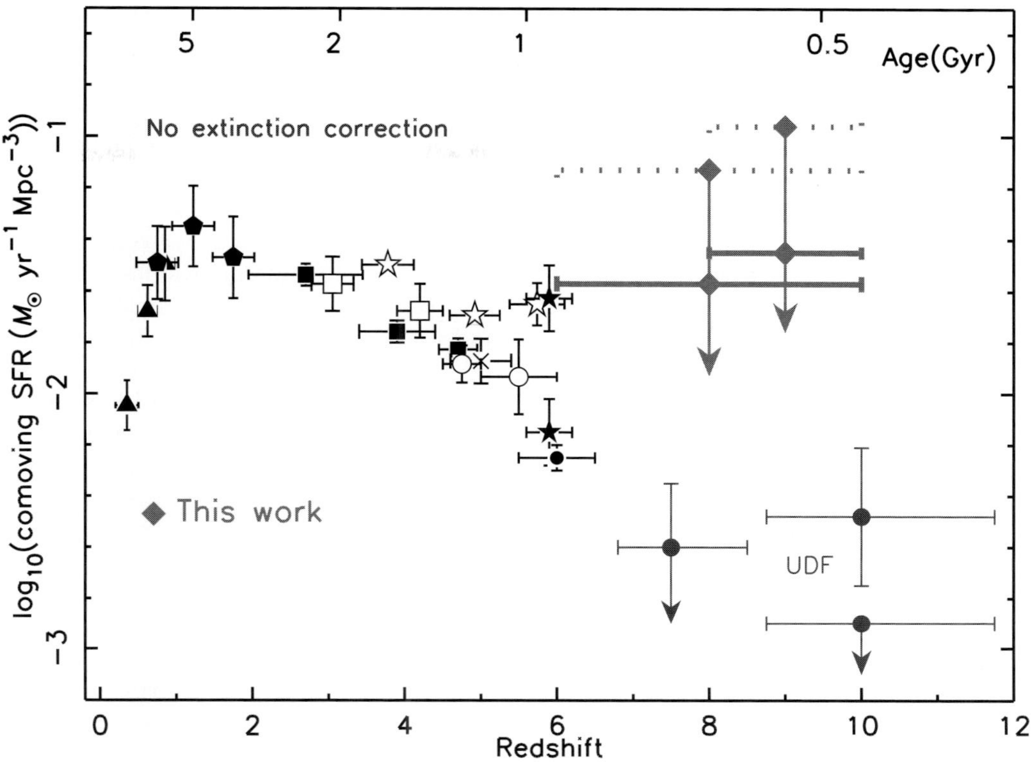

FIGURE 4.12. Evolution of the comoving SFR density as a function of redshift including a compilation of results at $z \lesssim 6$, estimates from the lensing-cluster survey of Richard *et al.* (2006) for the redshift ranges 6–10 and 8–10, and the values derived by Bouwens and collaborators from the Hubble Ultra-Deep Field (labelled "UDF"). Thick solid lines: SFR density obtained from integrating the LF of the first-category candidates of Richard *et al.* down to $L_{1500} = 0.3L^*_{z=3}$; dotted lines: the same but including also second-category candidates with a detection threshold of $<2.5\sigma$ in H. From Richard *et al.* (2006).

going from $z \sim 4$ to $z \sim 6$, whereas beyond this the results are quite controversial, as we will discuss; see e.g. a recent update by Hopkins (2006).

The properties of individual galaxies will be discussed in Section 4.4.3.3.

4.4.3.2 Optical-dropout samples ($z \gtrsim 7$)

Going beyond redshift 7 requires the use of near-IR observations, as the Lyα break of such objects moves out of the optical window. Given the different characteristics of such detectors and imagers (lower sensitivity and smaller field of view), progress has been less rapid than for lower-redshift observations.

In the NICMOS Ultra-Deep Field, Bouwens *et al.* (2004, 2006) have found 1–4 z-dropouts detected in J and K, compatible with redshift $z \sim 7$ starbursts. From this small number of objects and from the non-detection of J-dropouts by Bouwens *et al.* (2005) they deduce a low SFRD between $z \sim 7$ and 10, corresponding to a significant decrease of the SFRD with respect to lower redshift (see Figure 4.12, symbols labelled "UDF"). The properties of these and other $z \gtrsim 7$ galaxies will be discussed below.

As an alternative to the "blank fields" usually chosen for "classical" deep surveys, the use of gravitational-lensing clusters – i.e. galaxy clusters acting as strong gravitational lenses for background sources – has over the last decade or so proven very efficient at finding distant galaxies (e.g. Hu *et al.* 2002; Kneib *et al.* 2004). Using this method, and applying the Lyman-break technique plus a selection for blue UV-rest-frame spectra (i.e. starbursts), our group has undertaken very-deep near-IR imaging of several clusters to search for $z \sim 6$–10 galaxy candidates (Schaerer *et al.* 2006). Thirteen candidates whose SEDs are compatible with that of star-forming galaxies at $z > 6$ have been found (Richard *et al.* 2006). After taking into account the detailed lensing geometry and sample incompleteness, correcting for false-positive detections, and assuming a fixed slope taken from observations at $z \sim 3$, their LF was computed. Within the errors the resulting LF is compatible with that of $z \sim 3$ Lyman-break galaxies. At low luminosities it is also compatible with the LF derived by Bouwens *et al.* (2006) for their sample of $z \sim 6$ candidates in the Hubble Ultra Deep Field and related fields. However, the turnover observed by these authors towards the bright end relative to the $z \sim 3$ LF is not observed in the Richard *et al.* sample. The UV SFRD at $z \sim 6$–10 determined from this LF is shown in Figure 4.12. These values indicate a similar SFRD to that for $z \sim 3$–6, in contrast to the drop found from the deep NICMOS fields (Bouwens *et al.* 2006).[†] The origin of these differences concerning the LF and SFRD remains unknown. In any case, recent follow-up observations with the HST and Spitzer telescope undertaken to constrain the SEDs of these candidates better or to exclude some of them as intermediate-z contaminants show that data for the bulk of our candidates are compatible with their being truly at high-z (Schaerer *et al.* 2007a).

One of the main ways to clarify these differences is by improving the statistics, in particular by increasing the size (field of view) of the surveys. Both surveys of more lensing clusters and wide blank-field near-IR surveys, such as UKIDSS, are going on. The first $z \sim 7$ candidates have recently been found by UKIDSS (McLure 2007, private communication).

In this context it should also be remembered that not all optical dropout galaxies are at high-z, since a simple “dropout” criterion relies exclusively on a very red colour between two adjacent filters. As discussed for the i-dropouts above, extremely red objects (such as EROs) at $z \sim 1$–3 can be selected by such criteria. See Dunlop *et al.* (2007) and Schaerer *et al.* (2007b) for such examples. This warning is also of concern for searches for possible massive (evolved) galaxies at high redshift as undertaken by Mobasher *et al.* (2005) and McLure *et al.* (2006).

4.4.3.3 Properties of $z \gtrsim 6$ LBGs

Let us now review the main properties of individual $z \gtrsim 6$ LBGs, i.e. continuum-selected galaxies and discuss implications thereof. Lyα emitters (LAEs), such as the $z = 6.56$ lensed galaxy found by Hu *et al.* (2002), have already been discussed (Section 4.4.2). Determinations of stellar populations (ages, SF history), extinction and related properties of such distant galaxies have really been possible only recently with the advent of the Spitzer space telescope providing sensitive-enough imaging at 3.6 and 4.5 μm. These wavelengths, longwards of the K-band and hence not available for sensitive observations from the ground, correspond to the rest-frame optical domain, which is crucial to constrain properly stellar ages and stellar masses.

A triply lensed high-z galaxy magnified by a factor of ∼25 by the cluster Abell 2218 has been found by Kneib *et al.* (2004). Follow-up observations with the Spitzer telescope allowed the authors to constrain its SED up to 4.5 μm and revealed a significant Balmer

[†] The SFRD values of Bouwens have been revised upwards, reducing the differences with our study (Hopkins 2007).

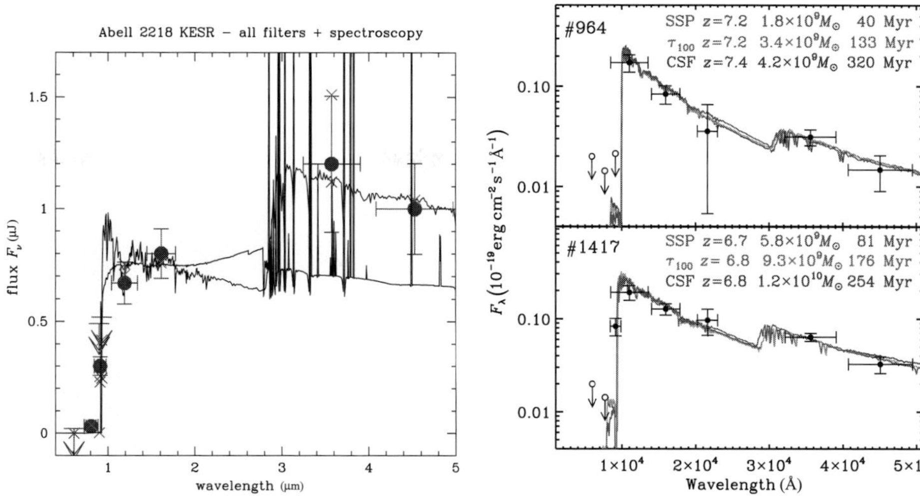

FIGURE 4.13. Left: Observed SED of the $z \sim 7$ lensed galaxy from Egami *et al.* (2005) and model fits from Schaerer & Pelló (2005) showing possible solutions with young ages (~15 Myr, solid line) or with a template of a metal-poor galaxy showing strong emission lines. Right: SEDs of two IRAC-detected $z \sim 7$ galaxies from the Hubble Ultra Deep Field and best fits using three different SF histories. From Labbé *et al.* (2006). Note the different flux units (F_ν versus F_λ) used in the two plots.

break (Egami *et al.* 2005; see Figure 4.13). Their analysis suggests that this $z \sim 7$ galaxy is in the post-starburst stage with an age of at least ~50 Myr, possibly a few hundred million years. If true, this would indicate that a mature stellar population is already in place at such a high redshift. However, the apparent 4000-Å break can also be reproduced equally well with a template of a young (~3–5-Myr) burst, where strong rest-frame optical emission lines enhance the 3.6- and 4.5-μm fluxes (Schaerer & Pelló 2005, and Figure 4.13). The stellar mass is an order of magnitude smaller (~$10^9 M_\odot$) than that of a typical LBG, the extinction low, and its SFR $\sim M_\odot \mathrm{yr}^{-1}$.

Two to four of the four $z \sim 7$ candidates of Bouwens *et al.* (2004) discussed above have been detected in the very-deep 23.3-h exposures taken with the Spitzer telescope at 3.6 and 4.5 μm by Labbé *et al.* (2006). Their SED analysis indicates photometric redshifts in the range 6.7–7.4, stellar masses of (1–10) × $10^9 M_\odot$, stellar ages of 50–200 Myr, SFRs up to ~$25 M_\odot$ yr and low reddening, $A_V < 0.4$.

Evidence for mature stellar populations at $z \sim 6$ has also been found by Eyles *et al.* (2005, 2007). By "mature" or "old" we mean here populations with ages corresponding to a significant fraction of the Hubble time, which is just ~1 Gyr at this redshift. By combining HST and Spitzer-telescope data from the GOODS survey they found that 40% of 16 objects for which they had clean photometry exhibit evidence for substantial Balmer/4000 spectral breaks. For these objects, they find ages of ~200–700 Myr, implying formation redshifts of $7 \leq z_\mathrm{f} \leq 18$, and large stellar masses in the range ~(1–3) × $10^{10} M_\odot$. Inverting the SF histories of these objects leads them to suggest that the past global SFR may have been much higher than that observed for the $z \sim 6$ epoch, as shown in Figure 4.14. This could support the finding of a relatively high SFRD at $z \gtrsim 7$, such as was found by Richard *et al.* (2006).

In short, although the samples of $z > 6$ Lyman-break galaxies for which detailed information is available are still very small, several interesting results concerning their properties have emerged already: mature stellar populations in possibly many galaxies indicating

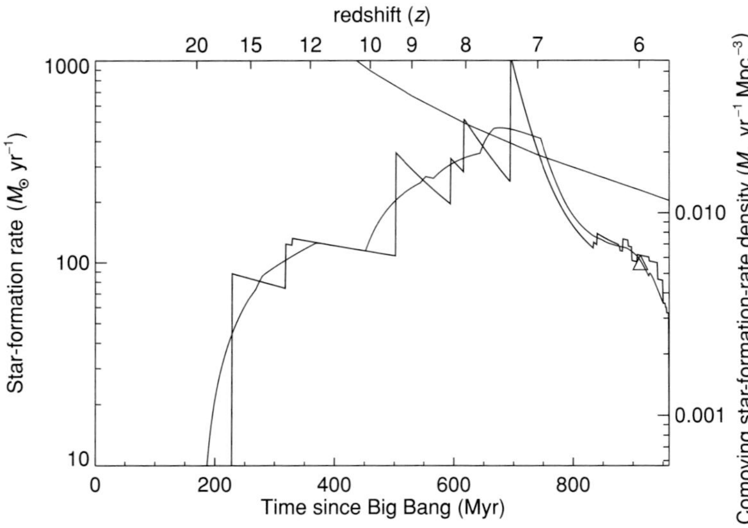

FIGURE 4.14. History of the star-formation-rate density determined by inversion from the observed i-dropout galaxies analysed by Eyles *et al.* (2007). The dotted curve is the sum of the past star-formation rates for our i'-dropout sample (left axis), with the corresponding star-formation-rate density shown on the right axis, corrected for incompleteness including a factor of 3.2 for galaxies below the flux threshold. The dashed curve is this star-formation history smoothed on a timescale of 100 Myr. The triangle is the estimate of the unobscured (rest-frame UV) star-formation-rate density at $z \approx 6$ from i'-dropouts in the HUDF from Bunker *et al.* (2004). The solid curve shows the condition for reionization from star formation, as a function of time (bottom axis) and redshift (top axis), assuming an escape fraction of unity for the Lyman-continuum photons. From Eyles *et al.* (2007).

a high formation redshift, stellar masses of the order of $(10^9\text{–}10^{10})M_\odot$ and generally low extinction. However, a fraction of these galaxies appears also to be young and less massive (Eyles *et al.* 2007) forming a different "group". Similar properties and two similar groups are also found among the high-z LAEs (Schaerer & Pelló 2005; Lai *et al.* 2007; Pirzkal *et al.* 2006) already discussed above. Whether such separate "groups" really exist and, if so, why, remains to be seen.

In a recent analysis Verma *et al.* (2007) find that ~70% of $z \sim 5$ LBGs have typical ages of $\lesssim$100 Myr and stellar masses of $\sim 10^9 M_\odot$, namely are younger and less massive than typical LBGs at $z \sim 3$. They also find indications for a relatively low extinction, lower than at $z \sim 3$. The trend of a decreasing extinction in LBGs with increasing redshift has been found in many studies, and is in agreement with the results discussed above for $z \sim 6$ and higher. However, the differences in age and mass e.g. compared with the objects of Eyles *et al.* (2007) may be surprising, especially given the short time (~200 Myr) between redshifts 5 and 6. Several factors, such as selection effects and the representativeness of the small $z \sim 6$ samples studied in detail, may contribute to such differences. Reaching a more-complete and coherent understanding of the primeval galaxy types, their evolution and their relation with galaxies at lower redshift will need more time and further observations.

4.4.4 *What next?*

It has been possible during the last decade to push the observational limits out to very high redshift and to identify and study the first samples of galaxies observed barely ~1 Gyr after the Big Bang. The current limit is approximately at $z \sim$ 7–10, where just a few galaxies (or galaxy candidates) have been detected, and where spectroscopic confirmation remains extremely challenging.

Thanks to very deep imaging in the near-IR domain it is possible to estimate or constrain the stellar populations (age, SF history, mass, etc.) and dust properties (extinction) of such "primeval" galaxies, providing us with a first glimpse on galaxies in the early Universe. Despite this great progress and these exciting results, the global observational picture on primeval galaxies, and on their formation and evolution, remains to be drawn. Many important questions remain or, better said, are starting to be posed now, and can now or in the near future be addressed not only by theory and modelling but also observationally!

We have already seen some of the emerging questions, but others, sometimes more-general ones, have not been addressed. Among the important questions concerning primeval galaxies we can list the following.

- How do different high-z populations such as LAEs and LBGs fit together? Are there other currently unknown populations? What are the evolutionary links between these populations and galaxies at lower redshift?
- What is the metallicity of the high-z galaxies? Where is Population III?
- What is the star-formation history of the Universe during the first Gyr after the Big Bang?
- Are there dusty galaxies at $z \gtrsim 6$? How, where and when do they form? How much dust is produced at high redshift?
- Which are the sources of reionization? Are these currently detectable galaxies or very-faint low-mass objects? What is the history of cosmic reionization?

We, and especially young students, are fortunate to live during a period when theory, computing power and observational facilities are rapidly growing, enabling astronomers to peer even deeper into the Universe. It is probably a fair guess to say that within the next 10–20 years we should have observed the very first galaxies forming in the Universe, found Population III, etc. We will thus have reached the limits of the map in this exploration of the Universe. However, a lot of challenging and interesting work will remain if we are to reach a global and detailed understanding of the formation and evolution of stars and galaxies!

Acknowledgments

I thank Jordi Cepa for the invitation to lecture on this topic and for his patience with the manuscript. I would also like to thank him and the IAC team for the organization of this excellent and enjoyable Winter School. Over the last years I've appreciated many interesting and stimulating discussions with my collaborators and other colleagues. Among them I'd like to thank in particular Roser Pelló, Johan Richard, Jean-François Le Borgne, Jean-Paul Kneib, Angela Hempel and Eiichi Egami, representing the "high"-z Universe, and Daniel Kunth, Anne Verhamme, Hakim Atek, Matthew Hayes and Miguel Mas-Hesse for the nearby Universe, as well as Andrea Ferrara, Grażyna Stasińska and David Valls-Gabaud. Both the list of people thanked and the references are quite incomplete, though. Apologies!

REFERENCES

Afonso, J., Ferguson H., Norris, R. (eds.). 2007, *At the Edge of the Universe: Latest Results from the Deepest Astronomical Surveys* (ASP)

Ahn, S. H. 2004, ApJ, 601, L25

Ajiki, M. *et al.* 2003, AJ, 126, 2091

Baraffe, I., Heger, A., Woosley, S. E. 2001, ApJ, 552, 464

Barkana, R., Loeb, A. 2001, Phys. Rep., 349, 125

Boone, F., Schaerer, D. *et al.* 2007, A&A, in preparation

Bouwens, R. J. *et al.* 2003, ApJ, 595, 589

Bouwens, R. J., Illingworth, G. D. 2006, Nature, 443, 189

Bouwens, R. J., Illingworth, G. D., Blakeslee, J. P., Franx, M. 2006, ApJ, 653, 53

Bouwens, R. J., Illingworth, G. D., Thompson, R. I., Franx, M. 2005, ApJL, 624, L5

Bouwens R. J., Thompson R. I., Illingworth G. D., Franx M., van Dokkum P., Fan X., Dickinson M. E., Eisenstein D. J., Rieke M. J. 2004, ApJ, 616, L79

Bromm, V., Kudritzki, R. P., Loeb, A. 2001, ApJ, 552, 464

Bromm V., Larson, R. B. 2004, ARA&A, 42, 79

Bunker, A. J., Stanway, E. R., Ellis, R. S., McMahon, R. G. 2004, MNRAS, 355, 374

Cen, R., Haiman, Z. 2000, ApJ, 542, L75

Chapman, S. C., Blain, A. W., Ivison, R. J., Smail, I. R. 2003, Nature, 422, 695

Charlot, S., Fall, S. M. 1993, ApJ, 415, 580

Chary, R.-R., Stern, D. Eisenhardt, P. 2005, ApJL, 635, L5

Ciardi, B., Ferrara, A. 2005, Space Sci. Rev., 116, 625

Ciardi, B., Ferrara, A., Governato, F., Jenkins, A. 2000, MNRAS, 314, 611

Dawson, S., *et al.* 2004, ApJ, 617, 707

Dijkstra, M., Haiman, Z., Spaans, M. 2006a, ApJ, 649, 37

Dijkstra, M., Haiman, Z., Spaans, M. 2006b, ApJ, 649, 14

Dijkstra, M., Wyithe, J. S. B. 2007, MNRAS, submitted (astro-ph/0704.1671)

Dijkstra, M., Wyithe, J. S. B., Haiman, Z. 2006c, MNRAS, submitted (astro-ph/0611195)

Dopita, M. A., Sutherland, R. S. 2003, *Astrophysics of the Diffuse Universe* (Berlin: Springer-Verlag)

Dunlop, J. S., Cirasuolo, M., McLure, R. J. 2007, MNRAS, 376, 1054

Egami, E., *et al.* 2005, ApJL, 618, L5

Ekström, S., Meynet, G., Maeder, A. 2006, Cosmology, 353, 141

Ellis, R. S. 2007, in *First Light in the Universe*, ed. D. Schaerer, A. Hempel, & D. Puy (Berlin: Springer-Verlag), in press (astro-ph/0701024)

Eyles, L. P., Bunker, A. J., Ellis, R. S., Lacy, M., Stanway, E. R., Stark, D. P., Chiu, K. 2007, MNRAS, 374, 910

Eyles, L. P., Bunker, A. J., Stanway, E. R., Lacy, M., Ellis, R. S., Doherty, M. 2005, MNRAS, 364, 443

Fan, X. *et al.* 2003, AJ, 125, 1649

Fan, X., Carilli, C. L., Keating, B. 2006, ARAA, 44, 415

Ferrara, A. 2007, in *First Light in the Universe*, ed. D. Schaerer, A. Hempel, & D. Puy (Berlin: Springer-Verlag), in press (obswww.unige.ch/saas-fee2006/)

Ferrara, A., Ricotti, M. 2006, MNRAS, 373, 571

Finkelstein, S. L., Rhoads, J. E., Malhotra, S., Pirzkal, N., Wang, J. 2006, ApJ, submitted (astro-ph/0612511)

Furlanetto, S. R., Hernquist, L., Zaldarriaga, M. 2004, MNRAS, 354, 695

Furlanetto, S. R., Zaldarriaga, M., Hernquist, L. 2006, MNRAS, 365, 1012

Gawiser, E. *et al.* 2006, ApJL, 642, L13

Giavalisco, M. 2002, ARAA, 40, 579

Giavalisco, M., Koratkar, A., Calzetti, D. 1996, ApJ, 466, 831

Gnedin, N. Y., Prada, F. 2004, ApJL, 608, L77

Haiman, Z. 2002, ApJ, 576, L1

Haiman, Z., Cen, R. 2005, ApJ, 623, 627

Haiman, Z., Spaans, M. 1999, ApJ, 518, 138

Hansen, M., Oh, S. P. 2006, MNRAS, 367, 979

Hartmann, L. W., Huchra, J. P., Geller, M. J. 1984, ApJ, 287, 487

Hayes, M., Östlin, G. 2006, A&A, 460, 681

Hayes, M., Östlin, G., Mas-Hesse, J. M., Kunth, D., Leitherer, C., Petrosian, A. 2005, A&A, 438, 71

Heger, A., Woosley, S. E. 2002, ApJ, 567, 532

Heger, A., Fryer, C. L., Woosley, S. E., Langer, N., Hartmann, D. H. 2003, ApJ, 591, 288

Hernandez, X., Ferrara, A. 2001, MNRAS, 324, 484

Hopkins, A. M. 2006, in *At the Edge of the Universe: Latest Results from the Deepest Astronomical Surveys* (ASP), in press (astro-ph/0611283)

Hu, E. M., Cowie, L. L. 2006, Nature, 440, 1145

Hu, E. M., Cowie, L. L., McMahon, R. G., Capak, P., Iwamuro, F., Kneib, J.-P., Maihara, T., Motohara, K. 2002, ApJL, 568, L75

Hu, E. M., Cowie, L. L., Capak, P., McMahon, R. G., Hayashino, T., Komiyama, Y. 2004, AJ, 127, 563

Hu, E. M., Cowie, L. L., Capak, P., Kakazu, Y. 2005, *Probing Galaxies through Quasar Absorption Lines* (IAU), p. 363 (astro-ph/0509616)

Hummer, D. G. 1962, MNRAS, 125, 21

Iye, M. *et al.* 2006, Nature, 443, 186

Jimenez, R., Haiman, Z. 2006, Nature, 440, 501

Kashikawa, N. *et al.* 2006, ApJ, 648, 7

Kneib, J.-P., Ellis, R. S., Santos, M. R., Richard, J. 2004, ApJ, 607, 697

Kudritzki, R. P. 2002, ApJ, 577, 389

Kunth, D., Leitherer, C., Mas-Hesse, J. M., Östlin, G., Petrosian, A. 2003, ApJ, 597, 263

Kunth, D., Lequeux, J., Sargent, W. L. W., Viallefond, F. 1994, A&A, 282, 709

Kunth, D., Mas-Hesse, J. M., Terlevich, E., Terlevich, R., Lequeux, J., Fall, S. M. 1998, A&A, 334, 11

Labbé, I., Bouwens, R., Illingworth, G. D., Franx, M. 2006, ApJL, 649, L67

Lai, K., Huang, J.-S., Fazio, G., Cowie, L. L., Hu, E. M., Kakazu, Y. 2007, ApJ, 655, 704

Le Delliou, M., Lacey, C., Baugh, C. M., Guiderdoni, B., Bacon, R., Courtois, H., Sousbie, T., Morris, S. L. 2005, MNRAS, 357, L11

Le Delliou, M., Lacey, C. G., Baugh, C. M., Morris, S. L. 2006, MNRAS, 365, 712

Madau, P. 1995, ApJ, 441, 18

Maiolino, R., Schneider, R., Oliva, E., Bianchi, S., Ferrara, A., Mannucci, F., Pedani, M., Roca Sogorb, M. 2004, Nature, 431, 533

Malhotra, S. *et al.* 2005, ApJ, 626, 666

Malhotra, S., Rhoads, J. E. 2002, ApJ, 565, L71

Malhotra, S., Rhoads, J. E. 2004, ApJL, 617, L5

Mao, J., Lapi, A., Granato, G. L., De Zotti, G., Danese, L. 2006, ApJ, submitted (astro-ph/0611799)

Marigo, P., Girardi, L., Chiosi, C., Wood, R. 2001, A&A, 371, 152

Mas-Hesse, J. M., Kunth, D., Tenorio-Tagle, G., Leitherer, C., Terlevich, R. J., Terlevich, E. 2003, ApJ, 598, 858

McLure, R. J. *et al.* 2006, MNRAS, 372, 357

Meier, D., Terlevich, R. 1981, ApJ, 246, L109

Meynet, G., Ekström, S., Maeder, A. 2006, A&A, 447, 623

Miralda-Escude, J. 1998, ApJ, 501, 15

Nagao, T., Motohara, K., Maiolino, R., Marconi, A., Taniguchi, Y., Aoki, K., Ajiki, M., Shioya, Y. 2005, ApJL, 631, L5

Nagao, T. *et al.* 2007, A&A, submitted (astro-ph/0702377)

Neufeld, D. A. 1990, ApJ, 350, 216

Neufeld, D. A. 1991, ApJ, 370, 85

Osterbrock, D. E., Ferland, G. J. 2006, *Astrophysics of Gaseous Nebulae and Active Galactic Nuclei*, 2nd edn (Sausalito, CA: University Science Books)

Ouchi, M. *et al.* 2005, ApJ, 620, L1

Partridge, R. B., Peebles, J. E. 1967, ApJ, 147, 868

Pelló, R., Schaerer, D., Richard, J., Le Borgne, J.-F., Kneib, J.-P. 2004, A&A, 416, L35

Pettini, M., Steidel, C. C., Adelberger, K. L., Dickinson, M., Giavalisco, M. 2000, ApJ, 528, 96

Pirzkal, N., Malhotra, S., Rhoads, J. E., Xu, C. 2006, ApJ, submitted (astro-ph/0612513)

Pritchet, J. C. 1994, PASP, 106, 1052

Richard, J., Pelló, R., Schaerer, D., Le Borgne, J.-F., Kneib, J.-P. 2006, A&A, 456, 861

Santos, M. R. 2004, MNRAS, 349, 1137

Sawicki, M., Thompson, D. 2006, ApJ, 642, 653

Schaerer, D. 2002, A&A, 382, 28.

Schaerer, D. 2003, A&A, 397, 527.

Schaerer, D., Hempel, A., Egami, E., Pelló, R., Richard, J., Le Borgne, J.-F., Kneib, J.-P., Wise, M., Boone, F. 2007b, A&A, in press (astro-ph/0703387)

Schaerer, D., Pelló, R. 2005, MNRAS, 362, 1054

Schaerer, D., Pelló, R., Richard, J., Egami, E., Hempel, A., Le Borgne, J.-F., Kneib, J.-P., Wise, M., Boone, F., Combes, F. 2006, The Messenger, 125, 20

Schaerer, D., Pelló, R., Richard, J., Egami, E., Hempel, A., Le Borgne, J.-F., Kneib, J.-P., Wise, M., Boone, F., Combes, F. 2007a, in *At the Edge of the Universe: Latest Results from the Deepest Astronomical Surveys* (ASP), in press (astro-ph/0701195)

Scannapieco, E., Madau, P., Woosley, S., Heger, A., Ferrara, A. 2005, ApJ, 633, 1031

Scannapieco, E., Schneider, R., Ferrara, A. 2003, ApJ, 589, 35

Schneider, R., Ferrara, A., Natarajan, P., Omukai, K. 2002, ApJ, 579, 30

Schneider, R., Ferrara, A., Salvaterra, R. 2004, MNRAS, 351, 1379

Schneider, R., Salvaterra, R., Ferrara, A., Ciardi, B. 2006, MNRAS, 369, 825

Shapiro, P. R., Giroux, M. L. 1987, ApJL, 321, L107

Shapley, A., Steidel, C. C., Pettini, M., Adelberger, K. L. 2003, ApJ, 588, 65

Shimasaku, K. *et al.* 2006, PASJ, 58, 313

Spitzer, L. 1978, *Physical Processes in the Interstellar Medium* (New York: Wiley)

Stanway, E. R., Bunker, A. J., McMahon, R. G. 2003, MNRAS, 342, 439

Stanway, E. R., Bunker, A. J., McMahon, R. G., Ellis, R. S., Treu, T., McCarthy, P. J. 2004, ApJ, 607, 704

Stark, D. P., Ellis, R. S., Richard, J., Kneib, J.-P., Smith, G. P., Santos, M. R. 2007, ApJ, submitted (astro-ph/0701279)

Steidel, C. C., Giavalisco, M., Dickinson, M., Adelberger, K. L. 1996, AJ, 112, 352

Stern, D. J., Spinrad, H. 1999, PASP, 111, 1475

Stratta, G. *et al.* 2007, ApJ, submitted (astro-ph/0703349)

Taniguchi, Y. *et al.* 2005, PASJ, 57, 165

Tapken, C., Appenzeller, I., Noll, S., Richling, S., Heidt, J., Meinkoehn, E., Mehlert, D. 2007, A&A, in press (astro-ph/0702414)

Tegmark, M., Silk, J., Rees, M. J., Blanchard, A., Abel, T., Palla, F. 1997, ApJ, 474, 1

Tenorio-Tagle, G., Silich, S.A., Kunth, D. *et al.* 1999, MNRAS, 309, 332

Thuan, T. X., Izotov, Y. I. 1997, ApJ, 489, 623

Thuan, T. X., Sauvage, M., Madden, S. 1999, ApJ, 516, 783

Todini, P., Ferrara, A. 2001, MNRAS, 325, 726

Tumlinson, J. 2006, ApJ, 641, 1

Tumlinson, J., Giroux, M. L., Shull, J. M. 2001, ApJ, 550, L1

Valls-Gabaud, D. 1993, ApJ, 419, 7

Verhamme, A., Schaerer, D., Maselli, A. 2006, A&A, 460, 397

Verhamme, A., Schaerer, D., Atek, H., Tapken, C. 2007, A&A, in preparation

Verma, A., Lehnert, M. D., Förster Schreiber, N. M., Bremer, M. N., Douglas, L. 2007, MNRAS, in press (astro-ph/0701725)

Walter, F. *et al.* 2003, Nature, 424, 406

Wang, J. X. *et al.* 2004, ApJL, 608, L21

Webb, T. M. A., Tran, K.-V. H., Lilly, S. J., van der Werf, P. 2007, ApJ, 659, 76

Weinmann, S. M., Lilly, S. J. 2005, ApJ, 624, 526

Williams, R. E. *et al.* 1996, AJ, 112, 1335

Wilman, R. J., Gerssen, J., Bower, R. G., Morris, S. L., Bacon, R., de Zeeuw, P. T., Davies, R. L. 2005, Nature, 436, 227

Wise, J. H., Abel, T. 2005, ApJ, 629, 615

Wu, Y. *et al.* 2008, ApJ, in press (astro-ph/0703283)

Wyithe, J. S. B., Loeb, A. 2004, Nature, 432, 194

Yan, H., Dickinson, M., Giavalisco, M., Stern, D., Eisenhardt, P. R. M., Ferguson, H. C. 2006, ApJ, 651, 24

Yoshida, M. *et al.* 2006, ApJ, 653, 988

5. Active galactic nuclei

BRADLEY M. PETERSON

5.1. Introduction

In this contribution, we will take a broad view of active galactic nuclei (AGNs), but concentrating on emission-line phenomena. As a result, some interesting topics such as blazars and jets, connections with starbursts, and the environments of AGNs will receive little attention, although these are important topics and are discussed at length elsewhere.

It is useful to begin a discussion of AGNs with some history of the subject, partly because the history of how a scientific field is launched and develops over time is interesting and instructive, but primarily because it gives us some insight into the observed properties of AGNs, which is essential for distinguishing them from other objects in the sky. As we introduce the observational properties of these sources, at least the basis of the sometimes complicated taxonomy we use to describe active galaxies will become clear. An underlying theme throughout this discussion will be the principle of "AGN unification," which posits that the diverse taxonomy of AGNs has more to do with observational circumstances, such as inclination and obscuration effects, than with intrinsic physical differences among various types of AGN. Considerable effort has been expended in attempts to explain the broadest range of phenomena with the most-limited range of physical mechanisms and structures.

5.2. The discovery and nature of active galaxies

The word "activity" in connection with phenomena in the nuclei of galaxies appears to have originated with Ambartsumian (1968). An ADS search on the specific phrase "active galactic nuclei" reveals that it was first used in the title of a paper by Weedman (1974) in an invited presentation at a meeting of the American Astronomical Society and first used in a Ph.D. dissertation title by Eilek (1975, 1977). Initially, activity in galactic nuclei referred primarily to radio emission, although it was sometimes stated that at least some of the observed activity might be attributable to vigorous star formation and the resultant supernovae.[†] Over time, the phrase "active galactic nuclei" gradually evolved to refer more generally to phenomena that are not clearly attributable to normal stars. A more-modern, narrow definition might be that active nuclei are those that emit radiation that is fundamentally powered by accretion onto supermassive ($M > 10^6 M_\odot$) black holes.

From a phenomenological point of view, the characteristics of active galaxies include

- strong X-ray emission,
- non-stellar ultraviolet/optical continuum emission,
- relatively strong radio emission, and
- ultraviolet through infrared spectra dominated by strong, broad emission lines.

Not every active galaxy has all of these characteristics, as we shall see.

Two of the largest subclasses of AGN are Seyfert galaxies and quasars; indeed, the reason for the modern ubiquity of the term "AGN" is that it provides a common

[†] Ambartsumian (1971) includes in his definition of activity "the violent motions of gaseous clouds, considerable excess radiation in the ultraviolet, relatively rapid changes in brightness, expulsions of jets and of condensations."

The Emission-Line Universe, ed. J. Cepa. Published by Cambridge University Press.

designation for these closely related types of sources. The connection between lower-luminosity Seyferts and higher-luminosity quasars was recognized relatively late because the original members of these two subclasses were extreme examples of each type. In the first case, Seyfert (1943) originally identified a sample of relatively nearby galaxies with extraordinarily bright cores. Spectroscopy of the nuclei of these galaxies reveals the strong, broad emission lines that are now widely regarded as the defining characteristic of the class. Woltjer (1959) offered the first physical analysis of these extraordinary sources, but they otherwise received scant attention for nearly a decade.

In the second case, quasars were discovered as a result of technology opening up a new observational window in the electromagnetic spectrum. In the period following the Second World War, some scientists who had spent the war years on radar research turned their attention to celestial sources, some of which had been detected serendipitously while the sky was being scanned for hostile aircraft. Ultimately many extraterrestrial radio sources were identified with bright galaxies and supernova remnants, but some radio sources seemed to have no obvious optical counterparts within the arcminute-size beams of single-dish antennas. As radio-source positions were made more accurate (by use of both interferometry and lunar occultations), it became clear that some of these radio sources were associated with star-like objects, i.e. spatially unresolved point sources. The optical counterparts of these radio sources tended to be blue (relative to most stars) and their spectra were found to be dominated by strong, but unidentified, emission lines. Despite their star-like appearance, these were clearly not stars, and were thus called "quasi-stellar radio sources," a term that was eventually shortened to "quasars." The extragalactic nature of quasars became clear when Schmidt (1963) recognized the strongest emission lines in 3C 273 as the hydrogen Balmer series redshifted to $z = 0.158$. While the high redshift was not unprecedented (clusters of galaxies at $z \approx 0.2$ were known at the time), the optical brightness of 3C 273, $B \approx 13.1$ mag, implied an intrinsic luminosity about 100 times larger than those of nearby giant galaxies. Further complicating the story was the soon-reported rapid variability of quasars (e.g. Smith & Hoffleit 1963), implying on the basis of coherence arguments that they could not be much larger than several light days in extent. Perhaps inevitably, this led to a nearly 20-year controversy as to whether or not quasars are in fact at the vast distances implied by their redshifts.[†]

It is now generally accepted that Seyfert galaxies and quasars are closely related phenomena. It is probably fair to say that most astronomers believe that these are distinguished from one another only by luminosity, though, of course, luminosity may directly affect other observable properties, leading to additional apparent differences.[‡] A powerful argument for this first "unification" of AGN phenomenology was made by Weedman (1976), and earlier failures to connect the two subclasses were because the archetypes of the subclasses were extreme, as explained in the following list.

- The original Seyfert galaxies had nuclei that were of about the same luminosity as their host galaxies. In contrast, the first quasars discovered were 100 times as luminous as normal giant galaxies; indeed, they are so bright compared with their host galaxies that it was not until the advent of CCD detectors in the early 1980s that the host-galaxy starlight features were observed in quasar spectra (Boroson & Oke 1982).
- The original Seyferts are well-resolved galaxies, but the first quasars discovered appeared to be star-like. Part of this, as indicated above, is simply the contrast between the nucleus and the host galaxy, which is much larger for quasars. But it is also a classic example of "Malmquist bias" – more-luminous objects are rarer than

† Remarkably, there are still a few hold-outs on this issue.

‡ An example of this might be the small number of "Type 2" quasars, as discussed below.

less-luminous objects and are thus more likely to be found at greater distances, and fainter objects at large distance are more likely to escape detection. The large distances and concomitant small angular diameters of quasar host galaxies make the starlight harder to detect. So, unlike Seyfert galaxies, quasars appear to be point-like, high-redshift sources.

- The first quasars discovered were powerful radio sources, which are now understood to constitute the minority of AGNs, around 5%–10% of the total population. None of the objects identified by Seyfert are luminous radio sources.
- The optical flux of the first known quasars was found to be highly variable. The variability of Seyfert galaxies was not detected until the late 1960s (Fitch *et al.* 1967), but apparently only because no one thought to look for it. Emission-line variability was first reported by Andrillat & Souffrin (1968).

5.3. Identifying active galaxies

Once a new class of object has been identified, we want to ask questions about the space density, luminosity function, and environment of these objects. This requires obtaining a large sample of these sources, and that requires an efficient means of separating them from other objects in the sky.

In the case of AGNs, all of the characteristics mentioned previously can be used to isolate them. Strategies for obtaining large samples of AGNs would thus include (a) radio surveys, (b) searches based on photometric color ("ultraviolet excess"), (c) searches for objects with strong emission lines, and (d) searches for variable sources. Indeed, all of these methods have been used to identify AGN candidates.

As already noted, the first quasars were discovered as a result of radio surveys of the sky. A serious difficulty with this approach was that the angular resolution of single-dish antennas is quite low (typically arcminutes) compared with ground-based resolution of the sky at optical wavelengths (typically around an arcsecond). Even at high Galactic latitude, there are usually multiple candidate objects that might be associated with the radio sources. However, once a few quasars with particularly accurate radio-source positions had been identified, it became clear that the quasars stand out in the $U - B$ color (Figures 5.1 and 5.2), greatly simplifying their identification among the candidates. This led to the further realization that other UV-excess sources at high Galactic latitude might also be quasars, and one might by-pass the laborious radio observations altogether. Searching for quasars by detecting UV excess met with immediate success, despite some gross initial errors in estimating their numbers on account of confusion with white dwarfs and horizontal-branch stars.[†] It quickly became clear that these "radio-quiet" UV-excess objects significantly outnumbered the "radio-loud" quasars by a large factor. Since the word "quasar" derived in part from "radio source," the term "quasi-stellar object (QSO)" was adopted generically for all of these sources. Over time, usage of the term "quasar" has become more generic.

In addition to radio and multicolor optical surveys, low-resolution spectroscopic surveys have been used very effectively to search for AGNs (see Chapter 2 of this volume). Variability, primarily in the optical, has been used to a lesser extent than other methods, but microlensing campaigns, for example, produce large numbers of candidate AGNs as a by-product.

[†] The initial overestimate of the space density of quasars was so large that the paper reporting the discovery of quasars by UV excess was entitled "The existence of a new major constituent of the Universe: the quasistellar galaxies" (Sandage 1965).

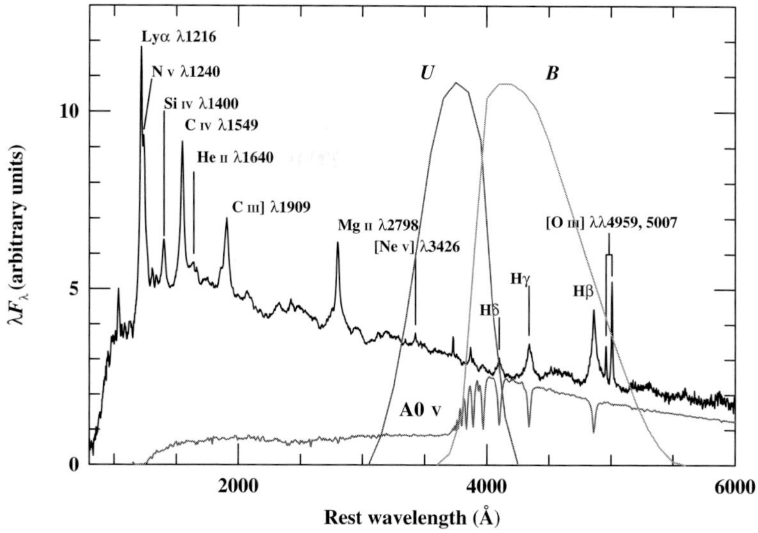

FIGURE 5.1. Relative energy flux λF_λ versus rest-frame wavelength for a composite quasar spectrum. Prominent emission lines are labeled. The lower spectrum, for comparison, is that of an A0 V star, in similar arbitrary flux units. The bandpasses for Johnson U and B are also shown. As one goes from longer to shorter wavelengths, note how the A-star spectrum drops dramatically at the Balmer limit, while that of the quasar continues to rise; this accounts for the very blue $U-B$ colors of quasars (see Figure 5.2). The composite spectrum is from the Large Bright Quasar Survey (Francis *et al.* 1991) and the A-star spectrum is from Pickles (1998). Based on a figure from Peterson (1997).

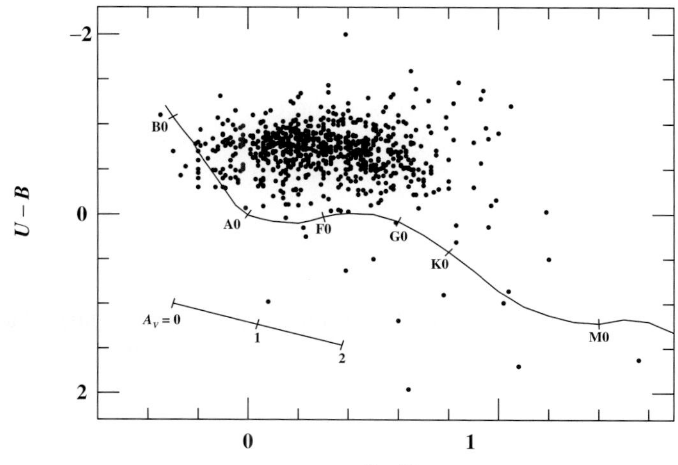

FIGURE 5.2. A two-color $U-B$ versus $B-V$ diagram for 788 quasars from the Hewitt & Burbidge (1993) catalog. The location of the zero-age main sequence is also shown, together with the extinction vector. Low-redshift AGNs stand out by having very small values of $U-B$ for a given $B-V$, for reasons shown in Figure 5.1. From Peterson (1997).

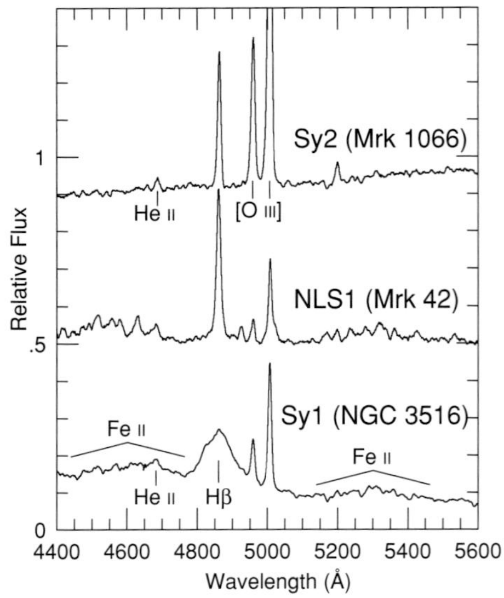

FIGURE 5.3. Optical spectra of a Seyfert 2 galaxy (top), a narrow-line Seyfert 1 galaxy (middle), and a Seyfert 1 galaxy (bottom). The most-prominent lines in each spectrum are Hβ λ4861 and [O III] λλ4959, 5007. From Pogge (2000), with permission from Elsevier.

In the years following the identification of the first quasars, an increasingly wide variety of AGN activity was identified, as discussed below. By the 1980s, however, it had become clear that the single best indicator of AGN activity is hard X-ray emission, since this is not subject to internal extinction in AGNs,† which often turn out to have fairly large columns of gas and dust along our sight lines to their centers, as we will see shortly.

5.4. A diverse taxonomy

Not all of the distinguishing characteristics of AGNs – variable non-stellar UV/optical continuum emission, strong emission lines, radio emission – are observed in every AGN. This has led to the categorization of a wide variety of AGN subclasses on the basis of the particular AGN characteristics that are observed in each case. One of the challenges of AGN astronomy is to determine *why* particular characteristics are or are not seen in each subclass, with the specific goal of accounting for AGN diversity with a minimum number of parameters. The premise of "AGN unification" is that the basic physics is the same in all AGNs, and that their apparent diversity is attributable primarily to how we observe them (e.g. some components may be hidden from us under certain circumstances). In addition to the "radio-loud" and "radio-quiet" subclasses already introduced, here we describe some of the more-important subclasses of AGNs.

Two types of Seyfert galaxies: Khachikian & Weedman (1974) were the first to separate Seyferts into two distinct spectroscopic classes (Figure 5.3). The spectra of Type 1 Seyfert galaxies have two distinct sets of emission lines that are superposed on one another. The "broad lines" or "broad-line components" arise in relatively high-density

† At energies above several keV, electron (Thompson) scattering is the sole remaining source of opacity since all of the abundant elements have been completely ionized. The electron column density necessary to make a source "Compton thick," i.e. with $\tau_e > 1$, is $N_e > 1/\sigma_e \approx 10^{24}\ \mathrm{cm}^{-2}$. Some AGNs do in fact have such large columns along the line of sight (e.g. Mathur *et al.* 2000).

gas ($n_e > 10^{10}\,\mathrm{cm}^{-3}$) and have Doppler widths that correspond to velocity dispersions of thousands of kilometers per second. At high nebular densities, forbidden-line emission is collisionally suppressed, so the broad components appear only in the permitted lines, and some intercombination lines such as C III] λ1909. The "narrow lines" or "narrow-line components" arise in gas of much lower density ($n_e \approx 10^3$–$10^6\,\mathrm{cm}^{-3}$) and have smaller Doppler widths, typically only a few hundred kilometers per second, which is comparable to or somewhat larger than stellar velocity dispersions in the cores of most galaxies. In Type 2 Seyfert galaxies, only the narrow-line components are observed. It is clear that at least some Type 2 Seyferts are intrinsically Type 1 sources whose broad-line-emitting regions are obscured along our line of sight.

Narrow-line Seyfert 1 galaxies: A subset of Seyfert 1 galaxies is notable for having unusually narrow permitted lines. As shown in Figure 5.3, they are true Type 1 objects in that their permitted lines are broader than their forbidden lines, even if only slightly, and their optical spectra exhibit Fe II emission. These objects, known as "narrow-line Seyfert 1" (NLS1) galaxies, are spectroscopically defined by the criteria that the [O III] λ5007/Hβ flux ratio is less than 3 (compare with the Seyfert 2 spectrum in Figure 5.3) and FWHM(Hβ) < 2000 km s^{-1} (Osterbrock & Pogge 1985). It was later discovered that they are very prominent soft X-ray sources with rapid, high-amplitude X-ray flux variability (Boller *et al.* 1996). As we discuss below, NLS1s are thought to be AGNs with high mass-accretion rates.

Blazars – lineless AGNs: By the early 1970s, certain radio-loud, variable, UV-excess objects had been found to have no emission lines or absorption lines at all. These were known initially as "BL Lacertae (or BL Lac) objects" after their archetype, which, as can be guessed from the name, had been previously, but erroneously, identified as a variable star. BL Lac objects were later recognized to share most properties of "optically violent variables," which differ from BL Lac objects only in that the broad emission lines are prominent spectral features. BL Lac objects and optically violent variables are thus often combined into a larger class of objects known as "blazars." Blazars are now thought to be AGNs with relativistic jets that are directed close to our line of sight. High signal-to-noise-ratio spectra of BL Lac objects often have weak emission or absorption lines that are simply swamped by the strong continuum emission.

LINERS – low-ionization AGNs: A final AGN subclass of which we should be aware is the sources known as LINERs, for "low-ionization nuclear emission region" galaxies (Heckman 1980). The LINER class is quite inhomogeneous, containing broad-line objects that are very much AGN-like as well as objects in which the emission-line spectrum appears to be driven by stars rather than by an AGN. In most cases, it now appears that LINERs are AGNs with low accretion rates and a comparatively weak ionizing spectrum.

5.5. The structure and physics of AGNs

5.5.1 *An overview*

The current paradigm for explaining the structure of AGNs is based on a combination of relatively straightforward physical considerations and some 40 years of sometimes ambiguous and hard-to-interpret observational data. While many important details are not understood, even some fundamental problems remain unsolved (e.g. how does mass

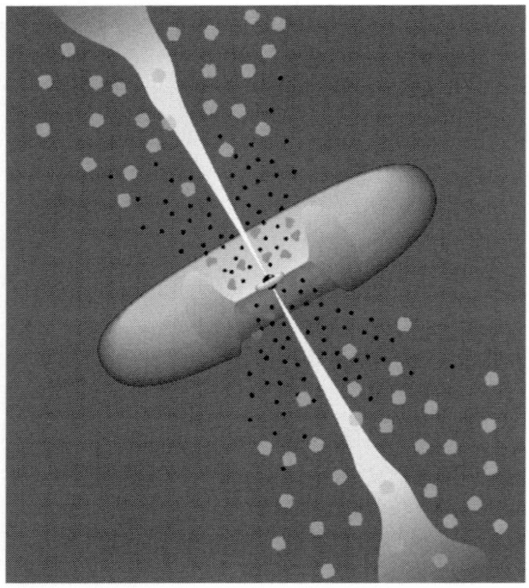

FIGURE 5.4. The classic unification model for (in this case, radio-loud) AGNs. Surrounding the black hole is a luminous accretion disk, with radio jets emitted along the axis of the system. On large scales, there is a thick dusty torus (the prominent feature in this diagram) that obscures the inner part of the torus along lines of sight near the midplane, thus blocking the direct view of the accretion disk and BLR, both of which are within the inner part of the torus. Narrow-line emission arises on larger scales and can be seen in all directions. Free electrons outside the torus can scatter light back towards an observer along the midplane; the inner regions are then visible in the scattered, polarized light. From Urry & Padovani (1995). Copyright PASP, with permission of the authors.

accrete onto the black hole, and how are jets formed?), our view of the basic structure of active nuclei, which we outline below, has been fairly secure and stable for at least the last two decades.

The major components of a classical active nucleus (Figure 5.4) are, from smallest to largest physical scale, a central supermassive black hole with a surrounding accretion disk that produces the X-ray through optical continuum, a broad-line region (BLR), and an extended narrow-line region (NLR) whose inner part is on the same spatial scale as an opaque disk-like structure, sometimes referred to as an "obscuring" (or "dusty") torus, which extends out to kiloparsec scales. On even larger scales, radio jets are sometimes seen along the axis of the system.

5.5.2 *The supermassive black hole*

5.5.2.1 Physical considerations

Supermassive black holes are thought to reside at the centers of all galaxies, at least of those with a nuclear bulge. About 5%–10% of bright galaxies harbor luminous active nuclei,[†] those black holes that are actively accreting surrounding matter, which is heated as it falls into the deep gravitational potential of the black hole.

It is easily demonstrated that the central black holes must be massive. Consider a quasar of black-hole mass $M_{\rm BH}$ and luminosity L that emits isotropically. At some distance r from the center, the energy flux is $F = L/(4\pi r^2)$. Since a photon's momentum is

[†] This is a rather luminosity-dependent statement. While only about 5% of bright galaxies are Seyfert galaxies, as many as $\sim$40% are LINERs.

related to its energy by $p = E/c$, the momentum flux, or radiation pressure, due to the quasar at r is thus

$$P_{\rm rad} = \frac{F}{c} = \frac{L}{4\pi r^2 c}. \tag{5.1}$$

The outward force felt by a particle will be $F_{\rm rad} = P_{\rm rad}\sigma$, where σ is the cross-section of the particle for interaction with a photon; for an ionized gas, this is the Thompson cross-section $\sigma_{\rm e} = 6.65 \times 10^{-25}\,{\rm cm}^2$, which is independent of wavelength. Counteracting this outward radiation pressure is the gravity of the central source, $F_{\rm grav} = G_{\rm BH}/r^2$. For a pure-hydrogen plasma, we can take the particle mass m to be that of a proton. For the quasar not to self-destruct due to radiation pressure, we require that $F_{\rm rad} < F_{\rm grav}$, which leads to the condition

$$L < \frac{4\pi G c m_{\rm p}}{\sigma_{\rm e}} M \approx 1.26 \times 10^{38} \frac{M_{\rm BH}}{M_\odot}\ {\rm erg\ s^{-1}}, \tag{5.2}$$

known as the Eddington limit. From this, we can infer a minimum viable mass for a source of a given luminosity; for a bright Seyfert galaxy, with $L \approx 10^{44}$ ergs s^{-1}, Equation (5.2) gives us $M_{\rm BH} > 10^6 M_\odot$.

To account for the luminous energy of quasars, we suppose that the process of mass accretion converts potential energy into radiant energy, without the need to specify the details of the process. For an infalling mass M, we suppose that the energy released is some fraction η of the rest energy, $E = \eta M c^2$. The *rate* at which energy is released globally is thus $L = {\rm d}E/{\rm d}t = \eta \dot{M} c^2$, where $\dot{M} = {\rm d}M/{\rm d}t$ is the mass-accretion rate (in units like g s^{-1}). Since the potential energy of the mass M at a distance r from the central mass $M_{\rm BH}$ is $U = GM_{\rm BH}M/r$, the rate at which energy is converted from potential energy to radiant energy is thus

$$L = \frac{{\rm d}U}{{\rm d}t} = \frac{GM_{\rm BH}}{r}\frac{{\rm d}M}{{\rm d}t} = \frac{GM_{\rm BH}\dot{M}}{r}. \tag{5.3}$$

The gravitational radius of the black hole is $R_{\rm g} = GM_{\rm BH}/c^2$, which provides a characteristic scale length. Most of the ultraviolet–optical radiation is emitted at $\sim 10R_{\rm g}$, as will be argued below, so the energy available is

$$U \approx \frac{GM_{\rm BH}M}{10R_{\rm g}} = 0.1Mc^2, \tag{5.4}$$

which suggests that $\eta \approx 0.1$ for gravitational accretion onto a collapsed object. Even without strong *observational* evidence supporting the existence of black holes in quasars, this high efficiency makes supermassive black holes an especially attractive solution for explaining the energetics of quasars (e.g. Lynden-Bell 1969) – even the proton–proton cycle in stars has an efficiency of only $\eta \approx 0.007$. To put specific numbers on this, a quasar with $L = 10^{46}$ erg s^{-1} needs to accrete only $\dot{M} \approx 2\,M_\odot$ yr^{-1} to account for its luminosity, assuming an efficiency of $\eta = 0.1$. This is the accretion rate necessary to reach the Eddington luminosity, and is thus known as the Eddington rate,

$$\dot{M}_{\rm Edd} = \frac{L_{\rm Edd}}{\eta c^2} = \frac{1.4 \times 10^{17}}{\eta}\frac{M_{\rm BH}}{M_\odot}\ {\rm g\ s^{-1}}. \tag{5.5}$$

Another useful quantity is the actual accretion rate relative to the Eddington accretion rate, which is simply called the "Eddington ratio,"

$$\dot{m} = \dot{M}/\dot{M}_{\rm Edd}. \tag{5.6}$$

5.5.2.2 Observational evidence

As noted above, supermassive black holes are now believed to reside in virtually all galaxies with central bulge components. In a formal sense, only our own Milky Way, M87 (a radio galaxy), and NGC 4258 (a Seyfert 2 galaxy with a megamaser disk) are so compact that a black hole is actually required. In the case of "quiescent galaxies" (those which do not show evidence of activity), modeling of stellar and gas dynamics indicates the presence of a supermassive central object in a few dozen galaxies that are close enough for us to resolve their black-hole-dynamical radius of influence, $GM_{\rm BH}/\sigma_*^2$, where σ_* is the stellar-bulge velocity dispersion of the host galaxy; within this radius of influence, the black-hole mass rather than the integrated mass of stars dominates the gravitational potential. In the case of AGNs, the masses of the central objects can be measured by "reverberation mapping" (described below). In both quiescent and active galaxies, there is a strong correlation between the black-hole mass and the stellar velocity dispersion, known as the "$M_{\rm BH}$–σ_* relationship" (Ferrarese & Merritt 2000; Gebhardt *et al.* 2000a, b; Ferrarese *et al.* 2001; Tremaine *et al.* 2002; Nelson *et al.* 2004; Onken *et al.* 2004). At the present time, it appears that the evidence for supermassive black holes in AGNs is quite good, but that the presence of a supermassive black hole in a galactic nucleus is a necessary, but not a sufficient, condition for the presence of activity.

5.5.3 *The accretion disk*

5.5.3.1 Physical considerations

From angular-momentum considerations alone, we expect that gas falling into a black hole will ultimately settle into a viscous, rotating structure that is called an accretion disk. Accretion disks have been studied observationally in detail in Galactic binary systems, which are less massive than AGNs by a factor of 10^6 or so (which is why such systems are sometimes referred to as "microquasars").

While we do not yet have a definitive accretion-disk theory for AGNs, we can arrive at some rather basic conclusions using classical thin-disk theory (e.g. Shakura & Sunyaev 1973). We assume that gravitational potential energy is released locally and that, from the virial theorem, half of the energy heats the disk and half of it is radiated away, and that the disk itself is optically thick, thus yielding

$$L = \frac{GM_{\rm BH}\dot{M}}{2r} = 2\pi r^2 \sigma T^4, \tag{5.7}$$

where σT^4 is the blackbody radiation per unit time per unit area and the factor of two on the right-hand side enters because the disk is two-sided. Rearranging this gives the radial temperature dependence,

$$T(r) = \left(\frac{GM_{\rm BH}\dot{M}}{4\pi\sigma r^3}\right)^{1/4}. \tag{5.8}$$

A more complete derivation gives

$$T(r) \approx \left\{\frac{3GM_{\rm BH}\dot{M}}{64\pi\sigma R_{\rm g}^3}\left[1 - \left(\frac{R_{\rm in}}{r}\right)^{1/2}\right]\right\}^{1/4}, \tag{5.9}$$

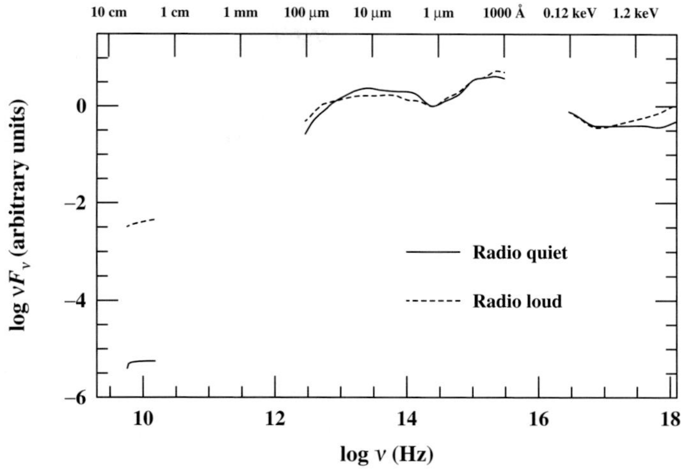

FIGURE 5.5. Mean spectral energy distributions for a sample of radio-quiet (solid line) and radio-loud (dashed line) quasars, from Elvis *et al.* (1994). The flux scale is arbitrarily normalized at 1 μm. The peak shortward of 1 μm is thought to be due to thermal emission from an accretion disk. The peak longward of 1 μm is attributable to thermal re-emission by dust. From Peterson (1997).

where we have expressed the distance from the black hole in terms of the gravitational radius and $R_{\rm in}$ is the inner radius of the disk. Some further substitution yields, for $r \gg R_{\rm in}$,

$$T(r) \approx 3.7 \times 10^5 \dot{m}^{1/4} \left(\frac{M_{\rm BH}}{10^8\, M_\odot} \right)^{-1/4} \left(\frac{r}{R_{\rm g}} \right)^{-3/4} \mathrm{K}. \tag{5.10}$$

Figure 5.5 shows that there is a peak in the spectral energy distribution of quasars at a wavelength $\lambda \approx 1000$ Å. If we ascribe this to thermal emission from the accretion disk, Wien's law indicates that the temperature is $T \approx 5 \times 10^5$ K. For a $10^8 M_\odot$ black hole, Equation (5.10) then tells us that the peak emission occurs at about $14R_{\rm g}$ for an AGN accreting at the Eddington rate or somewhat closer in for lower values of the Eddington ratio $\dot{m}$.

The continuous emission from a disk does not have a classic blackbody form because of the radial temperature gradient (Equation (5.10)). It is straightforward to show that the shape of the spectrum near its peak should have the approximate form $L_\nu \propto \nu^{1/3}$.

5.5.3.2 Observational evidence

Direct observational evidence for accretion disks in AGNs has been hard to come by. The slope of the continuum through the UV/optical actually seems to decrease rather than increase with increasing frequency. On the other hand, it has been argued that there could be other contributors to the continuum emission. Most simple models also predict that the Lyman edge (at 912 Å in the rest-frame) should be detectable in either emission or absorption, but no clear signature of either is seen. However, some lines of evidence suggest that the accretion disk should be embedded in a hot corona that scatters photons and could easily wash out the spectral features, as could relativistic broadening and non-LTE effects.

Despite the difficulties, it remains true that the broad-band spectral energy distribution of AGNs *does peak* in the spectral region expected for accretion disks around supermassive black holes (Figure 5.5), so, even though we cannot model the spectrum in detail at this time, the balance of evidence still supports their existence.

5.5.4 *The broad-line region*

Broad (FWHM $> 1000\,\mathrm{km\,s^{-1}}$) emission lines are almost uniquely an AGN phenomenon; such broad emission lines are seen nowhere else in nature other than in the spectra of supernova remnants. As shown below, the BLR is spatially unresolved – even in the nearest AGNs, the projected size of the BLR is of order tens of microarcseconds.

To some low order of approximation, the relative strengths of the various emission lines in AGN spectra are very similar to those emitted by a wide variety of astrophysical plasmas, such as planetary nebulae, H II regions, and supernova remnants, for the simple reason that photoionization equilibrium of all these gases occurs at the same temperature, $T \approx 10^4\,\mathrm{K}$.

Photoionization equilibrium is attained when the rate of photoionization is balanced by the rate of recombination. To determine the conditions under which photoionization equilibrium is achieved requires detailed calculation of models that produce predictions of the relative strengths of various emission lines and these are then compared with the observed relative strengths of these lines. Simple photoionization-equilibrium models are characterized by (1) the shape of the ionizing continuum, (2) the elemental abundances of the gas, (3) the particle density of the gas, and (4) an ionization parameter

$$U = \frac{Q_{\mathrm{ion}}(\mathrm{H})}{4\pi r^2 n_{\mathrm{H}} c}, \tag{5.11}$$

where the rate at which the central source produces ionizing photons is given by

$$Q_{\mathrm{ion}}(\mathrm{H}) = \int \frac{L_\nu}{h\nu}\,\mathrm{d}\nu, \tag{5.12}$$

where L_ν is the specific luminosity of the ionizing source and the integral is over ionizing photon energies. Thus the ionization parameter U essentially reflects the ratio of the rate at which photoionization occurs, which is proportional to $Q_{\mathrm{ion}}(\mathrm{H})$, to the recombination rate, which is proportional to the particle density n_{H}.

It is conventional to describe photoionization-equilibrium models in terms of the value of U at the face of a cloud exposed to the incident radiation from the central source. Over a rather broad range in values of U, photoionization equilibrium occurs at electron temperatures in the range 10 000–20 000 K, thus accounting for the similarities of their spectra. The *differences* in the spectra of the astrophysical plasmas in this temperature range are attributable primarily to the shape of the ionizing spectrum and gas density, and in some cases to the gas dynamics and elemental abundances.

Early photoionization-equilibrium models of the BLR were single-zone models, i.e. the emission was modeled in the context of multiple clouds with identical physical properties and ionization structure. An example of such a model is shown in Figure 5.6. Single-zone models suggest $n_{\mathrm{H}} \approx 3 \times 10^9\,\mathrm{cm^{-3}}$, on the basis of the presence of broad C III] λ1909 and the absence of broad [O III] $\lambda\lambda$4959, 5007. Acceptable matches to the C III] λ1909/C IV λ1549 and Lyα λ1215/C IV λ1549 flux ratios are obtained for $U \approx 0.01$.

The strong forbidden lines seen in the spectra of low-density plasmas (e.g. [N II] $\lambda\lambda$6548, 6584 and [O III] $\lambda\lambda$4363, 4959, 5007) are absent from BLR spectra. Thus, the physical diagnostics that can be used in lower-density gases (see Chapter 1 of this volume) cannot be used to determine the particle density or temperature of the BLR gas. The absence of forbidden lines, however, can be used to place a lower limit on the particle density in the gas. In the first generation of AGN photoionization-equilibrium models, the *presence* of

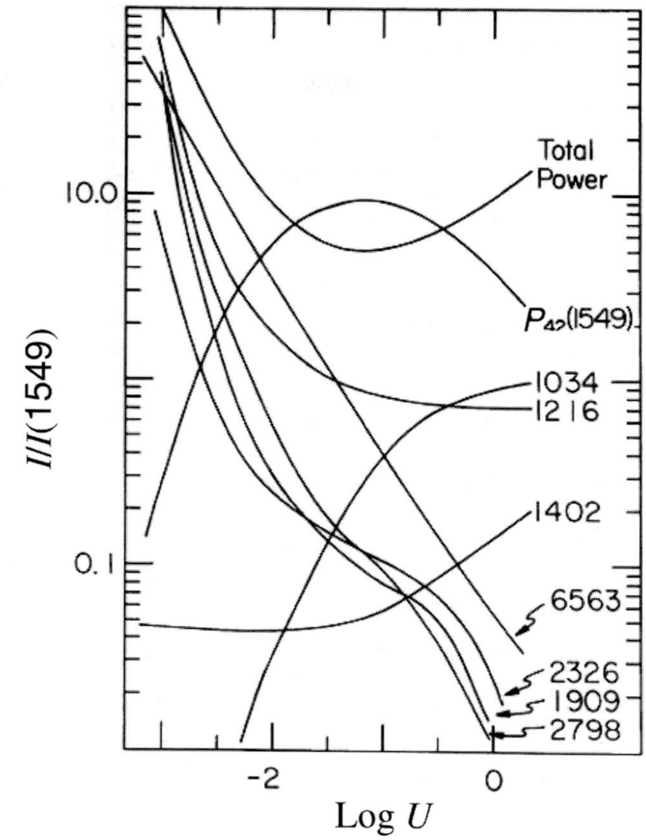

FIGURE 5.6. Results of a single-cloud photoionization equilibrium calculation, for an assumed particle density of 3×10^9 cm^{-3}. The predicted strengths of the broad components of O VI λ1034, Lyα λ1215, O IV λ1402, Hα λ6563, C II] λ2326, C III] λ1909, and Mg II λ2798 are shown relative to C IV λ1549, as a function of the ionization parameter U. The top line shows the total power emitted by the entire BLR (line and diffuse continuum), relative to C IV λ1549. The total power radiated in C IV λ1549 in units of 10^{42} erg s^{-1} is shown for the Seyfert galaxy NGC 4151 under the assumption that all the ionizing flux from the central source is absorbed by the BLR. From Ferland & Mushotzky (1982).

the semi-forbidden line C III] λ1909 in AGN broad-line spectra suggested an upper limit on the particle density of about 3×10^9 cm^{-3}. However, with the advent of reverberation mapping (see below), it became clear that the other strong UV lines arise in a different part of the BLR than where the C III] λ1909 line arises, so that a particle density of 3×10^9 cm^{-3} is in fact a *lower limit* on the particle density for the C IV-emitting part of the BLR.

Photoionization modeling is a powerful tool of long standing (Davidson & Netzer 1979) for trying to understand the physical nature of the line-emitting regions in AGNs. While progress was made in determining some of the physical parameters that characterize the BLR (e.g. relatively high nebular densities and electron temperatures), it became clear early that single-zone models could not simultaneously reproduce the relative strengths of both the low- and the high-ionization lines (Collin-Souffrin *et al.* 1986). Furthermore, an energy-budget problem was identified: the total amount of line emission observed exceeds the amount of continuum radiation available to power the lines (Netzer 1985; Collin-Souffrin 1986).

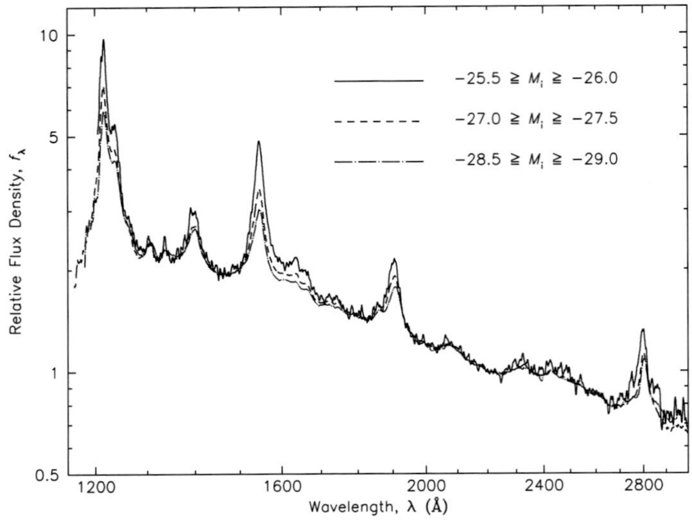

FIGURE 5.7. Composite spectra from the Sloan Digital Sky Survey, binned by luminosity. Note the similarity of the spectra, with the exception of the "Baldwin effect" in C IV λ1549; i.e., relative to the continuum and the other lines, C IV is stronger in lower-luminosity objects. Courtesy of D. Vanden Berk, based on an original from Vanden Berk *et al.* (2004).

Temperatures of around 10^4 K correspond to thermal line widths of order $10\,\mathrm{km\,s^{-1}}$, so the BLR gas moves supersonically. If the lines are Doppler-broadened, either the gas is moving in a deep gravitational potential or it is accelerated to high velocities, perhaps hydromagnetically or by radiation pressure. How the BLR gas is moving, whether in gravitational infall, in radiation-pressure-driven or hydromagnetically accelerated outflow, or in orbital motion, remains unknown, though the evidence (described below) points to gravitationally bound motions.

Perhaps surprisingly, emission-line profiles alone only weakly constrain the BLR kinematics. Numerous profoundly different kinematic models yield similar "logarithmic" profiles in which the flux at some displacement $\Delta\lambda$ from line center is proportional to $-\ln \Delta\lambda$, for $\Delta\lambda$ not too close to line center. This represents a rather idealized case since in many instances, especially among lower-luminosity AGNs, the line profiles have some structure, variously described as "bumps," "shelves," or "asymmetric wings." Emission-line profiles can change, in some cases drastically, on timescales comparable to the BLR dynamical timescale $\tau_{\rm dyn} \approx R_{\rm BLR}/{\rm FWHM}$, which is typically a few years for Seyfert galaxies of moderate luminosity.

Figure 5.7 shows several composite spectra of AGNs for various luminosity ranges. Their similarity suggests that not only the temperature of the line-emitting gas, but also the particle density and ionization parameter, are quite similar for AGNs of different luminosities. The only strong luminosity-dependence of AGN emission-line spectra is the equivalent width (i.e. the ratio of line flux to underlying continuum flux) in the C IV λ1549 emission line; relative to the continuum, C IV is weaker in more-luminous objects, a well-known anticorrelation known as the "Baldwin effect" (Baldwin 1977). The physical origin of the Baldwin effect is not known.

There are additional correlations among spectroscopic properties, some of which have been identified through principal-component analysis (PCA), such as that undertaken by Boroson & Green (1992). Given a set of input values (e.g. ratios of line fluxes, equivalent widths of lines, Doppler widths of lines), which are treated as vectors, PCA identifies the parameters that cause the greatest variation within a population. These are expressed as eigenvectors (also known as principal components), with a coefficient for each of the input vectors. We are thus able to identify correlated parameters, bearing in mind always that correlation does not necessarily imply causality – but it does give us a useful place to start. For Type 1 AGNs, Eigenvector 1 shows a strong anticorrelation between the strength of the optical Fe II blends on either side of the Hβ emission line and the flux in [O III] $\lambda5007$, while Eigenvector 2 shows a correlation between luminosity and the strength of He II $\lambda\lambda4686$ emission. Boroson (2002) argues that Eigenvector 1 is driven by the Eddington ratio (Equation (5.6)) and that Eigenvector 2 is driven by the accretion rate $\mathrm{d}M/\mathrm{d}t$.

5.5.5 *The narrow-line region*

Narrow (FWHM $< 1000\,\mathrm{km\,s^{-1}}$, typically) emission lines arise in a region that is spatially extended and is in fact at least partially resolvable in nearby AGNs. As in the BLR, the velocities are supersonic, indicating large-scale bulk motions of gas. The particle densities in the NLR are much lower than those in the BLR, so many forbidden lines over a wide range of ionization potential are present, and the emissivity per unit volume is much lower than that in the BLR, so the total amount of mass is large.

In many respects, understanding the NLR is much simpler than understanding the BLR.

(i) Since the NLR is spatially resolved, the kinematics are less ambiguous than in the BLR.

(ii) Forbidden-line emission allows use of temperature and density diagnostics, so the physical conditions are relatively better understood.

(iii) Forbidden lines are not self-absorbed, so the radiative-transfer problem is simpler than in the BLR.

Offsetting these tremendous advantages relative to the BLR is that dust, which extinguishes and reddens optical and ultraviolet radiation, is present in the NLR.

Relative line strengths for a typical NLR spectrum are shown in Table 5.1. The first column indicates the emission line and the second column gives the relative (unreddened) flux, normalized with respect to an Hβ value of unity, as is the usual convention. The third column gives the relevant ionization potential. For collisionally excited lines, the ionization potential is that necessary to achieve this particular ionization state. For recombination lines (e.g. all the hydrogen lines), the relevant ionization potential is that of the next-highest state of ionization, since it is recombination from these higher states that produces the observed emission lines. The fourth column gives the critical density for each particular line, as we describe below.

Table 5.1 reveals the presence of many of the lines seen also in other nebulae at temperatures of $\sim 10^4$ K, such as Galactic H II regions and planetary nebulae. However, AGN narrow-line spectra are distinguished by the presence of very-high-ionization lines, such as [Ne V], [Ar III], [Fe VII], and [Fe X]; the iron lines in particular are often referred to as "coronal lines" since they were first detected in the Solar corona, where the temperature is of order 10^6 K. The presence of these lines is due to the "hardness" of the AGN ionizing continuum, i.e. the relatively large fraction of very energetic ionizing photons compared with the Wien tail of a thermal spectrum of the type that powers Galactic nebulae.

The presence of forbidden and semi-forbidden lines is attributable to the low particle densities in the NLR. Consider for a moment a simple two-level atom with a ground state 1 and an excited state 2 at some higher energy ΔE. As discussed in Chapter 1 of this volume, in low-density gases in which the collision rate is low, each transition into the excited state 2 will ultimately lead to a radiative transition to the ground state, producing

TABLE 5.1. Strong narrow lines in Seyfert spectra

Line	Relative flux	Ionization potential (eV)	Critical density n_{crit} (cm^{-3})
Lyα λ1216	55	13.6	...
C IV λ1549	12	47.9	...
C III] λ1909	5.5	24.4	3.0×10^{10}
Mg II λ2798	1.8	7.6	...
[Ne V] λ3426	1.2	97.1	1.6×10^7
[O II] λ3727	3.2	13.6	4.5×10^3
[Ne III] λ3869	1.4	40.0	9.7×10^6
[O III] λ4363	0.21	35.1	3.3×10^7
He II λ4686	0.29	54.4	...
Hβ λ4861	1.00	13.6	...
[O III] λ4959	3.6	35.1	7.0×10^5
[O III] λ5007	11	35.1	7.0×10^5
[N I] λ5199	0.15	0.0	2.0×10^3
He I λ5876	0.13	24.6	...
[Fe VII] λ6087	0.10	100	3.6×10^7
[O I] λ6300	0.57	0.0	1.8×10^6
[Fe X] λ6375	0.04	235	4.8×10^9
[N II] λ6548	0.9	14.5	8.7×10^4
Hα λ6563	3.1	13.6	...
[N II] λ6583	2.9	14.5	8.7×10^4
[S II] λ6716	1.5	10.4	1.5×10^3
[S II] λ6731	...	10.4	3.9×10^3
[Ar III] λ7136	0.24	27.6	4.8×10^6

From Ferland & Osterbrock (1986) and Peterson (1997).

a photon of energy $h\nu_{21} = \Delta E$. In this case, the emissivity per unit volume j_{21} of the gas at frequency ν_{21} is proportional to the square of the particle density n^2. In high-density gases, the situation becomes slightly more complicated: as the density increases, the mean time between collisions approaches the lifetime of the upper level and the collisional de-excitation rate begins to compete with radiative de-excitation as a mechanism to depopulate the excited state. It is easy to show that in this case the emissivity per unit volume becomes a weaker function of the density, $j_{21} \propto n$. At the *critical density*, the rate of collisional de-excitation equals the rate of radiative de-excitation.

At low density, $j \propto n^2$ for all the lines in Table 5.1. If we increase only the gas density, as we reach densities $\sim 10^3$–10^4 cm^{-3}, some of the lines with low spontaneous radiative decay rates and thus low values of the critical density (e.g. [S II] $\lambda\lambda$6716, 6731) will become "collisionally suppressed" as they transition to the regime where $j \propto n$ only. There are two important things to keep in mind.

(i) Transitions with low spontaneous decay rates are known as "forbidden" lines only because they are non-electric-dipole transitions and have longer lifetimes. Transitions with somewhat higher critical densities, such as intercombination lines like C III] λ1909, are referred to as "semi-forbidden." Both of these labels are misnomers: the transitions *do* occur, but at lower rates than for "permitted" lines.

(ii) Even when lines are collisionally suppressed above the critical density, the volume emissivity is an increasing function of density, $j \propto n$. They are suppressed only relative to lines below their respective critical densities whose emissivity increases much more rapidly with density, $j \propto n^2$.

The size and mass of the NLR can be estimated in a straightforward fashion from the emissivity of the Balmer lines and the simplifying assumption that the NLR is entirely ionized hydrogen. The volume emissivity of Hβ is

$$j(\mathrm{H}\beta) = n_e^2 \alpha_{\rm eff}(\mathrm{H}\beta) \frac{h\nu}{4\pi} \ \mathrm{erg\ s^{-1}\ cm^{-3}\ sr^{-1}}, \tag{5.13}$$

where n_e is the electron density and $\alpha_{\rm eff}(\mathrm{H}\beta)$ is the effective recombination coefficient for Hβ, which has a value of $\sim 3 \times 10^{-14}\,\mathrm{cm^{-3}\ s^{-1}}$ at $T \approx 10^4$ K. The total luminosity in the Hβ narrow line is thus

$$L(\mathrm{H}\beta) = \iint j(\mathrm{H}\beta)\mathrm{d}\Omega\,\mathrm{d}V = \frac{4\pi\epsilon n_e^2}{3} \times 1.24 \times 10^{-25} r^3 \ \mathrm{erg\ s^{-1}}, \tag{5.14}$$

where we have introduced the filling factor ϵ: we make the simple assumption that the NLR is composed of N_c spherical clouds of radius R and particle density n_e, in which case the total *line-emitting* volume of the NLR is $4\pi\epsilon r^3/3 \approx 4\pi N_c R^3/3$. For local Seyfert 2 galaxies, a typical Hβ luminosity might be around $10^{41}\,\mathrm{erg\ s^{-1}}$, and, for a particle density of $n_3 \approx 10^3\,\mathrm{cm^{-3}}$, we find that $r \approx 20\epsilon^{1/3}$ pc. As noted earlier, the NLR is often at least partially resolved in nearby AGNs. For our fiducial luminosity, the size is typically $\sim$100 pc, from which we infer $\epsilon \approx 10^{-2}$, i.e. only about 1% of the volume is filled with line-emitting clouds. It is now trivial to compute the mass of the line-emitting clouds as well,

$$M_{\rm NLR} \approx \frac{4\pi}{3}\epsilon r^3 n_e m_p, \tag{5.15}$$

where m_p is the proton mass. For the same parameters as were used to estimate the NLR radius, we find $M_{\rm NLR} \approx 10^6 M_\odot$.

In general, narrow-line profiles are at least slightly asymmetric, with a stronger blue wing than red wing. This is usually taken to be a signature of mass outflow, with the far (receding) side of the NLR more strongly obscured from us than the near (approaching) side. Not all narrow lines in a particular AGN spectrum have the same width: indeed, the line widths are found to be correlated with both critical density and ionization potential (Pelat *et al.* 1981; Filippenko & Halpern 1984; De Robertis & Osterbrock 1984; Espey *et al.* 1994); this is consistent with a picture in which the more highly ionized and/or denser clouds are found deeper in the gravitational potential of the central black hole.

Careful narrow-band imaging of the central regions of nearby Seyfert 2 galaxies shows that the extended parts of the NLRs are generally wedge-shaped, i.e. there are "ionization cones" (Pogge 1988) whose vertices point back towards the central source. Ionization cones have opening angles in the range $\sim 30°$ to $\sim 100°$; this type of structure is apparently due to obscuration in the equatorial midplane of the system, and the ionization cones we see are where both gas and radiation from the central regions can flow outwards.† The nature of the obscuring material is described below.

5.5.6 *The "obscuring torus"*

An obvious, but important, question to ask is why do Seyfert 2 galaxies not have broad lines? It was recognized early on that this probably has something to do with obscuration, at least along certain lines of sight. The difficulty with this basic concept is that it is

† This statement should not be misconstrued as implying that the NLR gas is entirely the result of outflow. An H I map of the Seyfert 2 galaxy NGC 5252 shows that there is a neutral halo surrounding the galaxy that avoids the ionization cones (Prieto & Freudling 1993); clearly, the ionization cones are there because that's where the ionizing radiation is.

hard to completely obscure the BLR without also completely blocking the light from the smaller continuum source. Also, if the BLR and continuum source are extinguished by dust along the line of sight, then Seyfert 2 galaxies should show strong evidence of reddening; in particular, they should not be bright in the UV, where the opacity of dust grains is high.

Osterbrock (1978) suggested that, in the case of Seyfert 2 galaxies, our direct view of the nucleus is blocked by a structure that is highly opaque along our line of sight to the nucleus, but which is transparent or nearly so in other directions. He proposed that the opaque structure is in the form of a dusty "torus" surrounding the central accretion disk and BLR, but inside or co-spatial with the NLR. Thus, if one views the AGN in the torus midplane, one sees only the NLR and weak, scattered nuclear emission, i.e. a Seyfert 2 galaxy. If one views the AGN along the axis of the torus, then one sees the continuum source and BLR with little or no obscuration, i.e. a Seyfert 1 galaxy. Since radio morphology reveals that at least radio-loud AGNs have roughly axial symmetry, it is not too much of a stretch to suppose that the same axial symmetry extends down to nuclear scales, where a toroidal structure blocks our direct view of the continuum source and the BLR, but not the NLR. A free parameter in this basic model is the opening angle of the torus, which can be chosen to yield the observationally determined relative space densities of Seyfert 1 and Seyfert 2 galaxies.

The key to testing this hypothesis is *scattering*: if there are particles (specifically electrons or dust) outside the torus, but exposed to radiation from the central regions, they might scatter light, making it possible to observe the inner components, namely the accretion disk and the BLR, indirectly in the scattered radiation. The scattered light can be isolated in a relatively straightforward way because the scattering process polarizes the light such that the **E** vector is perpendicular to the axis of the torus. The veracity of the general torus picture was strongly supported by the polarized flux spectrum of the archetypical Seyfert 2 galaxy NGC 1068 (Antonucci & Miller 1985), as shown in Figure 5.8. Whereas the unpolarized spectrum shows a characteristic Seyfert 2 spectrum, the polarized spectrum (which accounts for only a few percent of the total flux) reveals the broad Hβ and Fe II lines characteristic of type 1 Seyferts. It is thus clear that at least NGC 1068 is intrinsically a Seyfert 1, but it appears to be a Seyfert 2 to us because we cannot view the inner AGN components directly.

Certainly this is a compelling argument for AGN unification. However, it should be noted that not all Seyfert 2 galaxies show broad lines in their polarized flux. Of course, in these cases it is unknown whether this is due to the absence of a BLR or the absence of a scattering medium. Whether or not there are "true Seyfert 2s" without BLRs remains unknown. A recent X-ray study of some of the best "true Seyfert 2" candidates reveals that in all cases there is evidence for line-of-sight obscuration, which argues against the existence of such objects (Ghosh *et al.* 2006).

It should also be noted that the notion of a dusty torus as originally conceived is not physically viable; we will return to this issue later.

5.6. Distinguishing AGNs from other emission-line galaxies

We have previously noted the general similarity of the emission-line spectra produced in a variety of astrophysical sources. There are, however, distinct differences in detail that depend on the nature of the ionizing source. This is important in distinguishing Seyfert 2 galaxies from star-forming galaxies where the emission is powered by early-type stars. Certain emission-line flux ratios are sensitive to the shape of the ionizing continuum or to the ionization parameter, and these can be used to distinguish AGNs from either

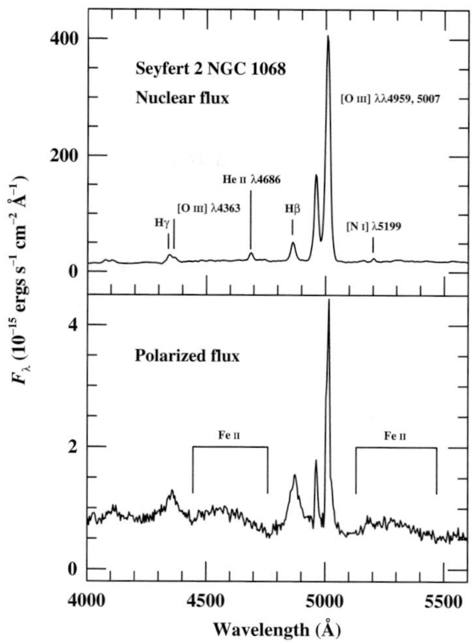

FIGURE 5.8. The upper panel shows the total-flux nuclear spectrum of the Seyfert 2 galaxy NGC 1068 and the lower panel shows the linearly polarized flux (Miller *et al.* 1991). Note the presence of Seyfert 1 features, specifically broad Hβ and Fe II emission lines, in the polarized spectrum; we do not observe the BLR and continuum source directly in this object, but only in scattered (and thus polarized) light. Data courtesy of R. W. Goodrich, from Peterson (1997).

star-forming galaxies or LINERs. We note that the most-robust line ratios are based on lines that are close in wavelength so that the observed ratios are fairly insensitive to reddening. Diagnostic diagrams can be constructed by plotting pairs of flux ratios that identify the type of object from its emission-line spectrum; these are known generically as "BPT diagrams" after Baldwin, Phillips, and Terlevich (1981) who introduced them (see Chapter 1 of this volume). In Figure 5.9, we show a BPT diagram for emission-line galaxies from the Sloan Digital Sky Survey (SDSS) that is based on [O III] $\lambda5007/\mathrm{H}\beta$ versus [N II] $\lambda6584/\mathrm{H}\alpha$. The band of points running from the upper left to the lower right represents star-forming galaxies, where the gas is ionized by the Wien tail of hot stars. The sequence from left to right is one of metallicity; the [O III] line is relatively stronger at low metallicity because of the greater importance of this particular transition as a coolant. In this diagram, the Seyfert galaxies (upper right) separate fairly cleanly from LINERs (lower right).

5.7. The theory of reverberation mapping

Because of the small angular size of central regions of even the nearest AGNs, the structure and kinematics of the BLR remain a mystery. Fundamentally, the problem is that the line profiles and line-flux ratios are insufficient to constrain what is certainly a complex region; indeed, the line profiles are highly degenerate and tell us virtually nothing about the BLR kinematics. Moreover, we have already seen that, whereas photoionization-equilibrium modeling can give us some insight into the physics of the BLR, simple single-cloud models run into problems regarding duplicating the strength

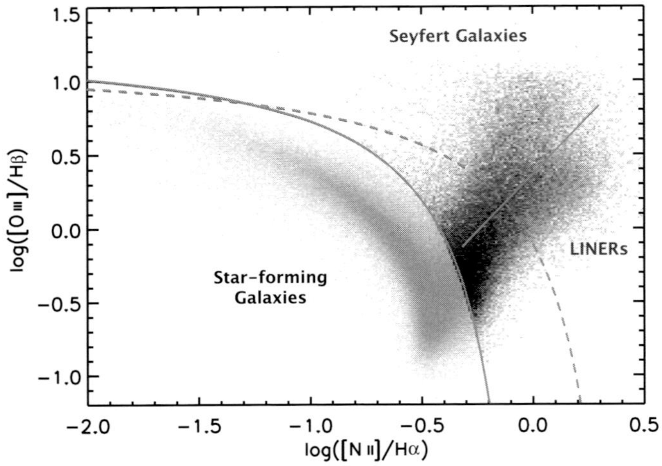

FIGURE 5.9. A BPT diagram for emission-line galaxies from the Sloan Digital Sky Survey. The points to the left are star-forming galaxies where the line emission arises in H II regions associated with hot stars. The upper-right points are Seyfert galaxies and the lower-right points are LINERs; these are actually two distinct populations, although with some overlap in their emission-line flux ratios. The two curves are delimiters from Kauffmann *et al.* (2003) (solid line) and Kewley *et al.* (2006) (dashed line). Courtesy of B. Groves, based on an original from Groves *et al.* (2006).

of low-ionization relative to high-ionization lines and accounting for the total energy in the emission lines.

In principle, many of the ambiguities can be resolved by using the technique of emission-line reverberation mapping (Blandford & McKee 1982; Peterson 1993), which makes use of the natural continuum variability of AGNs to probe the BLR. For reasons still not understood, the continuum emission from AGNs varies with time, and the broad emission-line fluxes change in response to these variations, but with a delay due to the time taken for light to travel across the BLR. We thus begin this section with a brief phenomenological discussion of AGN continuum-variability characteristics.

5.7.1 *The nature of continuum variability*

The continuum emission from AGNs is variable at every wavelength at which it has been observed, from gamma rays to long-wavelength radio. Blazars exhibit the most extreme variations and, unlike radio-quiet quasars and Seyfert galaxies, significant variations in the radio and long-wavelength infrared. Blazar variability is fundamentally tied to jet physics and relativistic beaming, and, since the jets are too well-collimated to have much influence on the line-emitting regions, we will restrict our discussion of variability to the continuum emission that arises in the accretion disk and its immediate vicinity.

In all cases, the observed continuum-flux variations in AGNs are aperiodic and of unpredictable amplitude. It has been found that a useful way to characterize AGN variability is in terms of the power-density spectrum (PDS), which is the product of the Fourier transform of the light curve and its complex conjugate. The common parameterization of the PDS for AGNs is as a power law in frequency f of the form $P(f) \propto f^{-\alpha}$. Depending on the observed frequency and on the particular AGN, the power-law index is generally in the range $1 < \alpha < 2.5$, with X-ray variations tending towards the lower values and UV/optical variations tending towards the larger values. On shorter timescales (days, for Seyfert galaxies), X-ray variations are poorly correlated with UV/optical variations, but on longer timescales (months to years) the X-ray variations appear to be well corrrelated with the UV/optical variations (e.g. Uttley *et al.* 2003). This is *not* true in a few cases where the phenomenology, let alone the physics, remains baffling; in the case of NGC 3516, for example, the X-ray and optical variations seem oddly disconnected (Maoz *et al.* 2002).

5.7.2 *Reverberation-mapping assumptions*

On the basis of early spectral-monitoring campaigns that revealed the correlation between continuum and emission-line variations, we can draw some basic conclusions that will simplify mapping the BLR.

(i) The rapid response of the broad emission lines to continuum variations tells us that the BLR is small (because the light travel-time is short) and that the BLR gas is dense (so the recombination time is short).

(ii) The BLR gas is optically thick in the H-ionizing continuum, i.e. at $\lambda < 912\,\text{Å}$. If the gas were optically thin, the emission lines would not respond strongly to continuum variations.

(iii) The variations of the observable continuum in the UV/optical must trace those of the unobservable H-ionizing continuum. Specifically, any time delay between the variations in the ionizing continuum and the observable continuum must be small.

On the basis of these very basic conclusions, we can make some further assumptions that will aid us in developing a theoretical framework for reverberation mapping. These assumptions are justifiable *ex post facto* by the following observations.

(i) The continuum is postulated to originate in a single central source that is small compared with the BLR. As noted earlier, the size of the UV/optical continuum-emitting region in an accretion disk around a black hole of mass $10^8 M_\odot$ is $\sim 10R_{\text{g}}$, or about 1.5×10^{14} cm or 0.06 lt-day, much smaller than the typical radius of the BLR in such a system, i.e. $\sim 10^{16}$ cm. We can thus treat the continuum as a point source.†

(ii) The light-travel time across the BLR, $\tau_{\text{LT}} = R_{\text{BLR}}/c$, is the most-important timescale. Other potentially important timescales include the following.

  (a) The dynamical time (introduced earlier), τ_{dyn}, which is the timescale over which the structure of the BLR could change due to bulk motions of the line-emitting gas. Since $\tau_{\text{dyn}}/\tau_{\text{LT}} = c/\text{FWHM} \approx 100$, the BLR structure is stable over the time it takes to map the BLR by reverberation, which is typically at least a few times τ_{LT}.

  (b) The recombination time $\tau_{\text{rec}} = (n_{\text{e}}\alpha_{\text{B}})^{-1}$, where n_e is the particle density, which we previously estimated to be greater than $\sim 10^{10}\,\text{cm}^{-3}$, and α_{B} is the case-B recombination coefficient, is the timescale over which photoionization equilibrium is re-established when the incident ionizing flux, i.e. $Q_{\text{ion}}(\text{H})$, changes. For the high densities of the BLR, $\tau_{\text{rec}} < 400\,\text{s}$, which means that the BLR gas responds virtually instantaneously for the purposes of reverberation mapping.

(iii) There is a simple, though not necessarily linear, relationship between the observable UV/optical continuum and the ionizing continuum that drives the emission-line variations. There is evidence that the continuum variations at shorter wavelengths precede those at longer wavelengths and the variations are more rapid and have more structure, but these are comparatively small effects.

† In contrast, this simplifying assumption cannot be made in reverberation mapping of the X-ray Fe Kα line (Reynolds *et al.* 1999).

5.7.3 *The transfer equation*

The continuum variations are described by the light curve $C(t)$ and the emission-line behavior as a function of time t and line-of-sight (LOS), or Doppler, velocity $V_{\rm LOS}$ is $L(t, V_{\rm LOS})$. Over the duration of a reverberation experiment, the continuum and emission-line variations are generally small, so it is useful to write both light curves in terms of deviations from mean values $\langle C \rangle$ and $\langle L(V_{\rm LOS}) \rangle$, i.e. $C(t) = \langle C \rangle + \Delta C(t)$ and $L(t, V_{\rm LOS}) = \langle L(V_{\rm LOS}) \rangle + \Delta L(t, V_{\rm LOS})$. Thus, even if the emission-line response or the relationship between the ionizing continuum and the observable continuum is non-linear, we can use a linearized-response model to describe the relationship between the continuum and emission-line variations,

$$\Delta L(t, V_{\rm LOS}) = \int_{-\infty}^{\infty} \Psi(\tau, V_{\rm LOS}) \Delta C(t - \tau) d\tau, \tag{5.16}$$

where $\Psi(\tau, V_{\rm LOS})$ is the "transfer function" and Equation (5.16) is the "transfer equation." It is apparent by inspection of Equation (5.16) that $\Psi(\tau, V_{\rm LOS})$ is the response of the emission line as a function of time and Doppler velocity to a δ-function continuum outburst. As we will see below, it is the projection of the BLR responsivity† in the LOS velocity–time-delay plane; for this reason, we often refer to $\Psi(\tau, V_{\rm LOS})$ by the more descriptive name "velocity–delay map."

5.7.4 *Velocity–delay maps*

At this point, we can construct a velocity–delay map from first principles by considering how an extended BLR would respond to a δ-function continuum outburst, as seen by a distant observer. For illustrative purposes and simplicity, we consider the response of an edge-on (inclination $i = 90°$) ring of radius R that is in Keplerian rotation around a central black hole of mass $M_{\rm BH}$. Photons produced in the continuum outburst would stream outward and, after some time R/c, some of these photons encounter the gas in the orbiting ring, are absorbed, and are quickly reprocessed into emission-line photons. The emission-line photons would be emitted in some pattern (we can assume isotropy for simplicity, but this is not necessarily the case), and some of them would return to the central course after another interval R/c. A hypothetical observer at the center of the system would observe the continuum flare go off, and after a time delay $2R/c$ would observe the response of the gas in the ring. From this privileged location, the observer sees the entire ring respond simultaneously because the total light-travel time is the same in each direction. At any other location in the ring plane, the total light-travel time from the continuum source to the ring to the observer differs for different parts of the ring; the locus of points in space for which the light-travel time from the source to the observer is constant is obviously an ellipse, with the source and the observer at the respective foci. Given that the observer is essentially at infinity, this locus of constant time delay, or "isodelay surface," is a parabola, as illustrated in Figure 5.10, in which the ring-shaped BLR is intersected by several isodelay surfaces corresponding to different time delays as seen by the distant observer.

The velocity–delay map we wish to construct transforms the BLR from configuration space into the observable space of LOS velocity and time delay. The upper panel of Figure 5.11 shows our ring-shaped BLR intersected by a single, arbitrary isodelay surface that intersects the BLR at polar coordinates (R, θ) centered on the continuum source.

† By "responsivity," we mean the marginal change in line emissivity δL as the continuum changes by δC.

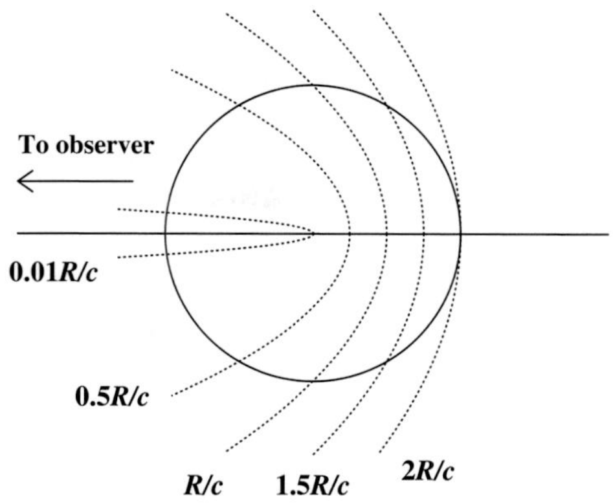

FIGURE 5.10. The circle represents a ring of gas orbiting around a central continuum source at a distance R. Following a continuum outburst, at any given time the observer far to the left sees the response of clouds along a surface of constant time delay, or isodelay surface. Here we show five isodelay surfaces, each one labeled with the time delay (in units of the shell radius R) we would observe relative to the continuum source. Points along the line of sight to the observer are seen to respond with zero time delay. The farthest point on the shell responds with a time delay $2R/c$.

Relative to continuum photons traveling from the central source to the distant observer, the time delay for this surface is shown by the dotted path whose two components have distances R and $R\cos\theta$, so the time delay for this isodelay surface is

$$\tau = (1 + \cos\theta)R/c, \tag{5.17}$$

which, as we anticipated earlier, is the equation for a parabola in polar coordinates. To transform the orbital velocity as a function of position into the observable line-of-sight velocity, we see by inspection that

$$V_{\rm LOS} = v_{\rm orb} \sin\theta, \tag{5.18}$$

where $v_{\rm orb} = (GM_{\rm BH}/R)^{1/2}$ is the orbital speed. In the lower frame of Figure 5.11, we show this transformation into velocity–delay space. A circular Keplerian orbit in configuration space transforms into an ellipse of semiaxes $v_{\rm orb}$ and R/c in velocity–time-delay space.

It is also instructive to "collapse" the velocity–delay map of Figure 5.11 in velocity to obtain the delay as a function of time for the entire emission-line flux (sometimes called the "one-dimensional transfer function" or the "delay map") and in time delay to obtain the emission-line profile. This case can be treated analytically in a straightforward fashion. We can assume that the azimuthal response of the ring is uniform, so $\Psi(\theta) = \epsilon$, where ϵ is a constant. To express the response as a function of time delay, we use the transformation

$$\Psi(\tau)\mathrm{d}\tau = \Psi(\theta)\frac{\mathrm{d}\theta}{\mathrm{d}\tau}\,\mathrm{d}\tau, \tag{5.19}$$

and from Equation (5.17) we have

$$\frac{\mathrm{d}\tau}{\mathrm{d}\theta} = -\frac{R}{c}\sin\theta. \tag{5.20}$$

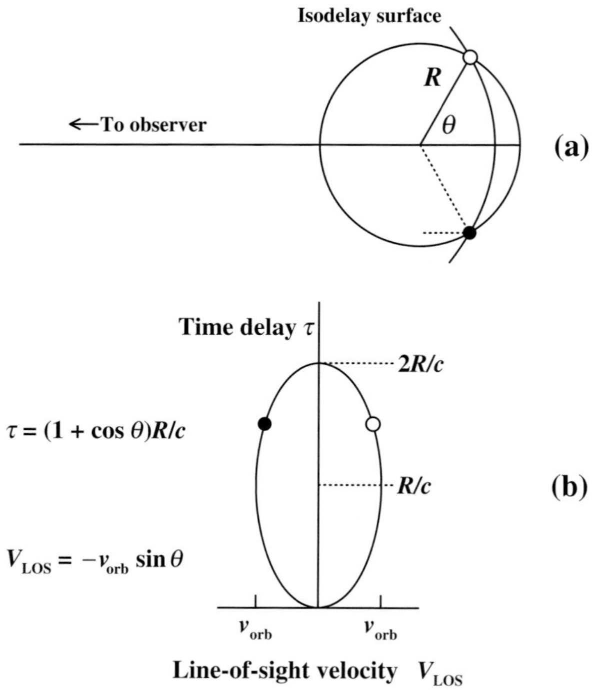

FIGURE 5.11. The upper diagram shows a ring that contains line-emitting clouds, as in Figure 5.10. An isodelay surface for an arbitrary time is given; the intersection of this surface and the ring shows the clouds that are observed to be responding at this particular time. The dotted line shows the additional light-travel time, relative to light from the continuum source, that signals reprocessed by the cloud into emission-line photons will incur (Equation (5.17)). In the lower diagram, we project the ring of clouds onto the line-of-sight velocity–time-delay $(V_{\rm LOS}, \tau)$ plane, assuming that the emission-line clouds in the upper diagram are orbiting in a clockwise direction (so that the cloud represented by a filled circle is blueshifted and the cloud represented by an open circle is redshifted).

By again making use of Equation (5.17) to express $\sin\theta$ in terms of τ, after some simple algebra, we find that the observed response of the BLR as a function of time delay is

$$\Psi(\tau)\mathrm{d}\tau = \frac{\epsilon}{R(2c\tau/R)^{1/2}\left[1 - c\tau/(2R)\right]^{1/2}}\,\mathrm{d}\tau. \tag{5.21}$$

We show the response of an edge-on ring in Figure 5.12. We leave it as an exercise for the reader to show that the mean response of the ring is

$$\langle\tau\rangle = \frac{\int \tau\Psi(\tau)\mathrm{d}\tau}{\int \Psi(\tau)\mathrm{d}\tau} = \frac{R}{c}, \tag{5.22}$$

as one would guess intuitively.

We can determine the emission-line profile for a ring using a similar transformation,

$$\Psi(V_{\rm LOS})\mathrm{d}V_{\rm LOS} = \Psi(\theta)\frac{\mathrm{d}\theta}{\mathrm{d}V_{\rm LOS}}\,\mathrm{d}V_{\rm LOS}. \tag{5.23}$$

Differentiating Equation (5.18) yields

$$\frac{\mathrm{d}V_{\rm LOS}}{\mathrm{d}\theta} = v_{\rm orb}\cos\theta, \tag{5.24}$$

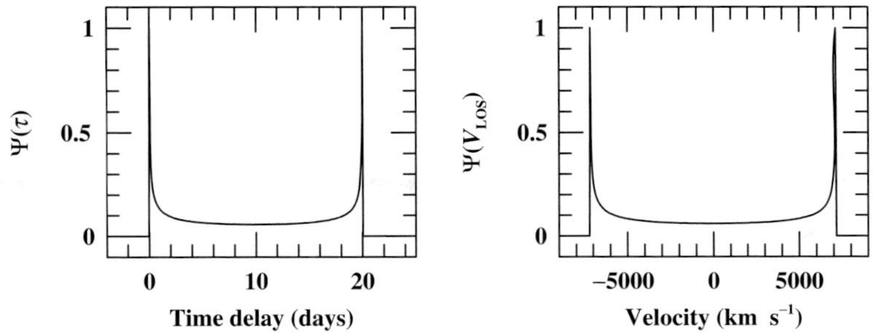

FIGURE 5.12. Left: a delay map $\Psi(\tau)$ for an edge-on circular Keplerian orbit, as described by Equation (5.21), at a distance of $R = 10$ light days from a black hole of mass $10^8 M_\odot$. Right: line profile for an edge-on circular Keplerian orbit, as described by Equation (5.25), at a distance of $R = 10$ light days from a black hole of mass $10^8 M_\odot$.

and after some algebra we can write the line profile as

$$\Psi(V_{\rm LOS})dV_{\rm LOS} = \frac{\epsilon}{v_{\rm orb}\left(1 - V_{\rm LOS}^2/v_{\rm orb}^2\right)^{1/2}}\, dV_{\rm LOS}, \tag{5.25}$$

which is plotted in Figure 5.12.

A useful way to characterize the line width is the line dispersion, defined as

$$\sigma_{\rm line} = \left(\langle V_{\rm LOS}^2 \rangle - \langle V_{\rm LOS}\rangle^2\right)^{1/2}. \tag{5.26}$$

By symmetry, $\langle V_{\rm LOS}\rangle = 0$. The first term is

$$\langle V_{\rm LOS}^2 \rangle = \frac{\int_{-v_{\rm orb}}^{v_{\rm orb}} V_{\rm LOS}^2 \Psi(V_{\rm LOS})dV_{\rm LOS}}{\int_{-v_{\rm orb}}^{v_{\rm orb}} \Psi(V_{\rm LOS})dV_{\rm LOS}} = \frac{v_{\rm orb}}{2}. \tag{5.27}$$

Thus, $\sigma_{\rm line} = v_{\rm orb}/2^{1/2}$. Inspection of Figure 5.12 shows that FWHM $= 2v_{\rm orb}$, so the ratio of these two line-width measures is FWHM$/\sigma_{\rm line} = 2\sqrt{2}$.

Generalization to a disk from a ring is obvious, though more parameters are required, specifically, inner and outer radii, the radial responsivity dependence, which is usually parameterized as a power law $\epsilon(r) \propto r^\alpha$, and the inclination. A example of a velocity–delay map for an inclined Keplerian disk is shown in Figure 5.13.

5.7.5 *Recovering velocity–delay maps from real data*

We have spent some effort transforming the BLR from configuration space into velocity–time-delay space. What we would really like to do is solve the inverse problem: we obtain experimentally a velocity–delay map from a spectroscopic monitoring program and transform this back into a unique solution for the BLR structure and kinematics. None of this is easy.

Inspection of the transfer equation (Equation (5.16)) would lead one immediately to think of Fourier inversion, i.e. the method of Fourier quotients. Indeed, this was the approach that was suggested in the seminal theoretical paper on reverberation mapping by Blandford & McKee (1982) nearly 25 years ago. The Fourier transform of the continuum light curve $C(t)$ is

$$C^*(f) = \int_{-\infty}^{+\infty} C(t)e^{-i2\pi f t}\, dt, \tag{5.28}$$

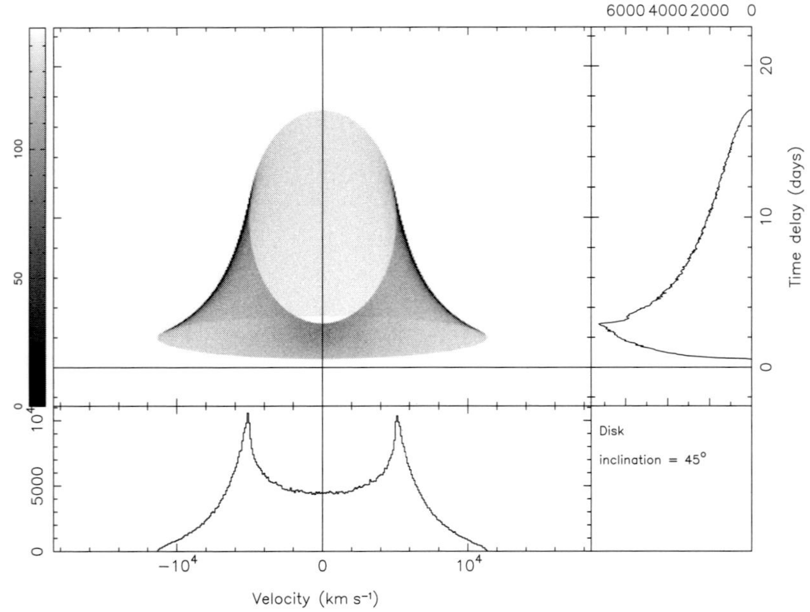

FIGURE 5.13. A velocity–delay map for a thin disk at inclination $i = 45°$. The upper left panel shows in gray scale the velocity–delay map, i.e. the transfer function or the observed emission-line response as a function of line-of-sight velocity $V_{\rm LOS}$ and time delay τ. The upper right panel shows the one-dimensional transfer function, i.e. the velocity–delay map integrated over $V_{\rm LOS}$, which is the response of the total emission line as a function of time (cf. Equation (5.21)). The lower left panel shows the emission-line response integrated over time delay; this is the profile of the variable part of the line (cf. Equation (5.25)). Adapted from Horne *et al.* (2004).

and a similar equation defines the Fourier transform of the emission-line light curve. Use of the convolution theorem (Bracewell 1965) gives us

$$L^* = \Psi^* C^*, \tag{5.29}$$

so $\Psi^* = L^*/C^*$. The velocity–delay map is then obtained by taking the inverse Fourier transform,

$$\Psi(\tau) = \int_{-\infty}^{+\infty} \Psi^*(f) e^{i2\pi f t}\, dt. \tag{5.30}$$

While this is straightforward in theory, it fails in practice simply because real astronomical data tend to be too sparse and too noisy. Other, more powerful methods, such as the maximum-entropy method (MEM) (Horne 1994) have been used, but the velocity–delay maps produced to date have been of poor quality, despite the effort and care involved (e.g. Ulrich & Horne 1996; Kollatschny 2003); it is important to understand, however, that the situation is far from hopeless. Extensive simulations show that recovery of a velocity–delay map will require only modest improvements over previous spectroscopic monitoring programs (Horne *et al.* 2004).

While the data requirements to obtain a fully resolved velocity–delay map are beyond what we can achieve with existing data, we can still learn much about the BLR by using simpler tools. In particular, it is possible to determine the mean response time for any particular emission line by cross-correlating the emission-line light curve with the continuum light curve. Under most circumstances, the centroid of the cross-correlation function is a good estimator of the centroid of the transfer function (Equation (5.22))

TABLE 5.2. *Reverberation results for NGC 5548*

Feature	$F_{\rm var}$	Lag (days)
UV continuum	0.321	—
Optical continuum	0.117	$0.6^{+1.5}_{-1.5}$
He II λ1640	0.344	$3.8^{+1.7}_{-1.8}$
N V λ1240	0.411	$4.6^{+3.2}_{-2.7}$
He II λ4686	0.052	$7.8^{+3.2}_{-3.0}$
C IV λ1549	0.136	$9.8^{+1.9}_{-1.5}$
Lyα λ1215	0.169	$10.5^{+2.1}_{-1.9}$
Si IV λ1400	0.185	$12.3^{+3.4}_{-3.0}$
Hβ λ4861	0.091	$19.7^{+1.5}_{-1.5}$
C III] λ1909	0.130	$27.9^{+5.5}_{-5.3}$

$\tau_{\rm cent} = \langle \tau \rangle$, which is the mean time delay, or lag, between continuum and emission-line variations.

The particular challenge of cross-correlating AGN continuum and emission-line light curves is that the light curves are often irregularly sampled, not necessarily by experimental design, of course, but more often because of uncontrollable variables such as weather. However, methodologies have been developed to deal with such data, either by interpolation of the light curves to obtain an evenly sampled grid (e.g. Gaskell & Sparke 1986; Gaskell & Peterson 1987; White & Peterson 1994) or by time-binning (Edelson & Krolik 1987; Alexander 1997). Error estimation is even more problematic, but some conceptually simple Monte Carlo methods seem to perform reasonably well under most circumstances (e.g. Peterson *et al.* 1998, 2004; Welsh 2000). A fairly complete tutorial on use of cross-corrrelation in reverberation mapping is provided by Peterson (2001).

5.8. Reverberation-mapping results

Thus far, spectroscopic monitoring programs undertaken for the purpose of reverberation mapping have yielded results for about three dozen galaxies. In most cases, emission-line lags have been measured for one or more of the Balmer lines, but in a few cases lags for multiple emission lines have been measured. In a few cases, lags for specific lines (usually Hβ) have been measured on more than one occasion. By far, the AGN that has been studied best using reverberation techniques is NGC 5548: it has been monitored in the ultraviolet with the International Ultraviolet Explorer (Clavel *et al.* 1991) and Hubble Space Telescope (Korista *et al.* 1995), and in the optical for 13 consecutive years (Peterson *et al.* 2002, and references therein), extended to 30 years if lower-quality archival data are included (Sergeev *et al.* 2007). In Table 5.2, we list results for several different time series obtained in 1988–89 (Clavel *et al.* 1991; Peterson *et al.* 1991, 2004). The first column gives the feature whose light curve is cross-correlated with the UV-continuum light curve, the second column gives the amplitude of variability (rms fractional variation, corrected for measurement uncertainties), and the third column gives the time delay, in days, relative to the variations in the UV continuum.

5.8.1 *The size of the broad-line region*

The first surprise from reverberation mapping was that the BLR is much smaller than had originally been supposed on the basis of photoionization equilibrium modeling. As noted earlier, single-zone models suggest $n_H \approx 10^{10}\,\text{cm}^{-3}$, on the basis of the presence of broad C III] λ1909 and the absence of broad [O III] λλ4959, 5007. Furthermore, the C III] λ1909/C IV λ1549 and Lyα λ1215/C IV λ1549 flux ratios suggest that $U \approx 0.01$. As a specific example, in NGC 5548, we can estimate the ionizing flux by interpolating between the observable UV at ∼1100 Å and the soft-X-ray flux at ∼2 keV and obtain $Q_{\text{ion}}(\text{H}) \approx 1.4 \times 10^{54}$ photons s^{-1}. Solving Equation (5.11) for the size of the line-emitting region yields $r \approx 3.3 \times 10^{17}\,\text{cm} \approx 130$ light days. However, comparison with Table 5.2 reveals that this is about an order of magnitude larger than the emission-line lags actually measured for NGC 5548.

The place where this calculation breaks down is in the single-zone assumption, i.e. a single representative gas cloud that produces each emission line with the flux ratios as observed. What we see instead, as described below, is that the BLR is extended and has a stratified ionization structure.

5.8.2 *Ionization stratification*

Inspection of Table 5.2 shows that the lines characteristic of highly ionized gas (e.g. He II and N V) have smaller lags than lines arising in less highly ionized gases (e.g. Hβ). In other words, the BLR, far from being homogeneous, exhibits ionization stratification – physical conditions in the BLR are functions of distance from the central source. There is also an indication that the higher-ionization lines vary with a larger amplitude than lower-ionization lines do, although part of this is certainly attributable to geometric dilution (i.e. the response of more distant gas is more spread out in time).

Note that C III] λ1909 and C IV λ1549 arise at very different distances from the central source, as mentioned above. The lack of strong response of C III] at distances of ∼10 lt-days indicates that the particle density at this distance is higher than $\sim 10^{11}\,\text{cm}^{-3}$ so that the C III] line is collisionally suppressed.

5.8.3 *The BLR radius–luminosity relationship*

We noted earlier that, at least to low order, AGN spectra are all very similar. Given the discussion of the previous section, we can conclude that the ionization parameter U and particle density n_H have approximately the same values in all AGNs. Because Q_{ion} is proportional to the ionizing luminosity L_{ion}, we can rearrange Equation (5.11) and obtain a simple prediction that

$$R \propto L^{1/2}, \tag{5.31}$$

where L can be any luminosity measure that is proportional to L_{ion}. This naïve expectation overlooks a number of subtleties such as (a) the expected dependence of the shape of the ionizing continuum on black-hole mass and (b) the Baldwin effect. Nevertheless, it serves as an interesting benchmark and does give us the idea that the BLR radius should depend on luminosity, and that the dependence is weak enough that it will take a reasonably large range of AGN luminosity to test it. Indeed, while a radius–luminosity (R–L) relationship had long been predicted (e.g. Koratkar & Gaskell 1991), it was only when reverberation experiments were extended from Seyfert galaxies to Palomar–Green quasars that this relationship became experimentally well defined. Kaspi *et al.* (2000), using a power-law parameterization of the form $R \propto L^{\alpha}$, found $\alpha = 0.70 \pm 0.03$, although the uncertainty is underestimated since it does not account for certain systematic effects. For example, the luminosity measure used is the optical flux at ∼5100 Å in the rest-frame, and some fraction of the luminosity is attributable to

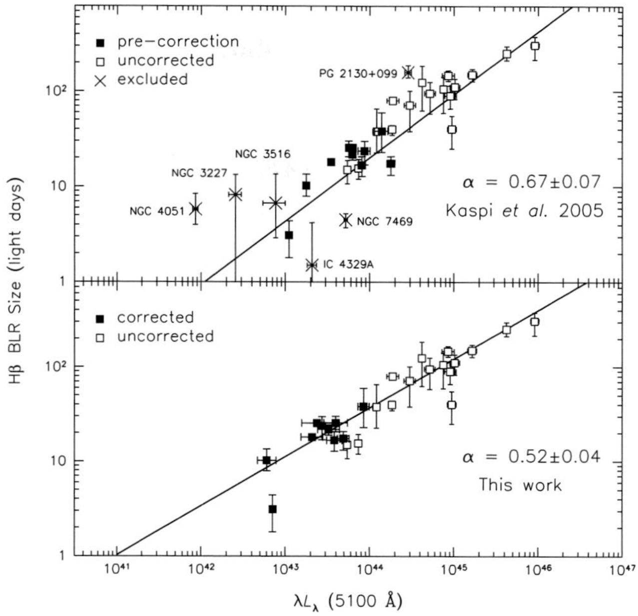

FIGURE 5.14. The relationship between the BLR radius, as determined from the reverberation lag of Hβ, and the optical continuum luminosity. The top panel shows the relationship without correction for the host-galaxy contribution to the luminosity. In the bottom panel, a correction for the host-galaxy contribution has been made for many of the lower-luminosity objects. The open squares are from Kaspi *et al.* (2005) and the filled squares are for the same AGNs, but after correction for starlight. Objects indicated by an × are not included in the fits, since host-galaxy models were not yet available. Adapted from Bentz *et al.* (2006a).

starlight arising from the host galaxy; this is significant for most reverberation-mapped AGNs because large spectrograph entrance apertures, which admit a lot of host-galaxy light, are used to mitigate against seeing effects that can result in photometric errors. Bentz *et al.* (2006a) have obtained images of the host galaxies of reverberation-mapped AGNs with the High Resolution Channel of the Advanced Camera for Surveys on the Hubble Space Telescope in order to accurately assess the host-galaxy contamination of the luminosity measurements for the reverberation-mapped AGNs. They find that, once the starlight has been accounted for, the slope of the R–L relationship reduces to $\alpha = 0.52 \pm 0.04$, as shown in Figure 5.14, in remarkable agreement with Equation (5.31).

It is also found that the emission-line lags for an individual AGN vary with the continuum flux. The best-studied case is NGC 5548, for which over a dozen independent measurements of the Hβ lag have been made (Figure 5.15); the Hβ lag varies between 6 and 26 days and depends on the continuum luminosity. The most-recent analysis finds that $\tau(\mathrm{H}\beta) \propto L_{\mathrm{opt}}^{0.66\pm0.13}$ (Bentz *et al.* 2007). However, we must at this point note another result from these spectrophotometric monitoring programs, namely that the continuum gets *harder* as it gets brighter, i.e. the shorter-wavelength continuum varies with a larger amplitude than does the longer-wavelength continuum. For NGC 5548, $L_{\mathrm{opt}}(5100\,\text{Å}) \propto L_{\mathrm{UV}}(1350\,\text{Å})^{0.84\pm0.05}$, so $\tau(\mathrm{H}\beta) \propto L_{\mathrm{UV}}(1350\,\text{Å})^{0.55\pm0.14}$ (Bentz *et al.* 2007). It should be noted, however, that this result is based on somewhat arbitrarily dividing up the NGC 5548 light curve by observing season (there are

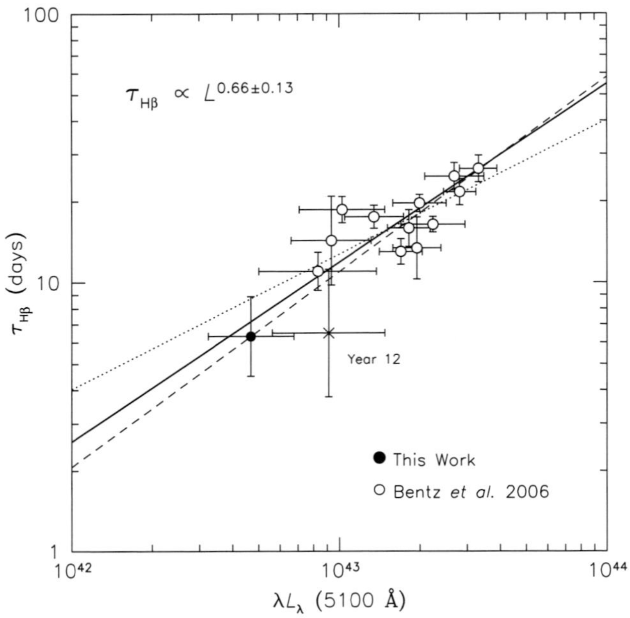

FIGURE 5.15. The relationship between the Hβ emission-line lag and the luminosity of the central source for NGC 5548. The dashed line is the best fit (slope 0.73 ± 0.14) to all the data, and the dotted line has a slope of 0.5 (Equation (5.31)). The solid line is the best fit excluding the suspicious data point labeled "Year 12." Adapted from Bentz *et al.* (2007).

~2-month gaps in coverage between the observing seasons when NGC 5548 is too close to the Sun to observe) and using the mean luminosity for each year. Cackett & Horne (2006) use a "dynamic" model in which the BLR "breathes," i.e. the velocity–delay map is a function of both time delay and continuum luminosity (Figure 5.16). They find a much shallower dependence on luminosity, $\tau(\mathrm{H}\beta) \propto L_{\mathrm{UV}}^{0.13}$. At least part of the difference is due to different assumptions about host-galaxy light, which would make the slope of the Cackett & Horne result somewhat steeper. This result is closer to the prediction from a more-detailed photoionization equilibrium model by Korista & Goad (2004) that yields $\tau(\mathrm{H}\beta) \propto L_{\mathrm{UV}}^{0.23}$.

5.8.4 *Black-hole masses*

To date, probably the most-important product of reverberation mapping is the mass of the central black holes in AGNs. Woltjer (1959) realized that, if the broad emission lines are Doppler-broadened on account of motion in a deep gravitational potential, then it would be possible to measure the mass of the nucleus if one could estimate the size of the line-emitting region. Since he had only a crude upper limit to the size from the fact that the nuclear regions are unresolved under seeing-limited conditions, the upper limits he obtained were extremely large and uninteresting, typically $\sim 10^{10} M_\odot$. Early emission-line-variability results suggested that the central masses in Seyfert galaxies were only of order $(10^7\text{–}10^8) M_\odot$ (e.g. Peterson 1987), making gravitationally bound motion an attractive alternative to the outflow models that had become prevalent by the 1980s. Reverberation mapping provided the first accurate estimates of the BLR size. The evidence that gravity is the most-important dynamical force in the BLR is the existence of a correlation between emission-line width ΔV and BLR size R of the form $R \propto \Delta V^{-2}$ (Figure 5.17), which has been found to hold in every case in which it can be critically tested (Peterson & Wandel

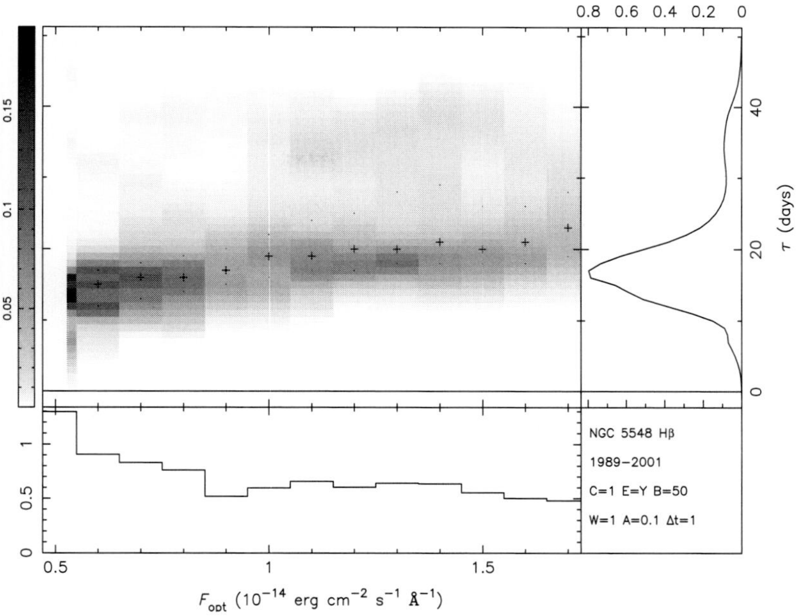

FIGURE 5.16. A luminosity-dependent delay map for Hβ in NGC 5548. This shows how the mean response time increases with the luminosity of the continuum source. Adapted from Cackett & Horne (2006).

1999, 2000; Onken & Peterson 2002; Kollatschny 2003). Given this result, the mass of the central source† is given by

$$M_{\rm BH} = f \frac{R\,\Delta V^2}{G}, \tag{5.32}$$

where $R = c\tau$ is the size of the BLR as measured for a particular emission line, ΔV is the width of the line, G is the gravitational constant, and f is a dimensionless scale factor of order unity, which we will discuss further below.

Additional evidence that the black-hole masses from Equation (5.32) have some validity is provided by the correlation between black-hole mass $M_{\rm BH}$ and host-galaxy bulge velocity dispersion σ_*, which is known as the "M–σ relationship" (Ferrarese & Merritt 2000; Gebhardt *et al.* 2000a). It was originally found in quiescent (non-active) galaxies, but holds also for AGNs (Gebhardt *et al.* 2000b; Ferrarese *et al.* 2001; Nelson *et al.* 2004; Onken *et al.* 2004). In Figure 5.18, we show the AGN M–σ relationship superposed on the best-fit lines to the relationship for quiescent galaxies. In this case, following Onken *et al.* (2004), we have used a value for the scale factor f that results in the best agreement between the relationship for AGNs and that for quiescent galaxies – in other words, we have scaled the AGN "virial products" (i.e. the observable quantities $R\,\Delta V^2/G$) to give the best global match to the quiescent-galaxy relationship.

An important point that we have glossed over is how the line width ΔV should be characterized. Two obvious candidates introduced earlier are the FWHM and the line

† Note that this is really the total mass enclosed by the radius R; in some models, the accretion disk also contains a significant amount of mass, which would be included in this calculation.

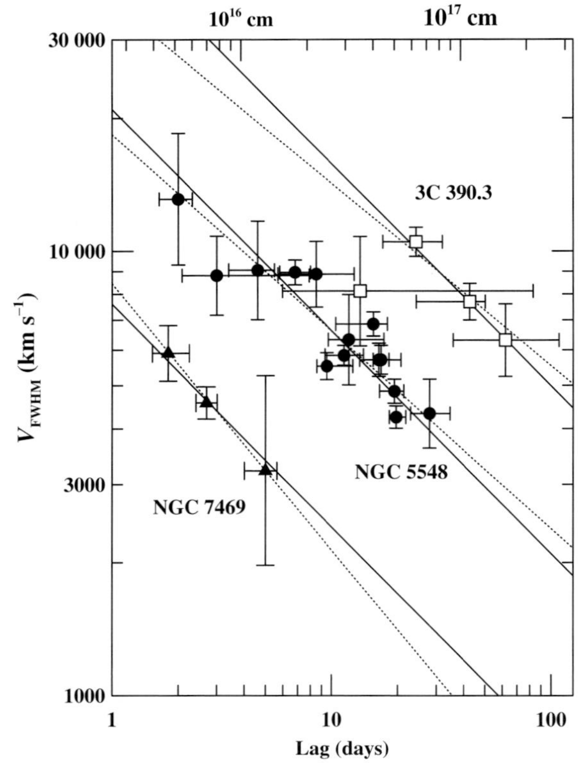

FIGURE 5.17. The relationship between the BLR line width and lag for three well-studied AGNs. The dotted lines are the best fits to each data set, and the solid lines show the best fit for a fixed virial slope, i.e. $\tau \propto V_{\rm FWHM}^{-2}$. From Peterson & Wandel (2000).

dispersion $\sigma_{\rm line}$, which is based on the second moment of the line profile $P(\lambda)$. The first moment of the line profile is the line centroid

$$\lambda_0 = \int \lambda P(\lambda) {\rm d}\lambda \bigg/ \int P(\lambda) {\rm d}\lambda, \tag{5.33}$$

and the second moment of the profile is used to define the variance or mean-square dispersion

$$\sigma_{\rm line}^2(\lambda) = \langle \lambda^2 \rangle - \lambda_0^2 = \left(\int \lambda^2 P(\lambda) {\rm d}\lambda \bigg/ \int P(\lambda) {\rm d}\lambda \right) - \lambda_0^2. \tag{5.34}$$

The square root of this equation is the line dispersion $\sigma_{\rm line}$ or root-mean square (rms) width of the line.

We can use the ratio of these two quantities, FWHM/$\sigma_{\rm line}$, as a low-order description of the line profile. For the ring of Figure 5.11, we have already noted that FWHM/$\sigma_{\rm line} = 2^{1/2}2 = 2.83$. Another well-known result is that, for a Gaussian, FWHM/$\sigma_{\rm line} = 2(2 \ln 2)^{1/2} = 2.35$. Observationally, the mean value of this ratio for the reverberation-mapped AGNs is FWHM/$\sigma_{\rm line} \approx 2$, which means that the typical AGN line profile is "peakier" than a Gaussian. Figure 5.19 shows two profiles with extreme values of the line-width ratio FWHM/$\sigma_{\rm line}$; this emphasizes the point that these two line-width

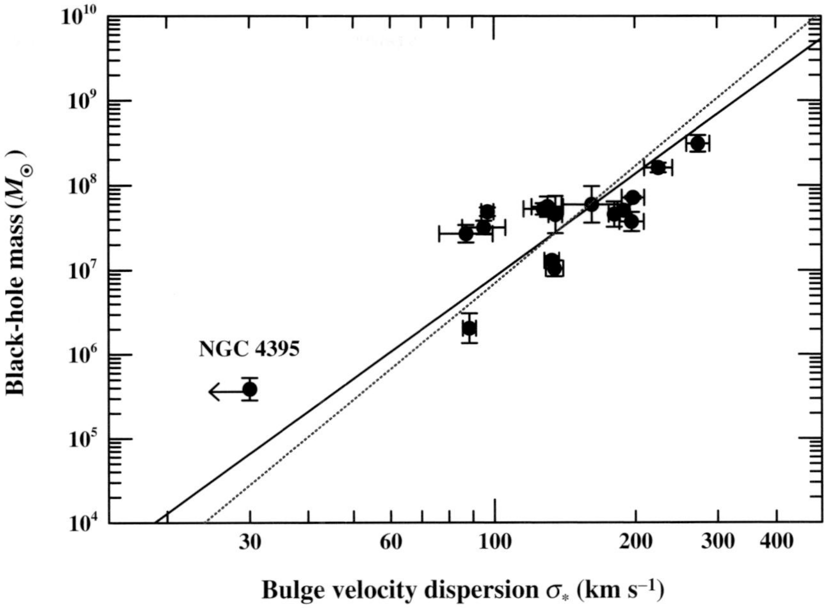

FIGURE 5.18. The relationship between the black-hole mass and host-galaxy stellar-bulge velocity dispersion for reverberation-mapped AGNs. The two lines are best-fits to the quiescent galaxy data from Tremaine *et al.* (2002) (shallower line) and Ferrarese (2002) (steeper line). This is an updated version of a similar plot from Onken *et al.* (2004) which incorporates additional data from Nelson *et al.* (2004), Bentz *et al.* (2006b, 2007), and Denney *et al.* (2006). For these data (excluding NGC 4395) the adopted scaling factor is $f = 5.9 \pm 1.8$.

measures are not simply interchangeable. Collin *et al.* (2006) argue that $\sigma_{\rm line}$ is the preferred line-width measure for black-hole-mass estimates because using the FWHM introduces a bias such that the scaling factor becomes a function of the line width, f(FWHM).

5.9. An evolving view of the BLR

The consensus view of AGN structure has remained fairly stable, if somewhat vague and lacking in details and self-consistency, for over 15 years. During that period, work has steadily progressed, and there have been some long-standing issues on which consensus is *beginning* to emerge. The following constitutes the author's personal view of some of the changes that are beginning to take hold.

5.9.1 *The BLR gas: discrete clouds or a continuous flow?*

The notion that the BLR is comprised of a large number of discrete gas clouds can be traced back to the early days of AGN research, when the best conceptual model of the BLR structure was a collection of optically thick clouds or filaments – in other words, something that might look like a large version of the Crab Nebula. The number of clouds or filaments required to account for the observed characteristics of the BLR turns out to be very large.

It was recognized long ago that the smoothness of broad-line profiles implies that there must be a lower limit ($\sim 10^5$ clouds) to the number of line-emitting clouds if the intrinsic line profile of each cloud is thermal (Capriotti *et al.* 1981; Atwood *et al.* 1982). This lower

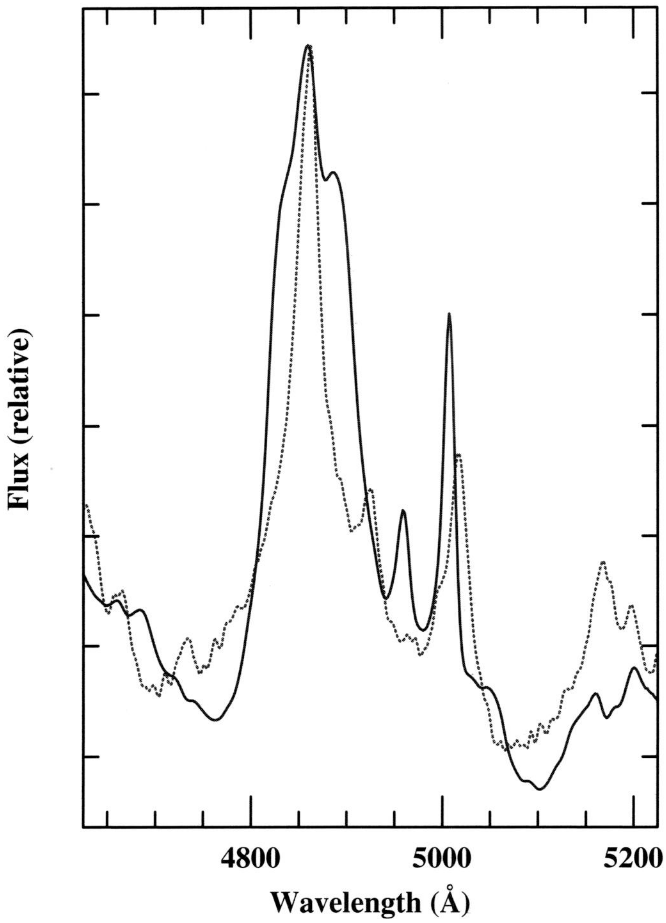

FIGURE 5.19. Two Hβ profiles representing extreme values of FWHM/$\sigma_{\rm line}$. PG 1700+518 (dotted line; FWHM/$\sigma_{\rm line} = 0.71$) has an Hβ profile that is narrower at the peak and broader at the base than a Gaussian of comparable width. Conversely, Akn 120 (solid line; FWHM/$\sigma_{\rm line} = 3.45$) has much stronger shoulders and is relatively more rectangular.

limit will depend on the spectral resolution and signal-to-noise ratio of the observations used in the analysis. Tight constraints are drawn from very-high-resolution observations of the least-luminous AGNs, with the most-recent limit based on Keck spectra of the dwarf Seyfert 1 galaxy NGC 4395 (Laor *et al.* 2006). This argument and other evidence are leading to complete abandonment of the notional discrete-cloud model in favor of some kind of a continuous flow.

5.9.2 *Physical conditions in the BLR*

As noted above, multiple reverberation measurements of the Hβ response in NGC 5548 reveal that the "BLR size" varies with continuum luminosity – when the nucleus is in a brighter state, the BLR is "bigger." The uncomfortable question that naturally arises is the following: how does the BLR know precisely where to be? What fine tunes the size of the BLR? Of course, the answer to this question is obvious from the BLR stratification results (Table 5.2): the BLR gas is distributed over a very wide range of distances, between at least $\sim$2 and $\sim$30 lt-days from the central source, in the case of NGC 5548. The

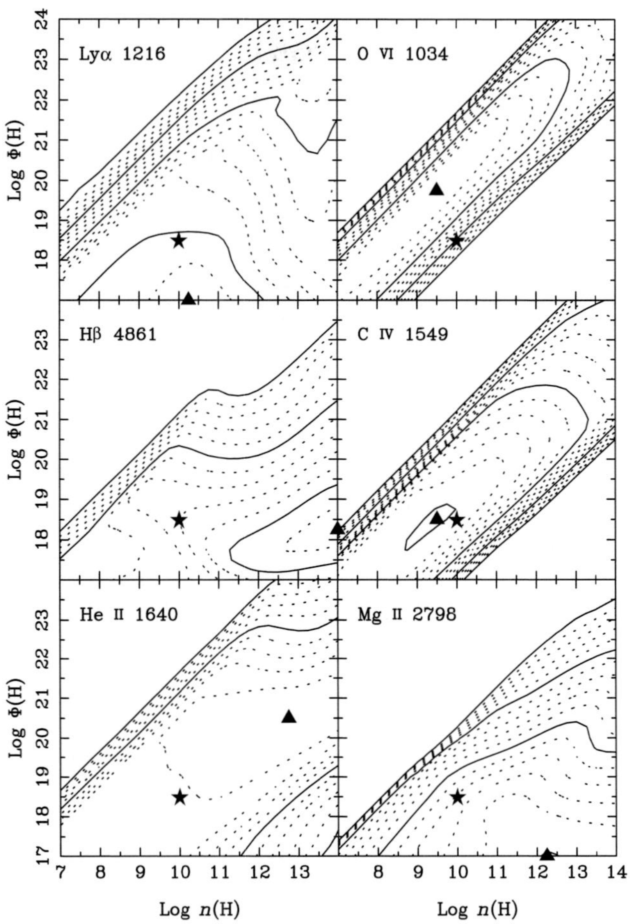

FIGURE 5.20. Contours of constant equivalent width for a grid of photoionization-equilibrium models, as a function of input ionizing flux ($\Phi(\mathrm{H}) = Q_{\mathrm{ion}}/(4\pi r^2)$) and particle density $n(\mathrm{H})$ for strong emission lines in AGN spectra. Dotted contours are separated by 0.2 dex and solid coutours by 1 dex. The solid star is a reference point to "standard BLR parameters," i.e. the best single-zone model. The triangle shows the location of the peak equivalent width for each line. Adapted from Korista *et al.* (1997).

very-well-defined emission-line responses that we measure indicate that the responsivity is well localized; the responsivity is clearly a very sensitive function of the physical conditions in the BLR. The dominant response we observe comes from where the physical conditions are optimal for a particular emission line. This is at least qualitatively consistent with the expectations of the "locally optimally emitting cloud (LOC)" model of Baldwin *et al.* (1995). In Figure 5.20, we show emissivity contours for various strong emission lines as a function of the ionizing photon flux $\Phi(\mathrm{H})$ and proton density $n(\mathrm{H})$ predicted by an LOC model.

It is perhaps worth mentioning that, in contrast to the other broad lines, the optical Fe II blends just longward and shortward of Hβ do not have a strongly localized reverberation signature, even though they do clearly vary (e.g. Vestergaard & Peterson 2005). This may be because the optimal responsivity for these blends is not sharply localized in U–n_{H} space.

5.9.3 *The mass of the BLR*

We used Equation (5.15) to estimate the mass of gas in the NLR. We neatly side-stepped this particular issue in our earlier preceding discussion of the BLR. We could in principle have carried out a nearly identical calculation for the BLR and found that the total amount of gas required to account for the Balmer-line emission is about $\sim M_{\odot}$ for moderate-luminosity AGNs. There is a hidden assumption in such calculations that the gas is emitting at high efficiency. The reverberation results clearly indicate that there is a lot of gas in the BLR that is radiating quite inefficiently at any given time. When this is taken into account, Baldwin *et al.* (2003) estimate that the total amount of gas in the BLR is closer to $(10^4\text{–}10^5)M_{\odot}$, indicating that the BLR gas flow is fairly massive.

5.9.4 *The kinematics of the BLR*

Members of a small subset of AGNs have the double-peaked profiles that are characteristic of disks in Keplerian rotation. The best examples (e.g. Storchi-Bergmann *et al.* 2003) are in the Balmer lines of lower-luminosity objects in cases where the line widths are very large. Double-peaked profiles sometimes become apparent in difference profiles (i.e. the spectrum formed by subtracting a low-flux state spectrum from a high-flux state spectrum), and in other cases we see consistent and similar weak signatures in the form of strong shoulders or secondary peaks in Balmer lines. It seems plausible that double-peak disk components are indeed present in virtually all AGNs, though they are sometimes masked by low inclination or by the presence of another emission-line component, namely a disk wind.

Disk winds have been invoked in order to describe a number of observed AGN features. Elvis (2000) presents a fairly comprehensive phenomenological disk-wind model that unifies several AGN emission- and absorption-line characteristics in a physically well-motivated way. A theoretical driver for disk winds is that hydromagnetically accelerated winds can serve to remove angular momentum that is carried in by infalling matter. Observational evidence supporting disk winds includes the following.

- Strong blueward asymmetries are present in the higher-ionization lines in NLS1s (e.g. Leighly 2000). This would occur if there were a strong outflow component with the far (receding) side partially obscured from our view (presumably by the torus).
- Peaks of high ionization lines tend to be blueshifted relative to systemic, and the maximum observed blueshift increases with source luminosity (e.g. Espey 1997, and references therein).
- In radio-loud quasars, the widths of the bases of the C IV lines are larger in edge-on sources than in face-on sources, implying that the wind has a strong radial (as opposed to polar) component to it (Vestergaard *et al.* 2000). Also, Rokaki *et al.* (2003) have shown that line width is anticorrelated with several beaming indicators, again arguing that the velocities are largest in the disk plane.

A natural explanation for at least some of the diversity in line profiles is that *both* components are present in Type 1 AGN spectra, in varying relative strength: the disk wind, however, is weak in low-accretion-rate objects and stronger in high-accretion-rate objects. This is hardly a radical suggestion. As noted earlier, at least two physical components are required to account for the strength of the low-ionization lines (which are strong in the dense rotating-disk component, which might in fact be thought of as an extension of the accretion-disk structure itself) versus the high-ionization lines (which are presumably strong in the wind component). This is consistent with double-peaked profiles being primarily a Balmer-line phenomenon and strong blueward-asymmetric (and sometimes self-absorbed) components appearing primarily in UV resonance lines.

Consider the following simple argument. Equation (5.5) tells us that the source luminosity depends on accretion rate as $L \propto \dot{M}$. We also have a relationship between luminosity and BLR radius, $R \propto L^{1/2}$ (Equation (5.31)). Now, using the virial equation (Equation (5.32)), we can write the dependence of the line width as

$$\Delta V \propto \left(\frac{GM}{R}\right)^{1/2} \propto \left(\frac{M}{L^{1/2}}\right)^{1/2} \propto \left(\frac{M}{\dot{M}^{1/2}}\right)^{1/2} \propto \left(\frac{M}{\dot{m}}\right)^{1/4}. \tag{5.35}$$

This suggests that at a fixed black-hole mass, the line width reflects the Eddington rate $\dot{m}$ (as defined in Equation (5.6)). Objects with high Eddington rates are those with narrower lines, such as NLS1 galaxies, which are also the objects with the strongest evidence for massive outflows. Certainly, in Galactic binary systems, high accretion rates and high outflow rates are closely coupled, and it stands to reason that this holds for larger-mass systems as well.

5.9.5 *Dust and the BLR*

Sanders *et al.* (1989) showed that spectral energy distributions of AGNs have a local minimum at a wavelength ~1 μm (see Figure 5.5). They argued that the IR emission longward of 1 μm is attributable to dust, as had others previously (e.g. Rieke 1976), at the sublimation temperature of ~1500 K. This corresponds to a minimum distance from the AGN, the sublimation radius, at which dust can exist. For graphite grains, this is approximately

$$r_{\mathrm{sub}} = 1.3\left(\frac{L_{\mathrm{UV}}}{10^{46}\ \mathrm{erg\ s^{-1}}}\right)^{1/2}\left(\frac{T}{1500\,\mathrm{K}}\right)^{-2.8}\ \mathrm{pc}, \tag{5.36}$$

where L_{UV} is the UV luminosity of the AGN and T is the grain sublimation temperature (Barvainis 1987). Clavel *et al.* (1989) first showed that IR-continuum variations followed those in the UV/optical with a time lag that is consistent with the IR emission arising at the sublimation radius. The IR-continuum emission we see is reprocessed UV/optical emission from the inner regions. This effect, sometimes referred to as "dust reverberation," has now been observed in several AGNs. Suganuma *et al.* (2006) show that in each case where measurements are available the emission-line lags are always slightly smaller than or comparable to the IR-continuum lags. This indicates that the outer boundary of the BLR is defined by the sublimation radius, as suggested by Laor (2003).

5.10. Unification issues and the NLR

5.10.1 *The geometry of the obscuring torus*

The notion of an obscuring torus surrounding the central source and BLR has been the key feature of AGN unification models for nearly two decades. The left panel of Figure 5.21 illustrates the torus more or less as it was envisaged by Antonucci & Miller (1985). The opening angle of the torus is a free parameter that can be selected to match the observed space densities of Seyfert 1 and Seyfert 2 galaxies, as noted earlier.

There are significant problems with the "doughnut" view of the torus (Elitzur & Shlosman 2006), notably in terms of stability and size; with regard to the latter point, Elitzur (2006) notes that a conventional torus model such as that in the left panel of Figure 5.21 is predicted to have a size of hundreds of parsecs for a galaxy like NGC 1068, although mid-IR observations show that the core is no larger than a few parsecs. By introducing a large number of smaller clouds, the size of the torus can be made much smaller because of the larger radiating surface area, and the clumpy medium better reproduces

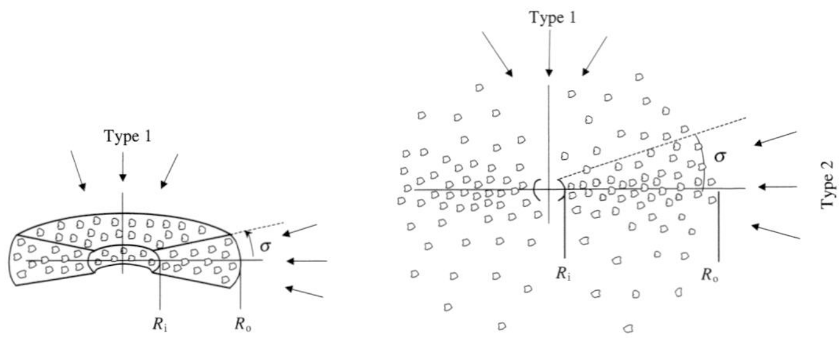

FIGURE 5.21. The illustration on the left shows the standard "doughnut" torus as originally envisaged by Antonucci & Miller (1985). The diagram on the right illustrates a more plausible arrangement of dusty obscuration in the vicinity of the AGN. Adapted from Elitzur *et al.* (2004).

the IR spectra of AGNs. Moreover, the clumpy medium has a natural origin in the disk wind.

5.10.2 *Type 2 quasars*

Among lower-luminosity AGNs, Seyfert 2 galaxies significantly outnumber Seyfert 1 galaxies, by a ratio of perhaps 3 : 1. However, there is a decreasing number of Type 2 AGNs at higher luminosity, i.e. "Type 2 quasars." Indeed, Type 2 quasars are rare and were not known in any reasonable numbers until the advent of the SDSS, which turned up a few hundred candidates (Zakamska *et al.* 2003). One possible explanation for this is that the inner edge of the obscuring torus is larger for more-luminous objects because the sublimation radius (Equation (5.36)) increases with luminosity. If the height of the torus remains the same or increases only slowly with luminosity (cf. Simpson 2005), the effective opening angle increases with luminosity, thus decreasing the numbers of obscured (Type 2) objects. This is sometimes referred to as the "receding-torus" model (Lawrence 1991).

5.10.3 *The NLRs in Type 1 and Type 2 AGNs*

A naïve expectation of AGN unification is that the narrow-line spectra of Type 1 and Type 2 AGNs should be the same. If not stated directly, this is implicit, for example, in AGN sample selection based on the luminosity of particular narrow emission lines, such as [O III] $\lambda5007$, where the underlying assumption is that the narrow lines are emitted isotropically. Careful comparison of the narrow-line components of Seyfert 1 galaxies with the narrow-line spectra of Seyfert 2 galaxies reveals that they are not identical (e.g. Schmitt 1998). For example, Seyfert 1 galaxies have relatively stronger high-ionization lines, which is probably because the most highly ionized NLR gas is at least partially within the throat of the obscuring torus and not visible in the directions from which the source would be viewed as a Seyfert 2. Also, it is observed that the size of the NLR scales differently with luminosity for the two types, with $r \propto L^{0.44}$ for Type 1 objects and $r \propto L^{0.29}$ for Type 2 objects. Bennert *et al.* (2006) suggest that this difference can be explained by invoking the differences in how the NLR projects onto the sky in a receding-torus model as a function of inclination.

5.11. Cosmological implications

Ever since their discovery, quasars have been recognized as potentially important cosmological probes.

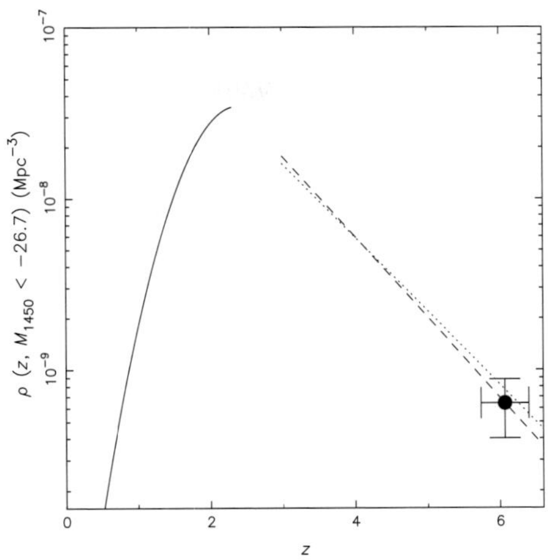

FIGURE 5.22. The comoving space density of QSOs with absolute magnitude $M_{1450} < -26.7$ mag. The lower-redshift result (full line) is from the 2DF survey (Croom *et al.* 2004). The dashed line is from Fan *et al.* (2001) and the dotted line is from Schmidt *et al.* (1995). The single point at high redshift is based on the ~19 quasars at $z > 5.7$ that were discovered in the SDSS. Adapted from Fan (2006), with permission of Elsevier.

(i) Because of their high luminosities, quasars can be seen at great distances. By determining the evolution of the luminosity function over cosmic time, we can trace the development of supermassive black holes and galaxies over the history of the Universe.
(ii) By using quasars simply as distant beacons, we can observe the cosmic evolution of the intergalactic medium through study of absorption lines that arise along our line of sight to quasars.
(iii) The appearance of the first quasars marks the appearance of discrete structures in the Universe and places strong constraints on models of galaxy formation (see Chapter 3 of this volume). Similarly, the presence of emission lines from elements more massive than helium places very stringent contraints on star formation in the early Universe (see Chapter 6 of this volume).

Determining the luminosity function over the history of the Universe obviously requires surveying large areas of the sky as deeply as possible using techniques that are designed to identify quasars efficiently, missing as few as possible while avoiding as many "false positives" as possible. It is also necessary to understand in detail the biases in selection methods so that survey selection functions can be used to correct statistically for biases. As discussed in Chapter 2 of this volume, emission-line surveys are an especially effective means of identifying quasars. Rather than repeat any of that discussion, we will concentrate on the results.

5.11.1 *The space density of AGNs*

Figure 5.22 shows the comoving† space density of the brightest quasars as a function of redshift. As soon as quasars were found in large numbers, it became apparent, even through such simple tests as V/V_{max} (Schmidt 1968) that quasars were much more common in the past than they are now. By the early 1980s, Osmer (1982) was able to conclude, however, that the density of quasars drops off dramatically by $z \approx 3.5$. The

† A "comoving volume" is one that expands with the Universe, in contrast to a "proper volume," which is fixed in an absolute sense.

"quasar era" corresponds to the redshift range $2 < z < 3$, when the comoving density of bright quasars was at a maximum.

Modern surveys, such as the Two-Degree Field (2DF) (Croom *et al.* 2004) and the SDSS (York *et al.* 2000), have been very successful at identifying large numbers of quasars out to very large redshifts. The space density of quasars remains very hard to constrain, however, above $z \approx 6$. At such large redshifts, Lyα, the shortest-wavelength strong emission line in AGN spectra, is shifted into the near infrared, and shortward of this quasar spectra are heavily absorbed by the "Lyα forest," i.e. absorption by neutral intergalactic hydrogen at lower redshift. In other words, at $z \approx 6$, quasars are extremely faint throughout the entire optical (observed frame) spectrum.

It is quite profound that quasars at $z > 5$ are found at all: this means that supermassive black holes must have been in place by the time the Universe was only a few hundred million years old. Furthermore, the spectra of these very-high-redshift quasars do not seem to be much different from those of relatively nearby AGNs; in particular, emission lines due to metals such as carbon, nitrogen, and iron are all present, which means that the material in the quasar BLRs must have been processed through at least one generation of massive stars within a few hundred million years of the Big Bang.

5.11.2 *The luminosity function*

It is customary to parameterize the quasar luminosity function as a double power law,

$$\Phi(L, z) = \frac{\Phi(L^*)}{(L/L^*)^{-\alpha} + (L/L^*)^{-\beta}}, \tag{5.37}$$

or, in magnitudes, as

$$\Phi(M, z) = \frac{\Phi(M^*)}{10^{0.4(\alpha+1)(M-M^*)} + 10^{0.4(\beta+1)(M-M^*)}}, \tag{5.38}$$

where M^* and L^* refer to a redshift-dependent fiducial luminosity,† and $\Phi(L^*)$ and $\Phi(M^*)$ represent the comoving density (normalized with respect to $z = 0$) of AGNs at the fiducial luminosity. Figure 5.23 shows the quasar luminosity function for several redshift ranges between $z \approx 0.5$ (when the Universe was about 80% its current age) and $z \approx 2$ (when the Universe was about 20% its current age) from the 2DF survey. For the lowest-redshift bin in Figure 5.23 (i.e. $0.40 < z < 0.68$), $\alpha \approx -3.3$ and $\beta \approx -1.1$ in Equations (5.37) and (5.38).

A major goal of quasar surveys is to understand how the luminosity function evolves with time. There are two simple descriptions of quasar evolution since $z \approx 2$.

- *Pure density evolution* describes a situation in which the number of AGNs per unit comoving volume decreases at a similar rate for all luminosities; in other words, the density of quasars decreases as individual quasars "turn off" or enter radiatively inefficient states such as the black hole at the center of the Milky Way (which is, in fact, probably a dead AGN). In this case, the luminosity function as shown in Figure 5.23 evolves by translating directly downward with time. There are many problems with this particular scenario, among the most important being that pure density evolution greatly overpredicts the number of very luminous quasars that should be found locally.
- *Pure luminosity evolution* supposes that all quasars simply become less luminous with time; in other words, the luminosity function shown in Figure 5.23 evolves by

† The transformation between the two equations above is shown explicitly by Peterson (1997).

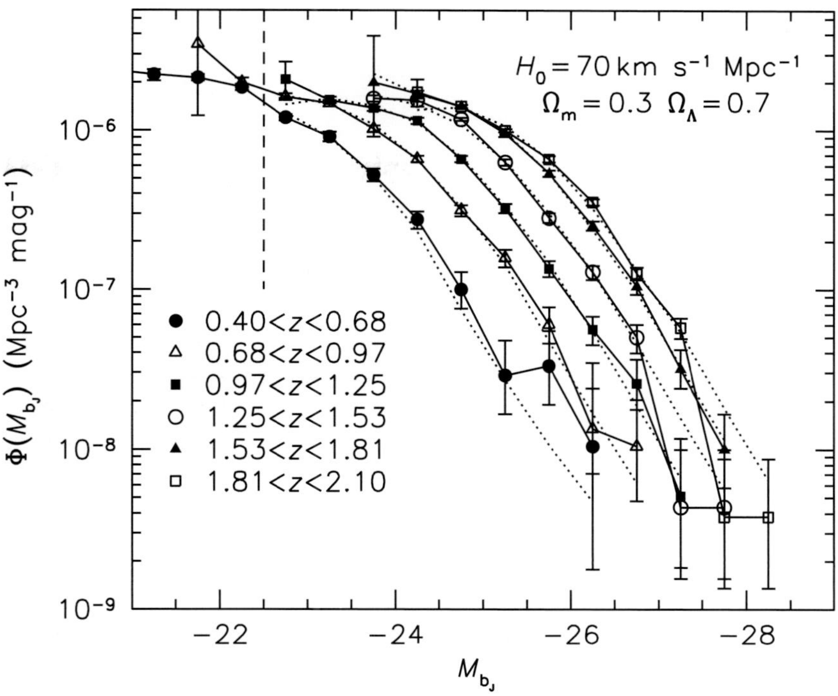

FIGURE 5.23. The quasar luminosity function from the 2DF for several redshift ranges. The luminosity function evolves in two ways: the overall number of quasars drops with time and the characteristic luminosities of the quasars also decrease with time. In the present epoch, quasars are much fewer in number and less luminous than were quasars in the past. From Croom *et al.* (2004).

translating directly to the left. There are also serious problems with this scenario, among them being that most local AGNs are currently radiating at Eddington ratios $\dot{m} \approx 0.1$, which would have required these AGNs to have been radiating at super-Eddington rates, $\dot{m} \gg 1$, at $z \approx 2$.

Close inspection of Figure 5.23 reveals two important things.

(i) The number of extremely bright quasars ($M_B < -26$ mag) decreases dramatically between $z \approx 2$ and $z \approx 0.5$.

(ii) There is a clear break in the slope of the luminosity function at some characteristic magnitude M^* that evolves towards lower luminosity between $z \approx 2$ and $z \approx 0.5$.

Since the quasar era at $z \approx 2$, the quasar population has diminished in number and in typical luminosity. This is shown even more dramatically in Figure 5.24, which clearly demonstrates that the epoch at which the maximum space density occurs is luminosity-dependent: more-luminous objects reach their maximum density earliest, at the highest redshifts. This phenomenon is often referred to as "cosmic downsizing," to indicate that the AGN population becomes dominated by lower-luminosity objects with time.

5.11.3 *Masses of distant quasars*

We have already noted the difficulty of measuring the black-hole masses of AGNs. In principle, reverberation mapping should still be viable even for distant AGNs since it does not depend on angular resolution. However, problems become apparent at higher redshift and luminosity. The BLR is larger in higher-luminosity objects, so the

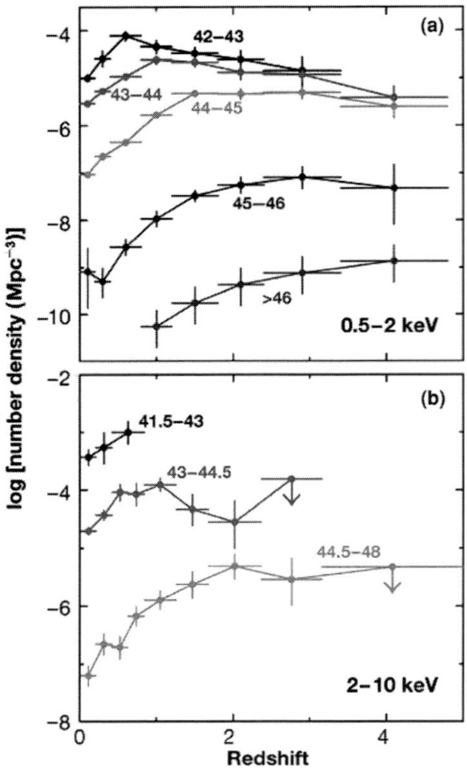

FIGURE 5.24. The evolution of the soft-X-ray (a) and hard-X-ray (b) luminosity functions for quasars. What this shows is that the higher-luminosity objects reach their highest space density at earlier epochs (higher redshifts) than do lower-luminosity objects. Labels are log of 2–10 keV flux in erg s^{-1}. Adapted from Brandt & Hasinger (2005).

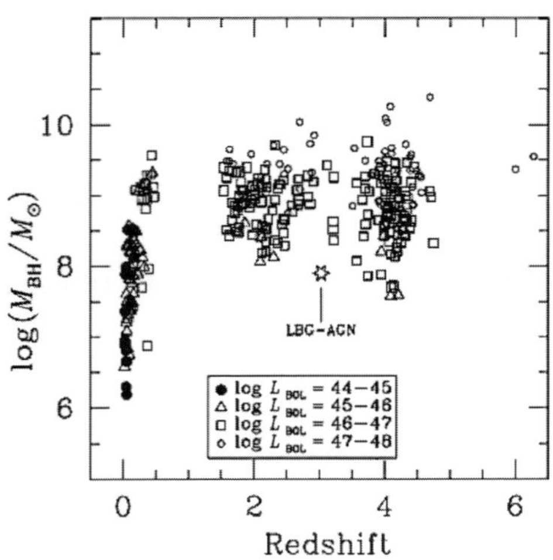

FIGURE 5.25. Distribution of black-hole masses with redshift. Note that black holes with $M > 10^9\,M_{\odot}$ are found at redshifts $z \approx 6$. Adapted from Vestergaard (2006).

emission-line response times are longer and are more geometrically diluted. Also, higher-luminosity sources have lower amplitudes of variability, so the variability signatures are weaker. Moreover, the light curves suffer $1 + z$ time dilation, requiring even longer reverberation programs.

Although direct measurement of the masses of black holes in distant quasars becomes very difficult, there are secondary methods that we can apply, notably by using the BLR radius–luminosity relationship discussed earlier. The BLR radius can be estimated from the quasar luminosity and the corresponding line width can be measured directly. From these two measurements it is possible to estimate the black-hole mass from a single spectrum. The great utility of this is that the masses of large numbers of AGNs can be estimated through measurements made from single spectra, thus by-passing the laborious reverberation-mapping process. Masses of large populations of AGNs have been measured by several investigators (Vestergaard 2002, 2004; McLure & Jarvis 2002; Kollmeier *et al.* 2006; Vestergaard & Peterson 2006) in this way. The example shown in Figure 5.25 shows the remarkable result that supermassive black holes with masses $M > 10^9\, M_\odot$ were already in place by $z \approx 4$–6.

Acknowledgments

I wish to thank the organizers, students, and other lecturers for making the XVIII Winter School a valuable and enjoyable experience. I am grateful to The Ohio State University for support of this work through NSF grant AST-0604066.

REFERENCES

Alexander, T. 1997, in *Astronomical Time Series*, ed. D. Maoz, A. Sternberg, & E. M. Leibowitz (Dordrecht: Kluwer), p. 163

Ambartsumian, V. A. 1968, in *Non-Stable Phenomena in Galaxies*, (Yerevan: Publishing House of the Armenian Academy of Sciences), p. 11

Ambartsumian, V. A. 1971, in *Nuclei of Galaxies*, ed. D. J. K. O'Connell (New York: American Elsevier), p. 1

Andrillat, Y., Souffrin, S. 1968, ApLett, 1, 111

Antonucci, R. R. J., Miller, J. S. 1984, ApJ, 297, 621

Atwood, B., Baldwin, J. A., Carswell, R. F. 1982, ApJ, 257, 559

Baldwin, J. A. 1977, ApJ, 214, 679

Baldwin, J. A., Ferland, G. J., Korista, K. T., Hamann, F., Dietrich, M. 2003, ApJ, 582, 590

Baldwin, J. A., Ferland, G. J., Korista, K. T., Verner, D. 1995, ApJ, 455, L119

Baldwin, J. A., Phillips, M. M., Terlevich, R. 1981, PASP, 93, 5

Barvainis, R. 1987, ApJ, 320, 537

Bennert, N. *et al.* 2006, New Astron. Rev., 50, 708

Bentz, M. C., Peterson, B. M., Pogge, R. W., Vestergaard, M., Onken, C. A. 2006a, ApJ, 644, 133

Bentz, M. C. *et al.* 2006b, ApJ, 651, 775

Bentz, M. C. *et al.* 2007, ApJ, in press

Blandford, R. D., McKee, C. F. 1982, ApJ, 255, 419

Boller, T, Brandt, W. N., Fink, H. 1996, A&A, 305, 53

Boroson, T. A. 2002, ApJ, 565, 78

Boroson, T. A., Green, R. F. 1992, ApJS, 80, 109

Bracewell, R. 1965, *The Fourier Transform and Its Applications* (New York: McGraw-Hill), p. 108

Brandt, W. N., Hasinger, G. 2005, ARAA, 43, 827

Cackett, E., Horne, K. 2006, MNRAS, 365, 1180

Capriotti, E. R., Foltz, C. B., Byard, P. L. 1981, ApJ, 245, 396

Clavel, J., Wamsteker, W., Glass, I. S. 1989, ApJ, 337, 236

Clavel, J. *et al.* 1991, ApJ, 366, 64

Collin, S., Kawaguchi, T., Peterson, B. M., Vestergaard, M. 2006, A&A, 456, 75

Collin-Souffrin, S. 1986, A&A, 166, 115

Collin-Souffrin, S., Joly, M., Pequignot, D., Dumont, S. 1986, A&A, 166, 27

Croom, S. M. *et al.* 2004, MNRAS, 349, 1397

Davidson, K., Netzer, H. 1979, Rev. Mod. Phys., 51, 715

Denney, K. D. *et al.* 2006, ApJ, 653, 152

DeRobertis, M., Osterbrock, D. E. 1984, ApJ, 286, 171

Edelson, R. A., Krolik, J. H. 1998, ApJ, 33, 646

Eilek, J. A. 1975, "Cosmic Ray Acceleration of Gas in Active Galactic Nuclei" PhD dissertation, University of British Columbia

Eilek, J. A. 1977, ApJ, 212, 278

Elitzur, M. 2006, New Astron. Rev., 50, 728

Elitzur, M., Nenkova, M., Ivezić, Z. 2004, in *The Neutral ISM in Starburst Galaxies*, ed. S. Aalto, S. Huttemeister, & A. Pedlar (San Francisco, CA: Astronomical Society of the Pacific), p. 242

Elitzur, M., Shlosman, I. 2006, ApJ, 648, 101

Elvis, M. 2000, ApJ, 545, 63

Elvis, M. *et al.* 1994, ApJS, 95, 1

Espey, B. R. *et al.* 1994, ApJ, 434, 484

Fan, X. 2006, New Astron. Rev., 50, 665

Fan, X. *et al.* 2001, AJ, 121, 54

Ferland, G. J., Mushotzky, R. F. 1982, ApJ, 262, 564

Ferland, G. J., Osterbrock, D. E. 1986, ApJ, 300, 658

Ferrarese, L. 2002, in *Current High-Energy Emission Around Black Holes*, ed. C.-H. Lee & H.-Y. Chang (Singapore: World Scientific), p. 3

Ferrarese, L., Merritt, D. 2000, ApJ, 539, L9

Ferrarese, L. *et al.* 2001, ApJ, 555, L79

Filippenko, A. V., Halpern, J. P. 1984, ApJ, 285, 458

Fitch, W. S., Pacholczyk, A. G., Weymann, R. J. 1967, ApJ, 150, L67

Francis, P. J. *et al.* 1991, ApJ, 373, 465

Fromerth, M. J., Melia, F. 2000, ApJ, 533, 172

Gaskell, C. M., Peterson, B. M. 1987, ApJS, 65, 1

Gaskell, C. M., Sparke, L. S. 1986, ApJ, 305, 175

Gebhardt, K. *et al.* 2000a, ApJ, 539, L13

Gebhardt, K. *et al.* 2000b, ApJ, 543, L5

Ghosh, H. *et al.* 2006, ApJ, 656, 105

Groves, B., Heckman, T. M., Kauffmann, G. 2006, MNRAS, 371, 1559

Heckman, T. M. 1980, A&A, 87, 152

Hewitt, A., Burbidge, G. 1993, ApJS, 87, 451

Horne, K. 1994, in *Reverberation Mapping of the Broad-Line Region in Active Galactic Nuclei*, ed. P. M. Gondhalekar, K. Horne, & B. M. Peterson (San Francisco, CA: Astronomical Society of the Pacific), p. 23

Horne, K., Peterson, B. M., Collier, S., Netzer, H. 2004, PASP, 116, 465

Kaspi, S. *et al.* 2000, ApJ, 533, 631

Kaspi, S. *et al.* 2005, ApJ, 629, 61

Kauffmann, G. *et al.* 2003, MNRAS, 346, 1055

Kewley, L. J., Groves, B., Kauffmann, G., Heckman, T. 2006, MNRAS, 372, 961

Khachikian, E. Ye., Weedman, D. W. 1974, ApJ, 192, 581

Kollatschny, W. 2003 A&A, 407, 461

Kollmeier, J. A. *et al.* 2006, ApJ, 648, 128

Koratkar, A. P., Gaskell, C. M. 1991 ApJ, 370, L61

Korista, K. T., Baldwin, J. A., Ferland, G., Verner, D. 1997, ApJS, 108, 401

Korista, K. T., Goad, M. R. 2004, ApJ, 606, 749

Korista, K. T. *et al.* 1995, ApJS, 97, 285

Laor, A. 2003, ApJ, 590, 86

Laor, A., Barth, A. J., Ho, L. C., Filippenko, A. V. 2006, ApJ, 636, 83

Lawrence, A. 1991, MNRAS, 252, 586

Leighly, K. M. 2000, New Astron. Rev., 44, 395

Lynden-Bell, D., 1969, Nature, 223, 690

Maoz, D., Markowitz, A., Edelson, R., Nandra, K. 2002, AJ, 124, 1988

Mathur, S. *et al.* 2000, ApJ, 533, L79

McLure, R. J., Jarvis, M. J. 2002, MNRAS, 337, 109

Miller, J. S., Goodrich, R. W., Mathews, W. G. 1991, ApJ, 378, 47

Nelson, C. H. *et al.* 2004, ApJ, 615, 652

Netzer, H. 1985, ApJ, 289, 451

Onken, C. A., Peterson, B. M. 2002 ApJ, 572, 746

Onken, C. A. *et al.* 2004, ApJ, 615, 645

Osmer, P. S. 1982, ApJ, 253, 28

Osterbrock, D. E. 1978, Proc. Natl. Acad. Sci. USA, 75, 540

Osterbrock, D. E., Pogge, R. W. 1985, ApJ, 297, 166

Pelat, D., Alloin, D., Fosbury, R. A. E. 1981, MNRAS, 195, 787

Peterson, B. M. 1987, ApJ, 312, 79

Peterson, B. M. 1993, PASP, 105, 247

Peterson, B. M. 1997, *An Introduction to Active Galactic Nuclei* (Cambridge: Cambridge University Press)

Peterson, B. M. 2001, in *Advanced Lectures on the Starburst–AGN Connection*, ed. I. Aretxaga, D. Kunth, & R. Mújica (Singapore: World Scientific), p. 3

Peterson, B. M., Wandel, A. 1999, ApJ, 521, L95

Peterson, B. M., Wandel, A. 2000 ApJ, 540, L13

Peterson, B. M., Wanders, I., Horne, K., Collier, S., Alexander, T., Kaspi, S. 1998, PASP, 110, 660

Peterson, B. M. *et al.* 1991, ApJ, 368, 119

Peterson, B. M. *et al.* 2002, ApJ, 581, 197

Peterson, B. M. *et al.* 2004, ApJ, 613, 682

Pogge, R. W. 1988, ApJ, 328, 519

Pogge, R. W. 2000, New Astron. Rev., 44, 381

Pickles, A. J. 1998, PASP, 110, 863

Prieto, M. A., Freundling, W. 1993, ApJ, 418, 668

Reynolds, C. S., Nowak, M. A. 2003, Phys. Rep., 377, 389

Reynolds, C. S., Young, A. J., Begelman, M. C., Fabian, A. C. 1999, ApJ, 514, 164

Sandage, A. 1965, ApJ, 141, 1560

Sanders, D. B. *et al.* 1989, ApJ, 347, 29

Schmidt, M. 1963, Nature, 197, 1040

Schmidt, M. 1968, ApJ, 151, 393

Schmidt, M., Schneider, D. P., Gunn, J. E. 1995, AJ, 110, 68

Schmitt, H. R. 1998, ApJ, 506, 347

Sergeev, S. G., Doroshenko, V. T., Dzyuba, S. A., Peterson, B. M., Pogge, R. W., Pronik, V. I. 2007, in preparation

Seyfert, C. 1943, ApJ, 97, 28

Shakura, N. I., Sunyaev, R. A. 1973, A&A, 24, 337

Simpson, C. 2005, MNRAS, 360, 565

Smith, H. J., Hoffleit, D. 1963, Nature, 198, 650

Storchi-Bergmann, T. *et al.* 2003, ApJ, 598, 956

Suganuma, M. *et al.* 2006, ApJ, 639, 46

Tremaine, S. *et al.* 2002, ApJ, 574, 740

Ulrich, M.-H., Horne, K. 1996, MNRAS, 283, 748

Urry, C. M., Padovani, P. 1995, PASP, 107, 803

Vanden Berk, D. E. *et al.* 2004, in *AGN Physics with the Sloan Digital Sky Survey*, ed. G. T. Richards and P. B. Hall (San Francisco, CA: Astronomical Society of the Pacific), p. 21.

Vestergaard, M. 2002, ApJ, 571, 733

Vestergaard, M. 2004, ApJ, 601, 676

Vestergaard, M., Peterson, B. M. 2005, ApJ, 625, 688

Vestergaard, M., Peterson, B. M. 2006, ApJ, 641, 133

Wandel, A., Peterson, B. M., Malkan, M. A. 1999 ApJ, 526, 579

Weedman, D. W. 1974, BAAS, 6, 441

Weedman, D. W. 1976, QJRAS, 16, 227

Welsh, W. F. 2000, PASP, 111, 1347

White, R. J., Peterson, B. M. 1994, PASP, 106, 879

Woltjer, L. 1959, ApJ, 130, 38

York, D. G. *et al.* 2000, AJ, 120, 1579

Zakamska, N. L. *et al.* 2003, AJ, 126, 2125

6. Chemical evolution

FRANCESCA MATTEUCCI

6.1. Basic assumptions and equations of chemical evolution

To build galaxy chemical-evolution models one needs to elucidate a number of hypotheses and make assumptions on the basic ingredients.

6.1.1 *The basic ingredients*

- Initial conditions: whether the mass of gas out of which stars will form is all present initially or will be accreted later on; the chemical composition of the initial gas (primordial or already enriched by a pregalactic stellar generation).
- The birthrate function:

$$B(M,t) = \psi(t)\varphi(M), \tag{6.1}$$

  where

$$\psi(t) = \text{SFR} \tag{6.2}$$

  is the star-formation (SF) rate (SFR) and

$$\varphi(M) = \text{IMF} \tag{6.3}$$

  is the initial mass function (IMF).
- Stellar evolution and nucleosynthesis: stellar yields, yields per stellar generation.
- Supplementary parameters: infall, outflow, radial flows.

6.1.2 *The star-formation rate*

Here we will summarize the most common parameterizations for the SFR in galaxies, as adopted by chemical-evolution models.

- Constant in space and time and equal to the estimated present-day SFR. For example, for the local disk, the present-day SFR is (2–5)$M_\odot$ pc^{-2} Gyr^{-1} (Boissier & Prantzos 1999).
- Exponentially decreasing:

$$\text{SFR} = \nu e^{-t/\tau_*} \tag{6.4}$$

  with $\tau_* =$ 5–15 Gyr (Tosi 1988). The quantity ν is a parameter that we call the efficiency of SF since it represents the SFR per unit mass of gas and is expressed in Gyr^{-1}.
- The most-used SFR is the Schmidt (1959) law, which assumes a dependence on the gas density, in particular

$$\text{SFR} = \nu \sigma_{\text{gas}}^k \tag{6.5}$$

  where $k = 1.4 \pm 0.15$, as suggested by Kennicutt (1998) in a study of local star-forming galaxies.
- Some variations of the Schmidt law with a dependence also on the total mass have been suggested, for example by Dopita & Ryder (1974). This formulation takes into account the feedback mechanism acting between supernovae (SNe) and stellar winds

The Emission-Line Universe, ed. J. Cepa. Published by Cambridge University Press.

injecting energy into the interstellar medium (ISM) and the galactic potential well. In other words, the SF process is regulated by the fact that, in a region of recent SF, the gas is too hot to form stars and is easily removed from that region. Before new stars could form, the gas needs to cool and collapse back into the star-forming region; this process depends on the potential well and therefore on the total mass density:

$$\mathrm{SFR} = \nu \sigma_{\mathrm{tot}}^{k_1} \sigma_{\mathrm{gas}}^{k_2} \tag{6.6}$$

with $k_1 = 0.5$ and $k_2 = 1.5$.

- Kennicutt (1998) also suggested, as an alternative to the Schmidt law to fit the data, the following relation:

$$\mathrm{SFR} = 0.017 \Omega_{\mathrm{gas}} \sigma_{\mathrm{gas}} \propto R^{-1} \sigma_{\mathrm{gas}} \tag{6.7}$$

with Ω_{gas} being the angular rotation speed of gas.

- Finally a SFR induced by spiral density waves was suggested by Wyse & Silk (1989):

$$\mathrm{SFR} = \nu V(R) R^{-1} \sigma_{\mathrm{gas}}^{1.5} \tag{6.8}$$

with R being the galactocentric distance and $V(R)$ the gas rotation velocity.

6.1.3 *The initial mass function*

The IMF is a probability function describing the distribution of stars as a function of mass. The present-day mass function is derived for the stars in the Solar vicinity by counting the main-sequence stars as a function of magnitude and then applying the mass–luminosity relation holding for main-sequence stars to derive the distribution of stars as a function of mass. In order to derive the IMF one has then to make assumptions on the past history of SF.

The derived IMF is normally approximated by a power law:

$$\varphi(M) \mathrm{d}M = a M^{-(1+x)} \mathrm{d}M, \tag{6.9}$$

where $\varphi(M)$ is the number of stars with masses in the interval M, $M + \mathrm{d}M$.

Salpeter (1955) proposed a one-slope IMF ($x = 1.35$) valid for stars with $M > 10 M_\odot$. Multi-slope (x_1, x_2, ...) IMFs were suggested later on always for the Solar vicinity (Scalo 1986, 1998; Kroupa *et al.* 1993; Chabrier 2003). The IMF is generally normalized as

$$a \int_{0.1}^{100} M \varphi(M) \mathrm{d}M = 1, \tag{6.10}$$

where a is the normalization constant and the assumed interval of integration is $(0.1\text{–}100) M_\odot$.

The IMF is generally considered constant in space and time with some exceptions such as the IMF suggested by Larson (1998) with

$$x = 1.35 (1 + m/m_1)^{-1}, \tag{6.11}$$

where m_1 is variable typical mass and is associated with the Jeans mass. This IMF predicts then that m_1 is a decreasing function of time.

6.1.4 *The infall rate*

For the rate of gas accretion there are in the literature several parameterizations.

- The infall rate is constant in space and time and equal to the present-day infall rate measured in the Galaxy ($\sim 1.0 M_\odot\,\mathrm{yr}^{-1}$).

- The infall rate is variable in space and time, and the most common assumption is an exponential law (Chiosi 1980; Lacey & Fall 1985):

$$\mathrm{IR} = A(R)\mathrm{e}^{-t/\tau(R)} \tag{6.12}$$

with $\tau(R)$ constant or varying with the galactocentric distance. The parameter $A(R)$ is derived by fitting the present-day total surface mass density, $\sigma_{\mathrm{tot}}(t_{\mathrm{G}})$, at any specific galactocentric radius R.
- For the formation of the Milky Way two episodes of infall have been suggested (Chiappini *et al.* 1997), such that during the first infall episode the stellar halo forms whereas during the second infall episode the disk forms. This particular infall law gives a good representation of the formation of the Milky Way. The proposed two-infall law is

$$\mathrm{IR} = A(R)\mathrm{e}^{-t/\tau_{\mathrm{H}}(R)} + B(R)\mathrm{e}^{-(t-t_{\mathrm{max}})/\tau_{\mathrm{D}}(R)}, \tag{6.13}$$

where $\tau_{\mathrm{H}}(R)$ is the timescale for the formation of the halo which could be constant or vary with galactocentric distance. The quantity $\tau_{\mathrm{D}}(R)$ is the timescale for the formation of the disk and is a function of the galactocentric distance; in most of the models it is assumed to increase with R (e.g. Matteucci & François 1989).
- More recently, Prantzos (2003) suggested a Gaussian law with a peak at 0.1 Gyr and a FWHM of 0.04 Gyr for the formation of the stellar halo.

6.1.5 *The outflow rate*

The so-called galactic winds occur when the thermal energy of the gas in galaxies exceeds its potential energy. Generally, gas outflows are called winds when the gas is lost forever from the galaxy. Only detailed dynamical simulations can suggest whether there is a wind or just an outflow of gas that will sooner or later fall back again into the galaxy. In chemical-evolution models, galactic winds can be sudden or continuous. If they are sudden, the mass is assumed to be lost in a very short interval of time and the galaxy is voided of all the gas; if they are continuous, one has to assume the rate of gas loss. Generally, in chemical-evolution models (Bradamante *et al.* 1998) and also in cosmological simulations (Springel & Hernquist, 2003) it is assumed that the rate of gas loss is several times the SFR:

$$W = -\lambda \mathrm{SFR}, \tag{6.14}$$

where λ is a free parameter with the meaning of wind efficiency. This particular formulation for the galactic wind rate is confirmed by observational findings (Martin 1999).

6.1.6 *Stellar evolution and nucleosynthesis: the stellar yields*

Here we summarize the various contributions to the element production by stars of all masses.

- Brown dwarfs ($M < M_{\mathrm{L}}$, $M_{\mathrm{L}} = (0.08\text{–}0.09)M_{\odot}$) are objects that never ignite H and their lifetimes are larger than the age of the Universe. They are contributing to the locking up of mass.
- Low-mass stars ($0.5 \leq M/M_{\odot} \leq M_{\mathrm{HeF}}$) ($(1.85\text{–}2.2)M_{\odot}$) ignite He explosively but without destroying themselves and then become C–O white dwarfs (WDs). If $M < 0.5M_{\odot}$, they become He WDs. Their lifetimes range from several times 10^9 years to several Hubble times!
- Intermediate-mass stars ($M_{\mathrm{HeF}} \leq M/M_{\odot} \leq M_{\mathrm{up}}$) ignite He quiescently. The mass M_{up} is the limiting mass for the formation of a C–O degenerate core and is in

the range (5–9)$M_\odot$, depending on which stellar-evolution calculations one believes. Lifetimes are from several times 10^7 to 10^9 years. They die as C–O WDs if not in binary systems. In binary systems they can give rise to cataclysmic variables such as novae and Type Ia SNe.

- Massive stars ($M > M_{\rm up}$). We distinguish here several cases.
  $M_{\rm up} \leq M/M_\odot \leq$ 10–12. Stars with main-sequence masses in this range end up as electron-capture SNe leaving neutron stars as remnants. These SNe will appear as Type II SNe with H lines in their spectra.
  10–12 $\leq M/M_\odot \leq M_{\rm WR}$, (with $M_{\rm WR} \sim$ (20–40)$M_\odot$ being the limiting mass for the formation of a Wolf–Rayet (WR) star). Stars in this mass range end their lives as core-collapse SNe (Type II) leaving a neutron star or a black hole as remnants.
  $M_{WR} \leq M/M_\odot \leq 100$. Stars in this mass range are probably exploding as Type Ib/c SNe, which lack H lines in their spectra. Their lifetimes are of the order of $\sim 10^6$ years.
- Very massive stars ($M > 100 M_\odot$) should explode by means of instability due to "pair creation" and are called *pair-creation* SNe. In fact, at $T \sim 2 \times 10^9$ K a large portion of the gravitational energy goes into creation of pairs (e^+, e^-), and the star becomes unstable and explodes. They leave no remnants and their lifetimes are $<10^6$ years. Probably these very massive stars formed only when the metal content was almost zero (Population III stars) (Schneider *et al.* 2004).

All the elements with mass number A from 12 to 60 have been formed in stars during quiescent burning. Stars transform H into He and then He into heavier elements until the Fe-peak elements, whereupon the binding energy per nucleon reaches a maximum and the nuclear fusion reactions stop.

H is transformed into He through the proton–proton chain or the CNO cycle, then ^{4}He is transformed into ^{12}C through the triple-α reaction.

Elements heavier than ^{12}C are then produced by synthesis of α-particles. They are called α-elements (O, Ne, Mg, Si and others).

The last main burning in stars is the ^{28}Si-burning which produces ^{56}Ni, which then decays into ^{56}Co and ^{56}Fe. Si-burning can be quiescent or explosive (depending on the temperature).

Explosive nucleosynthesis occurring during SN explosions mainly produces Fe-peak elements. Elements originating from s- and r-processes (with $A > 60$ up to Th and U) are formed by means of slow or rapid (relative to the β-decay) neutron capture by Fe seed nuclei; s-processing occurs during quiescent He-burning whereas r-processing occurs during SN explosions.

6.1.7 *Type Ia SN progenitors*

The Type Ia SNe, which lack H lines in their spectra, are believed to originate from WDs in binary systems and to be the major producers of Fe in the Universe. The models proposed are basically the following two.

- **The singly degenerate scenario (SDS)**, with a WD plus a main-sequence or red giant star, as originally suggested by Whelan and Iben (1973). The explosion (C-deflagration) occurs when the C–O WD reaches the Chandrasekhar mass, $M_{\rm Ch} = \sim 1.44 M_\odot$, after accreting material from the companion. In this model, the clock to the explosion is given by the lifetime of the companion of the WD (namely the less-massive star in the system). It is interesting to define the minimum timescale for the explosion which is given by the lifetime of a $8 M_\odot$ star, namely $t_{\rm SN\,Ia_{min}} = 0.03$ Gyr

(Greggio and Renzini 1983). Recent observations in radio galaxies by Mannucci *et al.* (2005, 2006) seem to confirm the existence of such prompt Type Ia SNe.

- **The doubly degenerate scenario (DDS)**, whereby the merging of two C–O WDs of mass $\sim 0.7 M_\odot$, due to loss of angular momentum as a consequence of gravitational-wave radiation, produces C-deflagration (Iben and Tutukov 1984). In this case the clock to the explosion is given by the lifetime of the secondary star, as above, plus the gravitational time delay, namely the time necessary for the two WDs to merge. The minimum time for the explosion is $t_{\mathrm{SN\,Ia_{min}}} = 0.03 + \Delta t_{\mathrm{grav}} = 0.04$ Gyr (Tornambè 1989).

Some variations of the above scenarios have been proposed, such as the model by Hachisu *et al.* (1996, 1999), which is based on the singly degenerate scenario whereby a wind from the WD is considered. Such a wind stabilizes the accretion from the companion and introduces a metallicity effect. In particular, the wind necessary to this model occurs only if the systems have metallicity ([Fe/H] < -1.0). This implies that the minimum time to the explosion is larger than in the previous cases. In particular, $t_{\mathrm{SN\,Ia_{min}}} = 0.33$ Gyr, which is the lifetime of the more-massive secondary considered ($2.3\,M_\odot$) plus the metallicity delay which depends on the chemical-evolution model assumed.

6.1.8 *Yields per stellar generation*

Under the assumption of the instantaneous-recycling approximation (IRA) which states that all stars more massive than $M_\odot$ die immediately, whereas all stars with masses lower than $M_\odot$ live forever, one can define the yield per stellar generation (Tinsley 1980);

$$y_i = \frac{1}{1-R} \int_1^\infty m p_{im} \varphi(m) \mathrm{d}m, \tag{6.15}$$

where p_{im} is the stellar yield of the element i, namely the newly formed and ejected amount of element i, produced by a star of mass m.

The quantity R is the so-called returned fraction:

$$R = \int_1^\infty (m - M_{\mathrm{rem}}) \varphi(m) \mathrm{d}m, \tag{6.16}$$

where M_{rem} is the mass of stellar remnants (white dwarfs, neutron stars, black holes). This is the total mass of gas restored into the ISM by an entire stellar generation.

6.1.9 *Analytical models*

The *simple model* for the chemical evolution of the Solar neighborhood is the simplest approach with which to model chemical evolution. The Solar neighborhood is assumed to be a cylinder of radius 1 kpc centered around the Sun.

The basic assumptions of the simple model are as follows:

- the system is one-zone and closed, with no inflows or outflows, with the total mass present since the beginning;
- the initial gas is primordial (i.e. there are no metals);
- the instantaneous-recycling approximation holds;
- the IMF, $\varphi(m)$, is assumed to be constant in time; and
- the gas is well mixed at any time (the IMA).

The simple model fails to describe the evolution of the Milky Way (G-dwarf metallicity distribution, elements produced on long timescales, and abundance ratios) because at least two of the above assumptions are manifestly wrong, especially if one intends to model the evolution of the abundances of elements produced on long timescales, such as Fe; in particular, the assumptions of the closed boxiness and the IRA. However, it is interesting to know the solution of the simple model and its implications. X_i is the abundance by mass of an element i. If $X_i \ll 1$, which is generally true for metals, we obtain the solution of the simple model. This solution is obtained analytically by ignoring the stellar lifetimes:

$$X_i = y_i \ln\left(\frac{1}{G}\right), \tag{6.17}$$

where $G = M_{\rm gas}/M_{\rm tot}$ and y_i is the yield per stellar generation, as defined above, otherwise called the *effective yield*. In particular, the effective yield is defined as

$$y_{i_{\rm eff}} = \frac{X_i}{\ln(1/G)}, \tag{6.18}$$

namely the yield that the system would have if it behaved as the simple closed-box model. This means that, if $y_{i_{\rm eff}} > y_i$, then the actual system has attained a higher abundance for the element i at a given gas fraction G. Generally, in the IRA, we can assume

$$\frac{X_i}{X_j} = \frac{y_i}{y_j}, \tag{6.19}$$

which means that the ratio of two element abundances is always equal to the ratio of their yields. This is no longer true when the IRA is relaxed. In fact, relaxing the IRA is necessary in order to study in detail the evolution of the abundances of single elements.

One can obtain analytical solutions also in the presence of infall and/or outflow but the necessary condition is to assume the IRA. Matteucci & Chiosi (1983) found solutions for models with outflow and infall and Matteucci (2001) found a solution for a model with infall and outflow acting at the same time. The main assumption in the model with outflow but no infall is that the outflow rate is

$$W(t) = \lambda(1 - R)\psi(t), \tag{6.20}$$

where $\lambda \geq 0$ is the wind parameter.

The solution of this model is

$$X_i = \frac{y_i}{(1 + \lambda)} \ln[(1 + \lambda)G^{-1} - \lambda]. \tag{6.21}$$

For $\lambda = 0$ the equation becomes that of the simple model (1.17).

The solution of the equation of metals for a model without wind but with a primordial infalling material ($X_{A_i} = 0$) at a rate

$$A(t) = \Lambda(1 - R)\psi(t) \tag{6.22}$$

and $\Lambda \neq 1$ is

$$X_i = \frac{y_i}{\Lambda}[1 - (\Lambda - (\Lambda - 1)G^{-1})^{-\Lambda/(1-\Lambda)}]. \tag{6.23}$$

For $\Lambda = 1$ one obtains the well-known case of *extreme infall* studied by Larson (1972), whose solution is

$$X_i = y_i[1 - e^{-(G^{-1}-1)}]. \tag{6.24}$$

This extreme-infall solution shows that when $G \to 0$ then $X_i \to y_i$.

6.1.10 *Numerical models*

Numerical models relax the IRA and closed-boxiness but generally retain the constancy of $\varphi(m)$ and the IMA.

If G_i is the mass fraction of gas in the form of an element i, we can write

$$\begin{aligned}
\dot{G}_i(t) = & -\psi(t)X_i(t) \\
& + \int_{M_{\rm L}}^{M_{\rm Bm}} \psi(t-\tau_m)Q_{mi}(t-\tau_m)\varphi(m)\mathrm{d}m \\
& + A\int_{M_{\rm Bm}}^{M_{\rm BM}} \phi(m) \\
& \times \left[\int_{\mu_{\rm min}}^{0.5} f(\gamma)\psi(t-\tau_{m2})Q_{mi}(t-\tau_{m2})\mathrm{d}\gamma\right]\mathrm{d}m \\
& + B\int_{M_{\rm Bm}}^{M_{\rm BM}} \psi(t-\tau_m)Q_{mi}(t-\tau_m)\varphi(m)\mathrm{d}m \\
& + \int_{M_{\rm BM}}^{M_{\rm U}} \psi(t-\tau_m)Q_{mi}(t-\tau_m)\varphi(m)\mathrm{d}m \\
& + X_{A_i}A(t) - X_i(t)W(t),
\end{aligned} \tag{6.25}$$

where $B = 1 - A$, $A = 0.05$–0.09. The meaning of the A parameter is the fraction in the IMF of binary systems with those specific features required to give rise to Type Ia SNe, whereas B is the fraction of all the single stars and binary systems in the mass range of definition of the progenitors of Type Ia SNe. The values of A indicated above are correct for the evolution of the Solar vicinity, where an IMF of Scalo (1986, 1989) or Kroupa *et al.* (1993) type is adopted. If one adopts a flatter IMF such as the Salpeter (1955) one, then A is different. In the above equations the contribution of Type Ia SNe is contained in the third term on the right-hand side. The integral is taken over a range of masses going from 3 to 16 times $M_\odot$, which represents the total mass of binary systems able to produce Type Ia SNe in the framework of the SDS. There is also an integration over the mass distribution of binary systems; in particular, one considers the function $f(\gamma)$ where $\gamma = M_2/(M_1 + M_2)$, with M_1 and M_2 being the primary and secondary masses of the binary system, respectively. For more details see Matteucci & Greggio (1986) and Matteucci (2001). The functions $A(t)$ and $W(t)$ are the infall and wind rates, respectively. Finally, the quantity Q_{mi} represents the stellar yields (both processed and unprocessed material).

6.2. The Milky Way and other spirals

The Milky Way galaxy has four main stellar populations: (1) the halo stars with low metallicities (the most-common metallicity indicator in stars is $[\mathrm{Fe/H}] = \log(\mathrm{Fe/H})_* - \log(\mathrm{Fe/H})_\odot$) and eccentric orbits; (2) the bulge population with a large range of metallicities, dominated by random motions; (3) the thin-disk stars with an average metallicity $\langle[\mathrm{Fe/H}]\rangle = -0.5\,\mathrm{dex}$ and circular orbits; and finally (4) the thick stars which possess chemical and kinematical properties intermediate between those of the halo and those of the thin disk. The halo stars have average metallicities of $\langle[\mathrm{Fe/H}]\rangle = -1.5$ dex and a maximum metallicity of ~ -1.0 dex, although stars with [Fe/H] as high as -0.6 dex and halo kinematics are observed. The average metallicity of thin-disk stars is ~ -0.6 dex, whereas that of bulge stars is $\sim -0.2\,\mathrm{dex}$.

6.2.1 *The Galactic formation timescales*

The kinematical and chemical properties of the various Galactic stellar populations can be interpreted in terms of the Galaxy-formation mechanism. Eggen *et al.* (1962), in a cornerstone paper, suggested a rapid collapse for the formation of the Galaxy lasting $\sim 3 \times 10^8$ yr. This suggestion was based on a kinematical and chemical study of Solar-neighborhood stars. Later on, Searle & Zinn (1979) proposed a central collapse like the one proposed by Eggen *et al.* but also that the outer halo formed by merging of large fragments taking place over a considerable timescale >1 Gyr. More recently, Berman & Suchov (1991) proposed the so-called hot-Galaxy picture, with an initial strong burst of SF that inhibited further SF for a few Gyr while a strong Galactic wind was created.

From an historical point of view, the modeling of the Galactic chemical evolution has passed through the phases that I summarize in the following.

- SERIAL FORMATION
  The Galaxy is modeled by means of one accretion episode lasting for the entire Galactic lifetime, whereby halo, thick disk, and thin disk form in sequence as a continuous process. The obvious limit of this approach is that it does not allow us to predict the observed overlapping in metallicity between halo and thick-disk stars and between thick- and thin-disk stars, but it gives a fair representation of our Galaxy (e.g. Matteucci & François 1989).
- PARALLEL FORMATION
  In this formulation, the various Galactic components start at the same time and from the same gas but evolve at different rates (e.g. Pardi *et al.* 1995). It predicts overlapping of stars belonging to the different components but implies that the thick disk formed out of gas shed by the halo and that the thin disk formed out of gas shed by the thick disk, and this is at variance with the distribution of the stellar angular momentum per unit mass (Wyse & Gilmore 1992), which indicates that the disk did not form out of gas shed by the halo.
- TWO-INFALL FORMATION
  In this scenario, halo and disk formed out of two separate infall episodes (overlapping in metallicity is also predicted) (e.g. Chiappini *et al.* 1997; Chang *et al.* 1999). The first infall episode lasted no more than 1–2 Gyr, whereas the second, during which the thin disk formed, lasted much longer, with a timescale for the formation of the Solar vicinity of 6–8 Gyr (Chiappini *et al.* 1997; Boissier & Prantzos 1999).
- THE STOCHASTIC APPROACH
  Here the hypothesis is that, during the early halo phases ([Fe/H] < -3.0 dex), mixing was not efficient and, as a consequence, one should observe in low-metallicity halo stars the effects of pollution from single SNe (e.g. Tsujimoto *et al.* 1999; Argast *et al.* 2000; Oey 2000). These models predict a large spread for [Fe/H] < -3.0 dex, which is not observed, as shown by recent data with metallicities down to -4.0 dex (Cayrel *et al.* 2004); see later.

6.2.2 *The two-infall model*

The SFR adopted (see Figure 6.1) is Equation (6.6) with different SF efficiencies for the halo and disk, in particular $\nu_{\rm H} = 2.0\ {\rm Gyr}^{-1}$ and $\nu_{\rm D} = 1.0\ {\rm Gyr}^{-1}$, respectively. A threshold density ($\sigma_{\rm th} = 7\,M_{\odot}\,{\rm pc}^{-2}$) for the SFR is also assumed, in agreement with results from Kennicutt (1989, 1998).

In Figure 6.2 we show the SN (II and Ia) rates predicted by the two-infall model. Note that the Type Ia SN rate is calculated according to the SDS (Greggio & Renzini 1983; Matteucci & Recchi 2001). There is a delay between the Type II SN rate and the Type

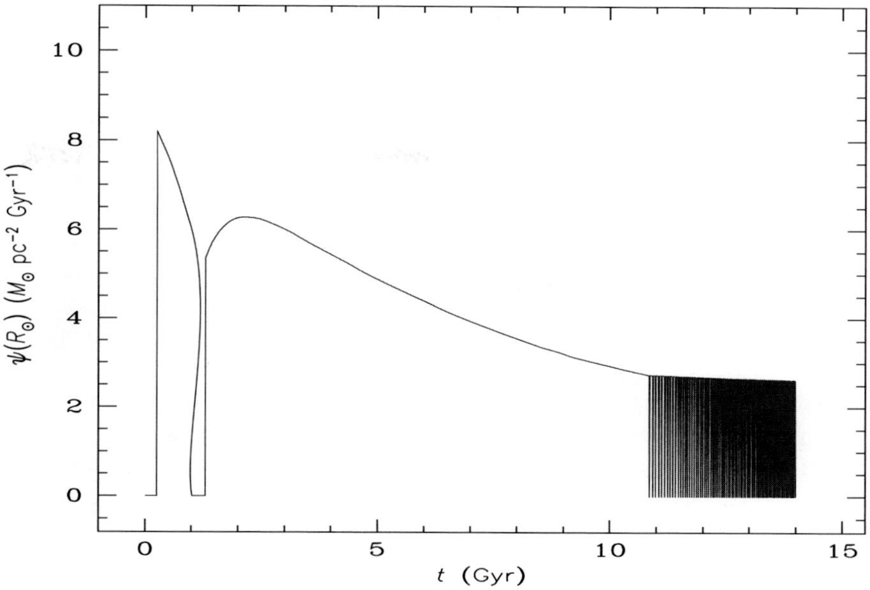

FIGURE 6.1. The predicted SFR in the Solar vicinity with the two-infall model, from Chiappini *et al.* (1997). The oscillating behavior at late times is due to the assumed threshold density for SF. The threshold gas density is also responsible for the gap in the SFR seen at around 1 Gyr.

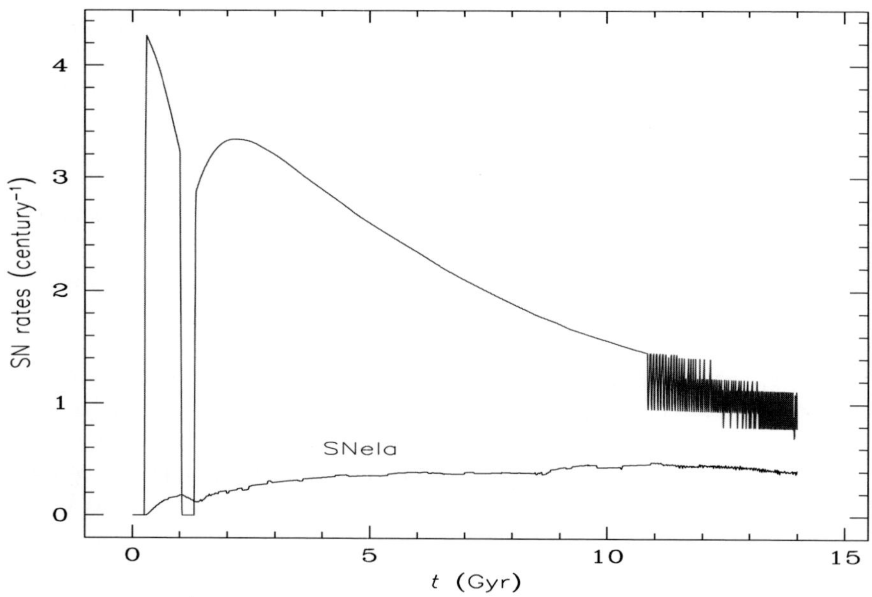

FIGURE 6.2. The predicted SN (Types II and Ia) rate in the Solar vicinity with the two-infall model. Figure from Chiappini *et al.* (1997).

Ia SN rate, and the Type II SN rate strictly follows the SFR, whereas the Type Ia SN rate is smoothly increasing.

François *et al.* (2004) compared the predictions of the two-infall model for the abundance ratios versus metallicity relations ([X/Fe] versus [Fe/H]), with the very recent and very accurate data of the project "First Stars" published by Cayrel *et al.* (2004). They adopted yields from the literature both for Type II and for Type Ia SNe and noticed that while for some elements (O, Fe, Si, Ca) the yields of Woosley & Weaver (1995) (hereafter WW95) reproduce the data fairly well, for the Fe-peak elements and heavier elements none of the available yields give a good agreement. Therefore, they varied empirically the yields of these elements in order to fit the data. In Figures 6.3 and 6.4 we show the predictions for α-elements (O, Mg, Si, Ca, Ti, K) plus some Fe-peak elements and Zn.

In Figure 6.4 we also show the ratios between the yields derived empirically by François *et al.* (2004), in order to obtain the excellent fits shown in the figures, and those of WW95 for massive stars. For some elements it was also necessary to change the yields from Type Ia SNe relative to the reference ones, which are those of Iwamoto *et al.* (1999) (hereafter I99).

In Figure 6.5 we show the predictions of chemical-evolution models for ^{12}C and ^{14}N compared with abundance data. The behavior of C gives a roughly constant [C/Fe] as a function of [Fe/H], although the amount of C seems to be slightly greater at very low metallicities, indicating that the bulk of these two elements comes from stars with the same lifetimes. The data in these figures, especially those for N, are old and do not concern very-metal-poor stars. Newer data concerning stars with [Fe/H] down to ~ -4.0 dex (Spite *et al.* 2005; Israelian *et al.* 2004) indicate that the [N/Fe] ratio continues to be high also at low metallicities, indicating a primary origin for N produced in massive stars. We recall here that we define as *primary* a chemical element that is produced in the stars starting from the H and He, whereas we define as *secondary* a chemical element that is formed from heavy elements already present in the star at its birth and not produced *in situ.* The model predictions shown in Figure 6.5 for C and N assume that the bulk of these elements is produced by low- and intermediate-mass stars (yields from van den Hoeck and Groenewegen (1997)) and that N is produced as a partly secondary and partly primary element. The N production from massive stars has only a secondary origin (yields from WW95). In Figure 6.5 we also show a model prediction in which N is considered to be a primary element in massive stars with the yields artificially increased. Recently, Chiappini *et al.* (2006) have shown that primary N produced by very-metal-poor fast-rotating massive stars can well reproduce the observations.

In summary, the comparison between model predictions and abundance data indicates the following scenario for the formation of heavy elements.

- ^{12}C and ^{14}N are mainly produced in low- and intermediate-mass stars ($0.8 \leq M/M_\odot \leq 8$). The amounts of primary and secondary N are still unknown and so is the fraction of C produced in massive stars. Primary N from massive stars seems to be required in order to reproduce the N abundance in low-metallicity halo stars.
- The α-elements originate in massive stars: the nucleosynthesis of O is rather well understood (there is agreement among authors); and the yields from WW95 as functions of metallicity produce excellent agreement with the observations for this particular element.
- Magnesium is generally underproduced by nucleosynthesis models. Taking the yields of WW95 as a reference, the Mg yields should be increased in stars with masses $M \leq 20 M_\odot$ and decreased in stars with $M > 20 M_\odot$ in order to fit the data. Silicon yields should be slightly increased in stars with masses $M > 40 M_\odot$.
- Fe originates mostly in Type Ia SNe. The Fe yields in massive stars are still unknown; WW95 metallicity-dependent yields overestimate Fe in stars of masses $< 30 M_\odot$. For this element, it is better to adopt the yields of WW95 for Solar metallicity.
- Fe-peak elements: the yields of Cr and Mn should be increased in stars of mass $(10\text{–}20) M_\odot$ relative to the yields of WW95, whereas the yield of Co should be

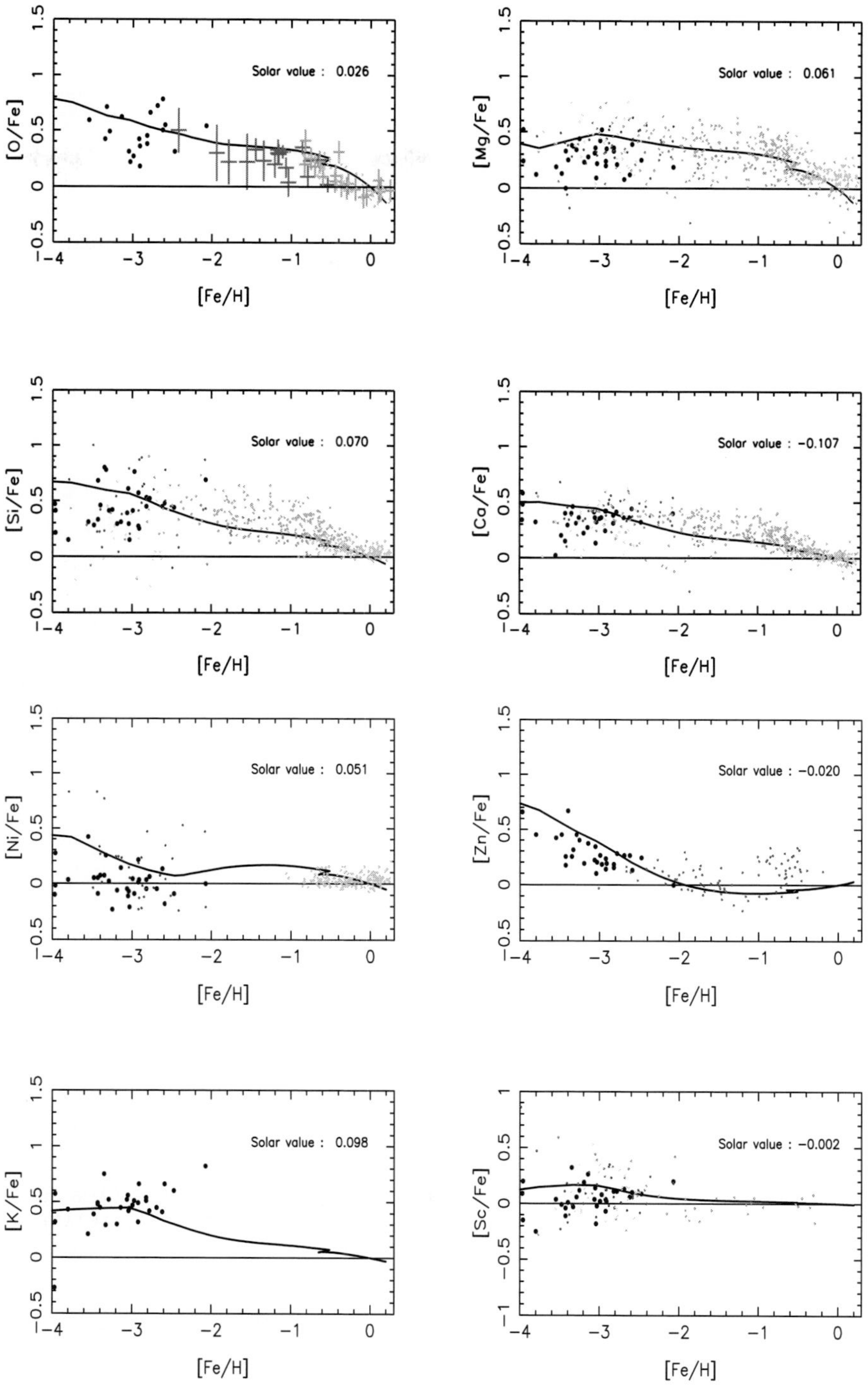

FIGURE 6.3. Predicted and observed [X/Fe] versus [Fe/H] for several α- and Fe-peak-elements plus Zn compared with a compilation of data. In particular the black dots are the recent high-resolution data from Cayrel *et al.* (2004). For the other data see references in François *et al.* (2004). The Solar value indicated in the upper right part of each figure is the predicted Solar value for the ratio [X/Fe]. The assumed Solar abundances are those of Grevesse & Sauval (1998) except that for oxygen, for which we take the value of Holweger (2001).

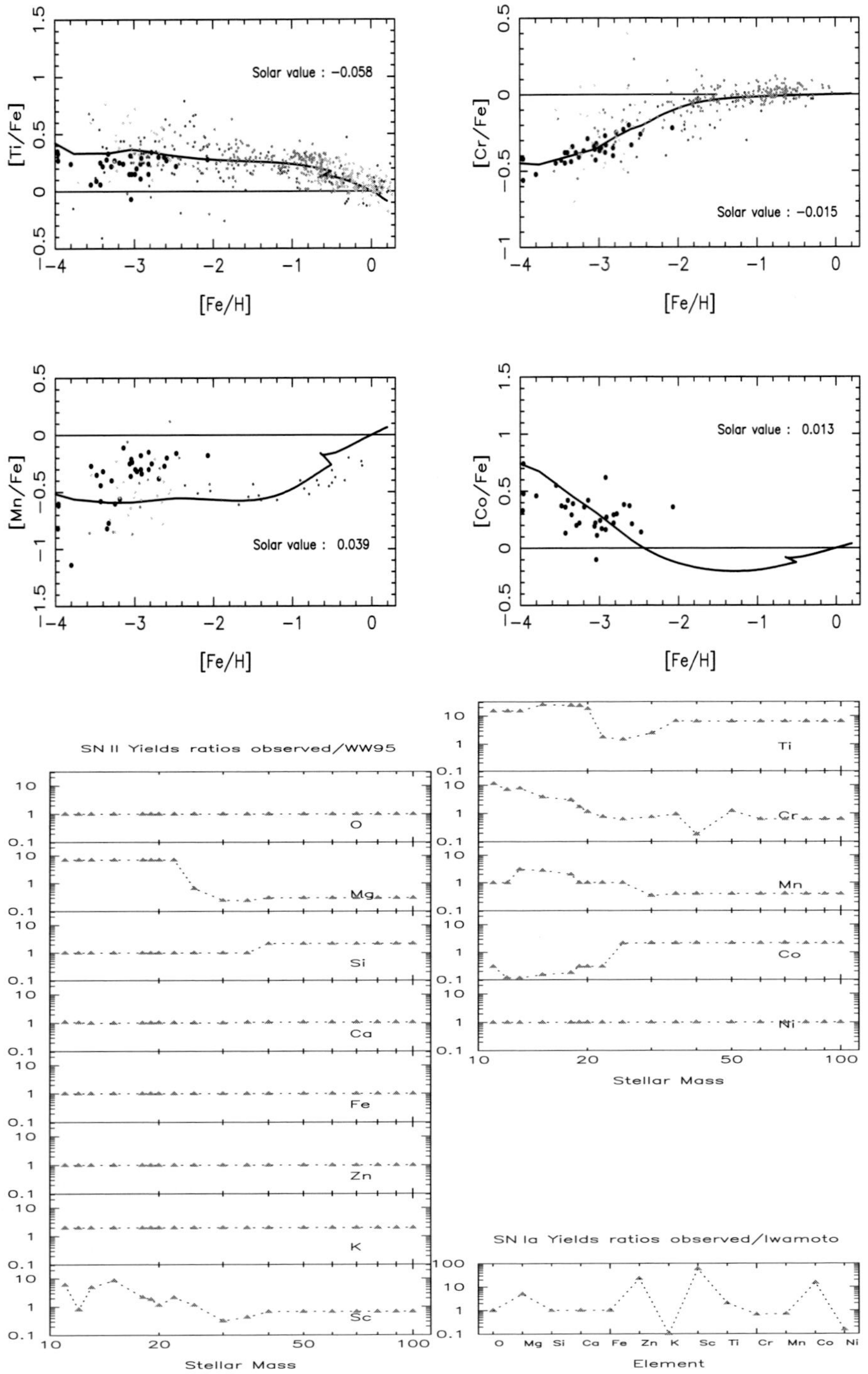

FIGURE 6.4. Upper panel: predicted and observed [X/Fe] versus [Fe/H] for several elements as in Figure 6.3. In the bottom part of this Figure are shown the ratios between the empirical yields and the yields published by WW95 for massive stars. Such empirical yields have been suggested by François *et al.* (2004) in order to fit best all the [X/Fe] versus [Fe/H] relations. In the small panel on the bottom right-hand side are shown also the ratios between the empirical yields for Type Ia SNe and the yields published by I99.

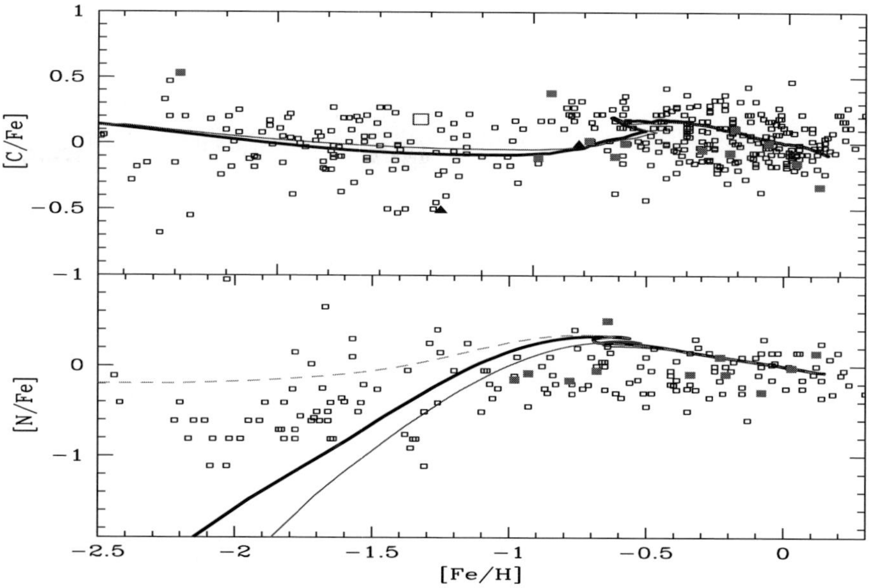

FIGURE 6.5. Upper panel: predicted and observed [C/Fe] versus [Fe/H]. Models from Chiappini *et al.* (2003a). Lower panel: predicted and observed [N/Fe] versus [Fe/H]. For references to the data see the original paper. The thin and thick continuous lines in both panels represent models with standard nucleosynthesis, as described in the text, whereas the dashed line represents the predictions of a model in which N in massive stars has been considered as a primary element with "ad-hoc" stellar yields.

increased in Type Ia SNe, relative to the yields of I99, and decreased in stars in the range (10–20)$M_{\odot}$, relative to the yields of WW95. Finally, the yield of Ni should be decreased in Type Ia SNe.

- The yields of Cu and Zn from Type Ia SNe should be larger, relative to the standard yields, as already suggested by Matteucci *et al.* (1993).

6.2.3 *Common conclusions from Milky Way models*

Most of the chemical-evolution models for the Milky Way in the literature lead to the following conclusions.

- The G-dwarf metallicity distribution can be reproduced only by assuming a slow formation of the local disk by infall. In particular, the timescale for the formation of the local disk should be in the range $\tau_{\rm d} \sim$ 6–8 Gyr (Chiappini *et al.* 1997; Boissier and Prantzos 1999; Chang *et al.* 1999; Chiappini *et al.* 2001; Alibès *et al.* 2001).
- The relative abundance ratios [X/Fe] versus [Fe/H], interpreted as a time delay between Type Ia and II SNe, suggest a timescale for the halo–thick-disk formation of $\tau_{\rm h} \sim$ 1.5–2.0 Gyr (Matteucci and Greggio 1986; Matteucci and François 1989; Chiappini *et al.* 1997). The external halo and thick disk probably formed more slowly or accreted (Chiappini *et al.* 2001).
- To fit the abundance gradients, SFR, and gas distribution along the Galactic thin disk we must assume that the disk formed *inside-out* (Matteucci & François 1989; Chiappini *et al.* 2001; Boissier & Prantzos 1999; Alibés *et al.* 2001). Radial flows can help in forming the gradients (Portinari & Chiosi 2000) but they are probably not the main cause for them. A variable IMF along the disk can in principle explain abundance gradients but it creates unrealistic situations: in fact, in order to reproduce

the negative gradients one should assume that in the external and less-metal-rich parts of the disk low-mass stars form preferentially. See Chiappini *et al.* (2000) for a discussion on this point.

- The SFR is a strongly varying function of the galactocentric distance (Matteucci & François 1989; Chiappini *et al.* 1997, 2001; Goswami & Prantzos 2000; Alibès *et al.* 2001).

6.2.4 *Abundance gradients from emission lines*

There are two types of abundance determinations in H II regions: one is based on recombination lines, which should have a weak temperature dependence (He, C, N, and O); the other is based on collisionally excited lines, for which a strong dependence is intrinsic to the method (C, N, O, Ne, Si, S, Cl, Ar, Fe, and Ni). The second method has predominated until now. A direct determination of the abundance gradients from H II regions in the Galaxy from optical lines is difficult because of extinction, so usually the abundances for distances larger than 3 kpc from the Sun are obtained from radio and infrared emission lines.

Abundance gradients can also be derived from optical emission lines in planetary nebulae (PNe). However, the abundances of He, C, and N in PNe are giving information only on the internal nucleosynthesis of the star. So, to derive gradients one should look at the abundances of O, S, and Ne, which are unaffected by stellar processes. In Figure 6.6 we show theoretical predictions of abundance gradients along the disk of the Milky Way compared with data from H II regions and B stars. The model adopted is from Chiappini *et al.* (2001, 2003a) and is based on an inside-out formation of the thin disk with the inner regions forming faster than the outer ones, in particular $\tau(R) = 0.875R - 0.75$ Gyr. Note that, to obtain a better fit for ^{12}C, the yields of this element have been increased artificially relative to those of WW95.

As already mentioned, most of the models agree on the inside-out scenario for the disk formation; however, not all models agree on the evolution of the gradients with time. In fact, some models predict a flattening with time (Boissier and Prantzos 1998; Alibès *et al.* 2001), whereas others, such as that of Chiappini *et al.* (2001), predict a steepening. The reason for the steepening is that in the model of Chiappini *et al.* there is a threshold density for SF, which induces the SF to stop when the density decreases below the threshold. This effect is particularly strong in the external regions of the disk, thus contributing to a slower evolution and therefore to a steepening of the gradients with time, as shown in the bottom panel of Figure 6.6.

6.2.5 *Abundance gradients in external galaxies*

Abundance gradients expressed in dex kpc^{-1} are found to be steeper in smaller disks but the correlation disappears if they are expressed in dex/R_d, which means that there is a universal slope per unit scale length (Garnett *et al.* 1997). The gradients are generally flatter in galaxies with central bars (Zaritsky *et al.* 1994). The SFR is measured mainly from H_α emission (Kennicutt 1998) and exhibits a correlation with the total surface gas density (H I plus H_2), in particular the suggested law is that of Equation (6.5).

In the observed gas distributions, differences between field and cluster spirals are found in the sense that cluster spirals have less gas, probably as a consequence of stronger interactions with the environment. Integrated colors of spiral galaxies (Josey & Arimoto 1992; Jimenez *et al.* 1998; Prantzos & Boissier 2000) indicate inside-out formation, as has also been found for the Milky Way.

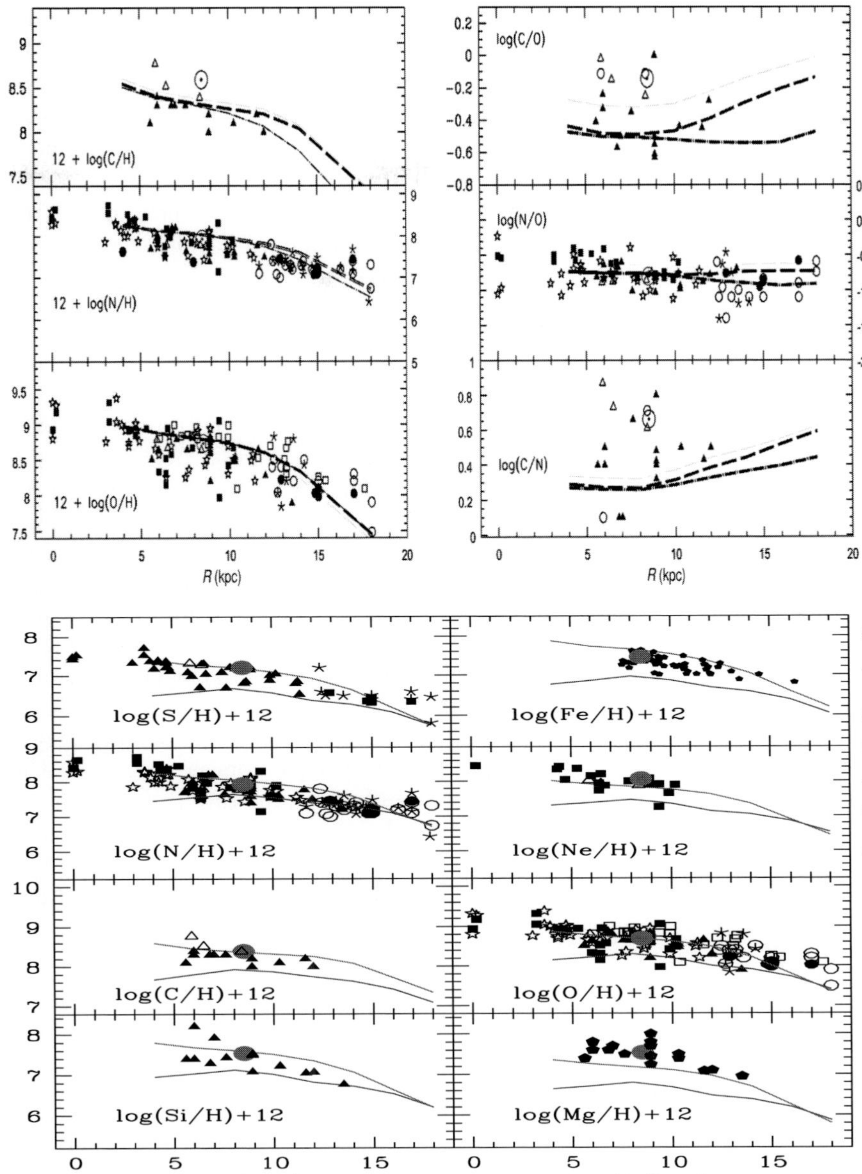

FIGURE 6.6. Upper panel: abundance gradients along the disk of the Milky Way. The lines are the models from Chiappini *et al.* (2003a): these models differ in terms of the nucleosynthesis prescriptions. In particular, the dash–dotted line represents a model with van den Hoeck & Groenewegen (1997) (hereafter HG97) yields for low- to intermediate-mass stars with η (the mass-loss parameter) constant and Thielemann *et al.*'s (1996) yields for massive stars, the long-dashed thick line has HG97 yields with variable η and Thielemann *et al.* yields, and the long-dashed thin line has HG97 yields with variable η but WW95 yields for massive stars. It is interesting to note that in all of these models the yields of ^{12}C in stars of masses $>40M_{\odot}$ have artificially been increased by a factor of three relative to the yields of WW95. Lower panel: the temporal behavior of abundance gradients along the disk as predicted by the best model of Chiappini *et al.* (2001). The upper lines in each panel represent the present-day gradient, whereas the lower ones represent the gradient a few Gyr ago. It is clear that the gradients tend to steepen with time, which is a still-controversial result.

As an example of abundance gradients in a spiral galaxy we show in Figure 6.7 the observed and predicted gas distribution and abundance gradients for the disk of M101. In this case the gas distribution and the abundance gradients are reproduced with systematically smaller timescales for disk formation relative to the Milky Way (M101 formed faster), and the difference between the timescales of formation of the internal and external regions is smaller, $\tau_{\mathrm{M101}} = 0.75R - 0.5$ Gyr (Chiappini *et al.* 2003a).

To conclude this section, we would like to recall a paper by Boissier *et al.* (2001), where a detailed study of the properties of disks is presented. They conclude that more-massive disks are redder, metal-richer, and gas-poorer than smaller ones. On the other hand, their estimated SF efficiencies for various (defined as the SFR per unit mass of gas) spirals seem to be similar: this leads them to conclude that more-massive disks are older than less-massive ones.

6.2.6 *How to model the Hubble sequence*

The Hubble sequence can be simply thought of as a sequence of objects for which SF proceeds faster in the early than in the late types. See also Sandage (1986).

We take the Milky Way galaxy, whose properties are best known, as a reference galaxy and we change the SFR relatively to the Galactic one, for which we adopt Equation (6.6). The quantity ν in Equation (6.6) is the efficiency of SF, which we assume to be characteristic of each Hubble type. In the two-infall model for the Milky Way we adopt $\nu_{\mathrm{halo}} = 2.0\ \mathrm{Gyr}^{-1}$ and $\nu_{\mathrm{disk}} = 1.0\ \mathrm{Gyr}^{-1}$ (see Figure 6.1). The choice of adopting a dependence on the total surface mass density for the Galactic disk is due to the fact that it helps in producing a SFR that is strongly varying with the galactocentric distance, as required by the observed SFR and gas-density distribution as well as by the abundance gradients. In fact, the inside-out scenario influences the rate at which the gas mass is accumulated by infall at each galactocentric distance and this in turn influences the SFR.

For bulges and ellipticals we assume that the SF proceeds as in a burst with very high SF efficiency, namely

$$\mathrm{SFR} = \nu \sigma^k \tag{6.26}$$

with $k = 1.0$ for the sake of simplicity; $\nu = 10$–$20\ \mathrm{Gyr}^{-1}$ (Matteucci 1994; Pipino & Matteucci 2004).

For irregular galaxies, on the other hand, we assume that SF proceeds more slowly and less efficiently than that in the Milky Way disk; in particular, we assume the same SF law as for spheroids but with $0.01 \leq \nu\,(\mathrm{Gyr}^{-1}) \leq 0.1$. Among irregular galaxies, a special position is taken by the blue compact galaxies (BCGs), namely galaxies that have blue colors as a consequence of the fact that they are forming stars at the present time, and have small masses, large amounts of gas, and low metallicities. For these galaxies, we assume that they suffered on average from one to seven short bursts, with the SF efficiency mentioned above (Bradamante *et al.* 1998).

Finally, dwarf spheroidals are also a special category, characterized by having old stars, no gas, and low metallicities. For these galaxies we assume that they suffered one long starburst lasting 7–8 Gyr or at most a couple of extended SF periods, in agreement with their measured color–magnitude diagram. It is worth noting that both ellipticals and dwarf spheroidals should lose most of their gas and therefore one may conclude that galactic winds should play an important role in their evolution, although ram pressure stripping cannot be excluded as a mechanism for gas removal. Also for these galaxies we assume the previous SF law with $k = 1$ and $\nu = 0.01$–$1.0\ \mathrm{Gyr}^{-1}$. Lanfranchi & Matteucci

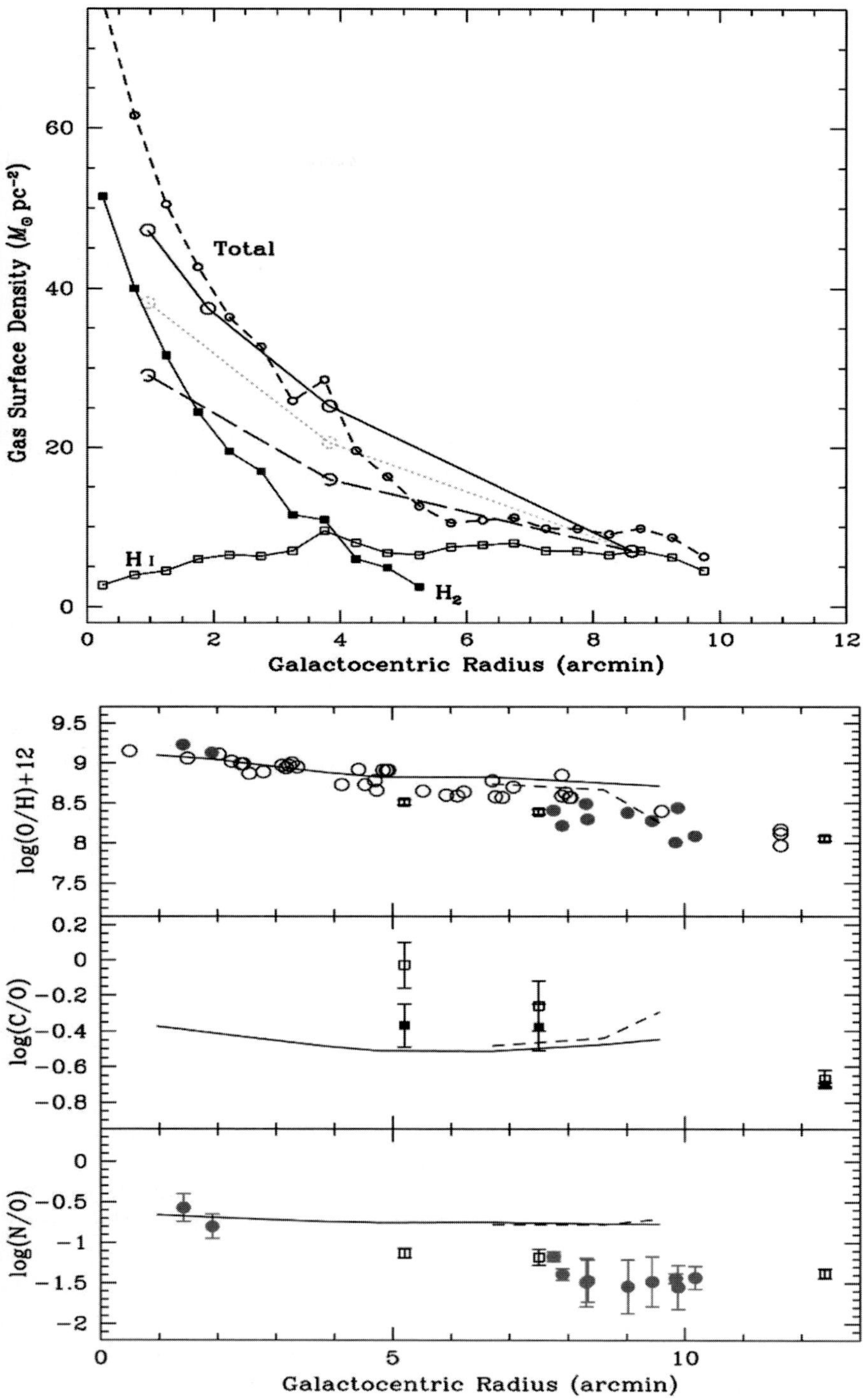

FIGURE 6.7. Upper panel: predicted and observed gas distribution along the disk of M101. The observed H I, H_2, and total gas are indicated. The large open circles indicate the models: in particular, the open circles connected by a continuous line refer to a model with central surface mass density of $1000 M_\odot\,pc^{-2}$, while the dotted line refers to a model with $800 M_\odot\,pc^{-2}$ and the dashed to a model with $600 M_\odot\,pc^{-2}$. Lower panel: predicted and observed abundance gradients of C, N, O elements along the disk of M101. The models are the lines and differ for a different threshold density for SF, being larger in the dashed model. All the models are by Chiappini *et al.* (2003a).

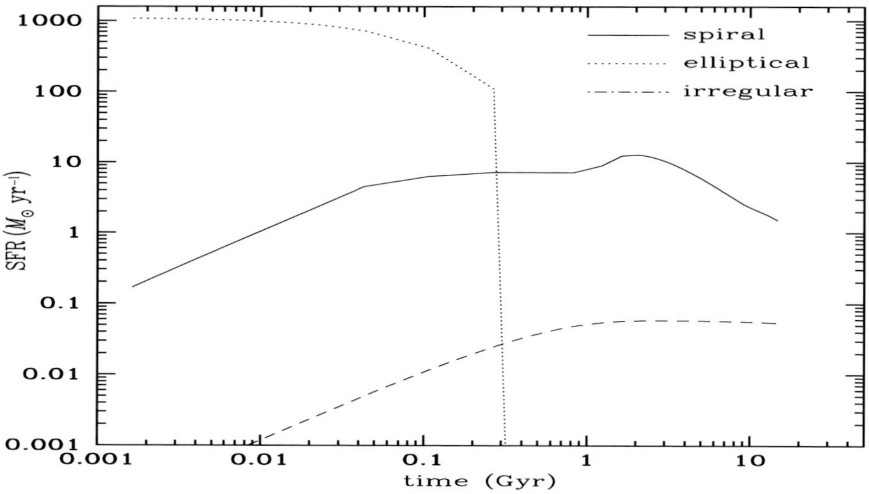

FIGURE 6.8. Predicted SFRs in galaxies of various morphological types, from Calura (2004). Note that for the elliptical galaxy the SF stops abruptly as a consequence of the galactic wind.

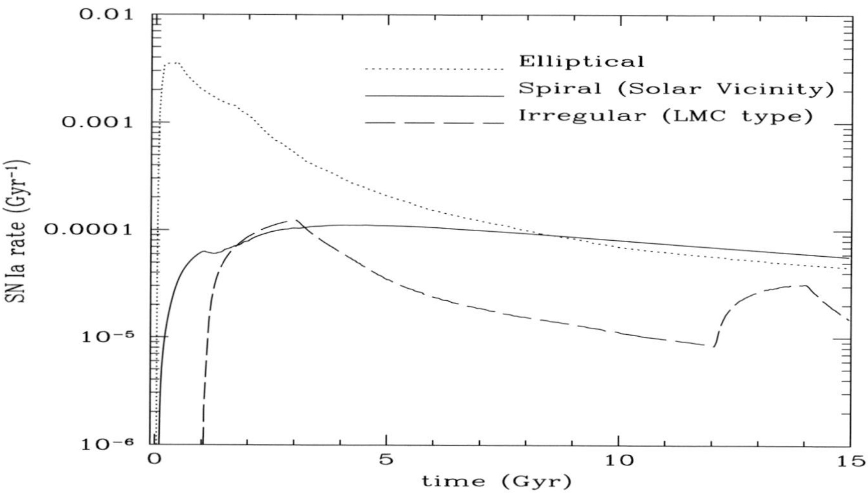

FIGURE 6.9. Predicted Type Ia SN rates for the SFRs of Figure 6.8. Figure from Calura (2004). Note that for the irregular galaxy here the predictions are for the LMC, where a recent SF burst is assumed.

(2003, 2004) developed more-detailed models for dwarf spheroidals by adopting the SF history suggested by the color–magnitude diagrams of single galaxies, with the same efficiency of SF as above. In Figure 6.8 we show the SFRs adopted for various types of galaxy and in Figure 6.9 the corresponding predicted Type Ia SN rates. For the irregular galaxy, the predicted Type Ia SN rate refers to a specific galaxy, the Lesser Magellanic Cloud (LMC), with a SFR taken from observations (Calura *et al.* 2003) with an early and a late burst of SF and low SF in between.

6.2.7 *Type Ia SN rates in various types of galaxy*

Following Matteucci & Recchi (2001), we define the typical timescale for Type Ia SN enrichment as *the time taken for the SN rate to reach the maximum.* In the following we will always adopt the SDS for the progenitors of Type Ia SNe. A point that is not often understood is that this timescale depends upon the progenitor lifetimes, IMF, and SFR and therefore is not universal. Sometimes in the literature the typical Type Ia SN timescale is quoted as being universal and equal to 1 Gyr, but this is just the timescale at which the Type Ia SNe start to be important in the process of Fe enrichment in the Solar vicinity.

Matteucci & Recchi (2001) showed that for an elliptical galaxy or a bulge of a spiral with a high SFR the timescale for Type Ia SN enrichment is quite short, in particular $t_{\mathrm{SN\,Ia}} = 0.3$–$0.5$ Gyr. For a spiral like the Milky Way, in the two-infall model, a first peak is reached at 1.0–1.5 Gyr (the time at which SNe Ia become important as Fe producers) (Matteucci and Greggio 1986) and a second, less-important, peak occurs at $t_{\mathrm{SN\,Ia}} = 4$–5 Gyr. For an irregular galaxy with a continuous but very low SFR the timescale is $t_{\mathrm{SN\,Ia}} > 5$ Gyr.

6.2.8 *A time-delay model for galaxies of various types*

As we have already seen, the time delay between the production of oxygen by Type II SNe and that of Fe by Type Ia SNe allows us to explain the [X/Fe] versus [Fe/H] relations in an elegant way. However, the [X/Fe] versus [Fe/H] plots depend not only on nucleosynthesis and IMF but also on other model assumptions, such as the SFR, through the absolute Fe abundance ([Fe/H]). Therefore, we should expect different behaviors in galaxies with different SF histories. In Figure 6.10 we show the predictions of the time-delay model for a spheroid like the Bulge, for the Solar vicinity, and for a typical irregular Magellanic galaxy.

As one can see in Figure 6.10, we predict a long plateau, well above the Solar value, for the [α/Fe] ratios in the Bulge (and ellipticals), owing to the fast Fe enrichment reached in these systems by means of Type II SNe: when the Type Ia SNe start enriching the ISM substantially, at 0.3–0.5 Gyr, the gas Fe abundance is already Solar. The opposite occurs in irregulars, where the Fe enrichment proceeds very slowly so that when Type Ia SNe start restoring the Fe in a substantial way (>3 Gyr) the Fe content in the gas is still well below Solar. Therefore, here we observe a steeper slope for the [α/Fe] ratio. In other words, we have below-Solar [α/Fe] ratios at below-Solar [Fe/H] ratios. This diagram is very important since it allows us to recognize a galaxy type solely by means of its abundances, and therefore it can be used to understand the nature of high-redshift objects.

6.3. Interpretation of abundances in dwarf irregulars

They are rather simple objects with low metallicity and large gas content, suggesting that they are either young or have undergone discontinuous SF activity (bursts) or a continuous but not efficient SF. They are very interesting objects for studying galaxy evolution. In fact, in "bottom-up" cosmological scenarios they should be the first self-gravitating systems to form and could also be important contributors to the population of systems giving rise to QSO absorption lines at high redshift (Matteucci *et al.* 1997; Calura *et al.* 2002).

6.3.1 *Properties of dwarf irregular galaxies*

Among local star-forming galaxies, sometimes referred to as H II galaxies, most are dwarfs. Dwarf irregular galaxies can be divided into two categories: dwarf irregular

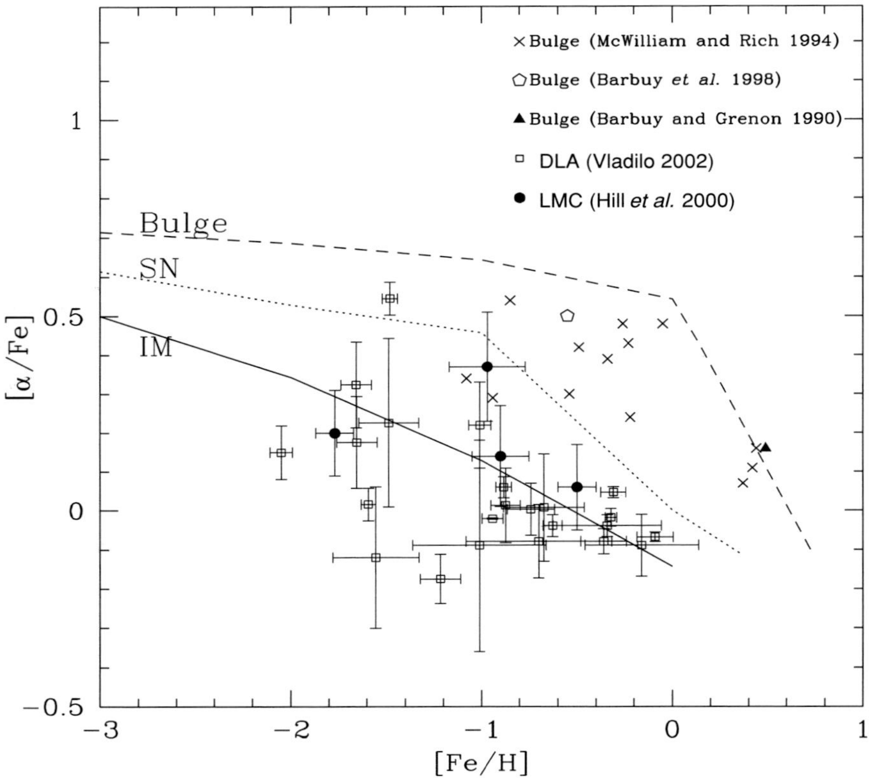

FIGURE 6.10. Predicted [α/Fe] ratios in galaxies with different SF histories. The top line represents the predictions for the Bulge or for an elliptical galaxy of the same mass ($\sim 10^{10} M_\odot$), the next line down represents the prediction for the Solar vicinity, and the lowest line shows the prediction for an irregular Magellanic galaxy. The differences among the various models are in the efficiency of star formation, this being quite high for spheroids ($\nu = 20$ Gyr^{-1}), moderate for the Milky Way ($\nu = 1$–2 Gyr^{-1}), and low for irregular galaxies ($\nu = 0.1$ Gyr^{-1}). The nucleosynthesis prescriptions are the same for all objects. The time delay between the production of α-elements and that of Fe coupled with the different SF histories produces the differences among the plots. Data for damped-Lyman-α systems (DLA), the LMC, and the Bulge are shown for comparison.

galaxies (DIGs) and blue compact galaxies (BCGs). The latter have very blue colors due to active SF at the present time.

Chemical abundances in these galaxies are derived from optical emission lines in H II regions. Both DIGs and BCGs show a distinctive spread in their chemical properties, although this spread is decreasing in the new, more-accurate, data, but also a definite mass–metallicity relation.

From the point of view of chemical evolution, Matteucci and Chiosi (1983) first studied the evolution of DIGs and BCGs by means of analytical chemical-evolution models including either outflow or infall and concluded that closed-box models cannot account for the Z–log G ($G = M_{\rm gas}/M_{\rm tot}$) distribution even if the number of bursts varies from galaxy to galaxy and suggested possible solutions to explain the observed spread. In other words, the data show a range of values of metallicity for a given G ratio, and this means that the effective yield is lower than that of the simple model and varies from galaxy to galaxy.

The possible solutions suggested to lower the effective yield were
(a) different IMFs,
(b) different amounts of galactic wind, and
(c) different amounts of infall.

In Figure 6.11 we show graphically the solutions (a), (b), and (c). Concerning the solution (a), one simply varies the IMF, whereas solutions (b) and (c) have already been described (Equations (1.21) and (1.23)).

Later on, Pilyugin (1993) put forward the idea that the spread observed also in other chemical properties of these galaxies such as in the He/H versus O/H and N/O versus O/H relations could be due to self-pollution of the H II regions, which do not mix efficiently with the surrounding medium, coupled with "enriched" or "differential" galactic winds, namely different chemical elements are lost at different rates. Other models (Marconi *et al.* 1994; Bradamante *et al.* 1998) followed the suggestions of differential winds and introduced the novelty of the contribution to the chemical enrichment and energetics of the ISM by SNe of different types (II, Ia, and Ib).

Another important feature of these galaxies is the mass–metallicity relation.

The existence of a luminosity–metallicity relation in irregulars and BCGs was suggested first by Lequeux *et al.* (1979), then confirmed by Skillman *et al.* (1989), and extended also to spirals by Garnett & Shields (1987). In particular, Lequeux *et al.* suggested the relation

$$M_{\mathrm{T}} = (8.5 \pm 0.4) + (190 \pm 60)Z \qquad (6.27)$$

with Z being the global metal content. Recently, Tremonti *et al.* (2004) analyzed 53 000 local star-forming galaxies (irregulars and spirals) in the SDSS. Metallicity was measured from the optical nebular emission lines. Masses were derived from fitting spectral-energy-distribution (SED) models. The strong optical nebular lines of elements other than H are produced by collisionally excited transitions. Metallicity was then determined by fitting simultaneously the most-prominent emission lines ([O III], H_β, [O II], H_α, [N II], and [S II]). Tremonti *et al.* (2004) derived a relation indicating that $12 + \log(\mathrm{O/H})$ is increasing steeply for M_* going from $10^{8.5}$ to $10^{10.5}$ but flattening for $M_* > 10^{10.5}$.

In particular, the Tremonti *et al.* relation is

$$12 + \log(O/H) = -1.492 + 1.847(\log M_*) - 0.080\,26(\log M_*)^2. \qquad (6.28)$$

This relation extends to higher masses the mass–metallicity relation found for star-forming dwarfs and contains very important information on the physics governing galactic evolution. Even more recently, Erb *et al.* (2006) found the same mass–metallicity relation for star-forming galaxies at redshift $z > 2$, with an offset from the local relation of $\sim$0.3 dex. They used H_α and [N II] spectra. In Figure 6.12 we show the figure from Erb *et al.* (2006) for the mass–metallicity relation at high redshift which includes the relation of Tremonti *et al.* (2004) for the local mass–metallicity relation.

The simplest interpretation of the mass–metallicity relation is that the effective yield increases with galactic mass. This can be achieved in several ways, as shown in Figure 6.11: by changing the IMF or the stellar yields as a function of galactic mass, or by assuming that the galactic wind is less efficient in more-massive systems, or that the infall rate is less efficient in more-massive systems. One of the commonest interpretations of the mass–metallicity relation is that the effective yield changes because of the occurrence of galactic winds, which should be more important in small systems. There is evidence that galactic winds exist for dwarf irregular galaxies, as we will see in the following section.

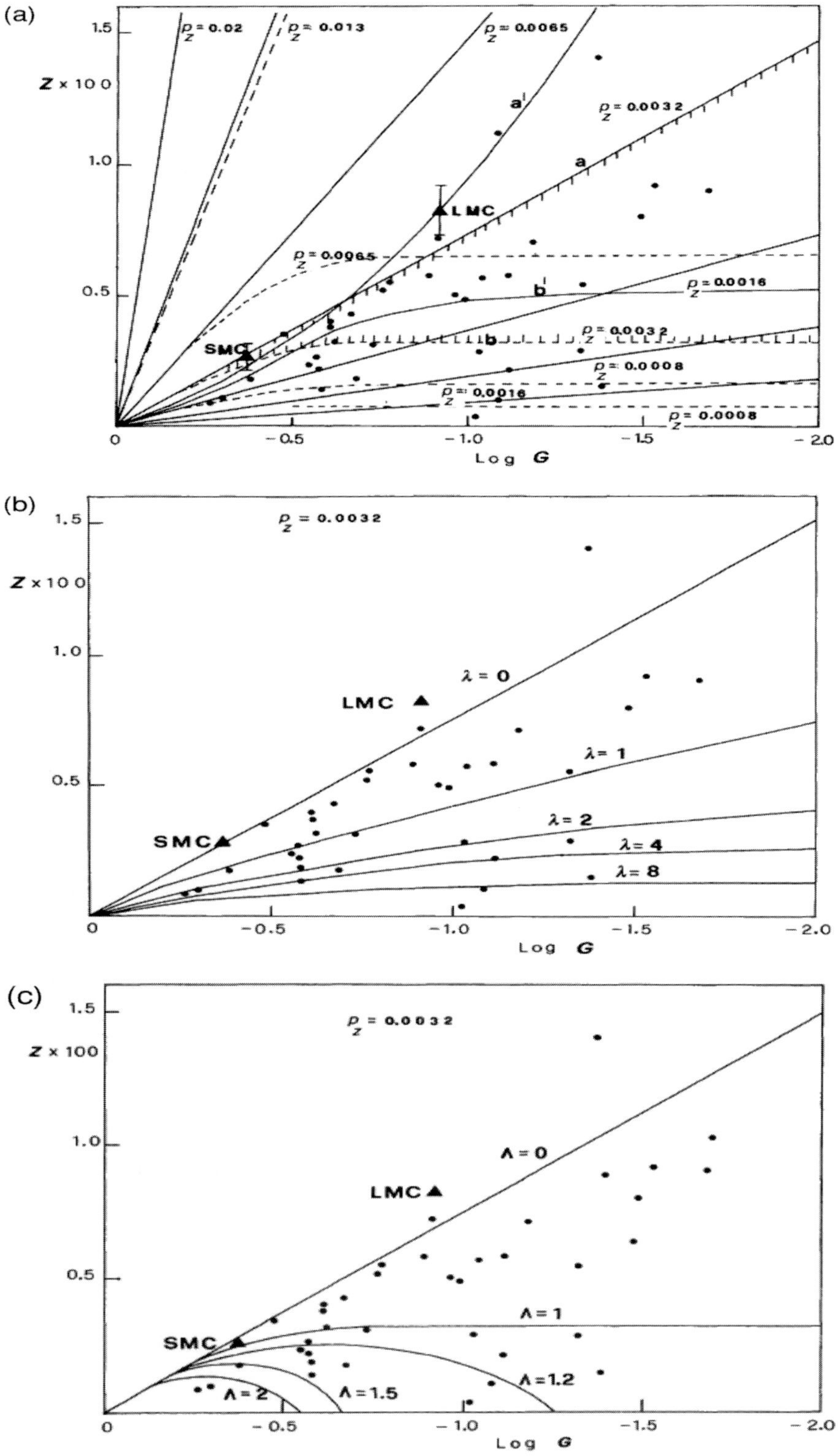

FIGURE 6.11. The Z–log G diagram. Solutions (a), (b), and (c) to lower the effective yield in DIGs and BCGs by Matteucci & Chiosi (1983). Solution (a) consists in varying the yield per stellar generation, here indicated by p_Z, just by changing the IMF. Solutions (b) and (c) correspond to Equations (6.21) and (6.23), respectively.

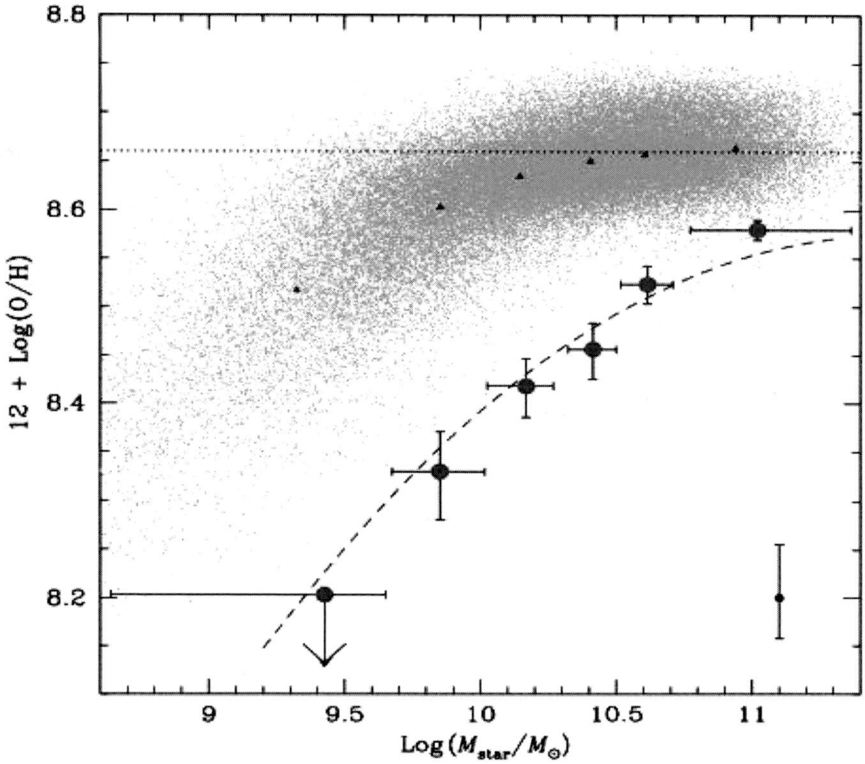

FIGURE 6.12. Figure 3 from Erb *et al.* (2006), showing the mass–metallicity relation for star-forming galaxies at high redshift. The data from Tremonti *et al.* (2004) are also shown.

6.3.2 *Galactic winds*

Papaderos *et al.* (1994) estimated a galactic wind flowing at a velocity of 1320 km s^{-1} for the irregular dwarf VIIZw403. The escape velocity estimated for this galaxy is $\simeq$50 km s^{-1}. Lequeux *et al.* (1995) suggested a galactic wind in Haro2=MKn33 flowing at a velocity of $\simeq$200 km s^{-1}, also larger than the escape velocity of this object. More recently, Martin (1996, 1998) found supershells as well in 12 dwarfs, including IZw18, which imply gas outflow. Martin (1999) concluded that the galactic-wind rates are several times the SFR. Finally, the presence of metals in the ICM (revealed by X-ray observations) and in the IGM (Ellison *et al.* 2000) constitutes a clear indication of the fact that galaxies lose their metals. However, we cannot exclude the possibility that the gas with metals is lost also by ram pressure stripping, especially in galaxy clusters.

In models of chemical evolution of dwarf irregulars (e.g. Bradamante *et al.* 1998) the feedback effects are taken into account and the condition for the development of a wind is

$$(E_{\rm th})_{\rm ISM} \geq E_{\rm Bgas}, \tag{6.29}$$

namely, that the thermal energy of the gas is larger than or equal to its binding energy. The thermal energy of gas due to SN and stellar-wind heating is

$$(E_{\rm th})_{\rm ISM} = E_{{\rm th}_{\rm SN}} + E_{{\rm th}_{\rm w}} \tag{6.30}$$

with the contribution of SNe being

$$E_{\mathrm{th_{SN}}} = \int_0^t \epsilon_{\mathrm{SN}} R_{\mathrm{SN}}(t) \mathrm{d}t, \tag{6.31}$$

while the contribution of stellar winds is

$$E_{\mathrm{th_w}} = \int_0^t \int_{12}^{100} \varphi(m)\psi(t)\epsilon_{\mathrm{w}}\, \mathrm{d}m\, \mathrm{d}t \tag{6.32}$$

with $\epsilon_{\mathrm{SN}} = \eta_{\mathrm{SN}}\epsilon_{\mathrm{o}}$ and $\epsilon_{\mathrm{o}} = 10^{51}\mathrm{erg}$ (typical SN energy) and $\epsilon_{\mathrm{w}} = \eta_{\mathrm{w}} E_{\mathrm{w}}$ with $E_{\mathrm{w}} = 10^{49}\mathrm{erg}$ (the typical energy injected by a $20 M_{\odot}$ star, taken as representative). η_{w} and η_{SN} are two free parameters and indicate the efficiency of energy transfer from stellar winds and SNe into the ISM, respectively, quantities that are still largely unknown. The total mass of the galaxy is expressed as $M_{\mathrm{tot}}(t) = M_*(t) + M_{\mathrm{gas}}(t) + M_{\mathrm{dark}}(t)$ with $M_{\mathrm{L}}(t) = M_*(t) + M_{\mathrm{gas}}(t)$ and the binding energy of gas is

$$E_{\mathrm{Bgas}}(t) = W_{\mathrm{L}}(t) + W_{\mathrm{LD}}(t) \tag{6.33}$$

with

$$W_{\mathrm{L}}(t) = -0.5G\frac{M_{\mathrm{gas}}(t) M_{\mathrm{L}}(t)}{r_{\mathrm{L}}}, \tag{6.34}$$

which is the potential well due to the luminous matter and with

$$W_{\mathrm{LD}}(t) = -G w_{\mathrm{LD}} \frac{M_{\mathrm{gas}}(t) M_{\mathrm{dark}}}{r_{\mathrm{L}}}, \tag{6.35}$$

which represents the potential well due to the interaction between dark and luminous matter, where $w_{\mathrm{LD}} \sim S(1 + 1.37S)/(2\pi)$, with $S = r_{\mathrm{L}}/r_{\mathrm{D}}$ being the ratio between the galaxy's effective radius and the radius of the dark-matter core. The typical model for a BCG has a luminous mass of $(10^8\text{–}10^9)\, M_{\odot}$, a dark-matter halo ten times larger than the luminous mass, and various values for the parameter S. The galactic wind in these galaxies develops easily but it carries out mainly metals, so the total mass lost in the wind is small.

6.3.3 *Results on DIGs and BCGs from purely chemical models*

Purely chemical models (Bradamante *et al.* 1998; Marconi *et al.* 1994) for DIGs and BCGs have been computed in the last few years by varying the number of bursts, the time of occurrence of bursts t_{burst}, the SF efficiency, the type of galactic wind (differential or normal), the IMF, and the nucleosynthesis prescriptions. The best model of Bradamante *et al.* (1998) suggests that the number of bursts should be $N_{\mathrm{bursts}} \leq 10$, so the SF efficiency should vary from 0.1 to 0.7 Gyr^{-1} for either a Salpeter or a Scalo (1986) IMF (the Salpeter IMF is favored). Metal-enriched winds are favored. The results of these models also suggest that SNe of Type II dominate the chemical evolution and energetics of these galaxies, whereas stellar winds are negligible. The predicted [O/Fe] ratios tend to correspond to overabundance relative to the Solar ratios, owing to the predominance of Type II SNe during the bursts, in agreement with observational data (see the upper panel of Figure 6.15 later). Models with strong differential winds and $N_{\mathrm{bursts}} = 10\text{–}15$ can, however, give rise to negative [O/Fe] ratios. The main difference between DIGs and BCGs, in these models, is that the BCGs suffer a present-day burst, whereas the DIGs are in a quiescent phase.

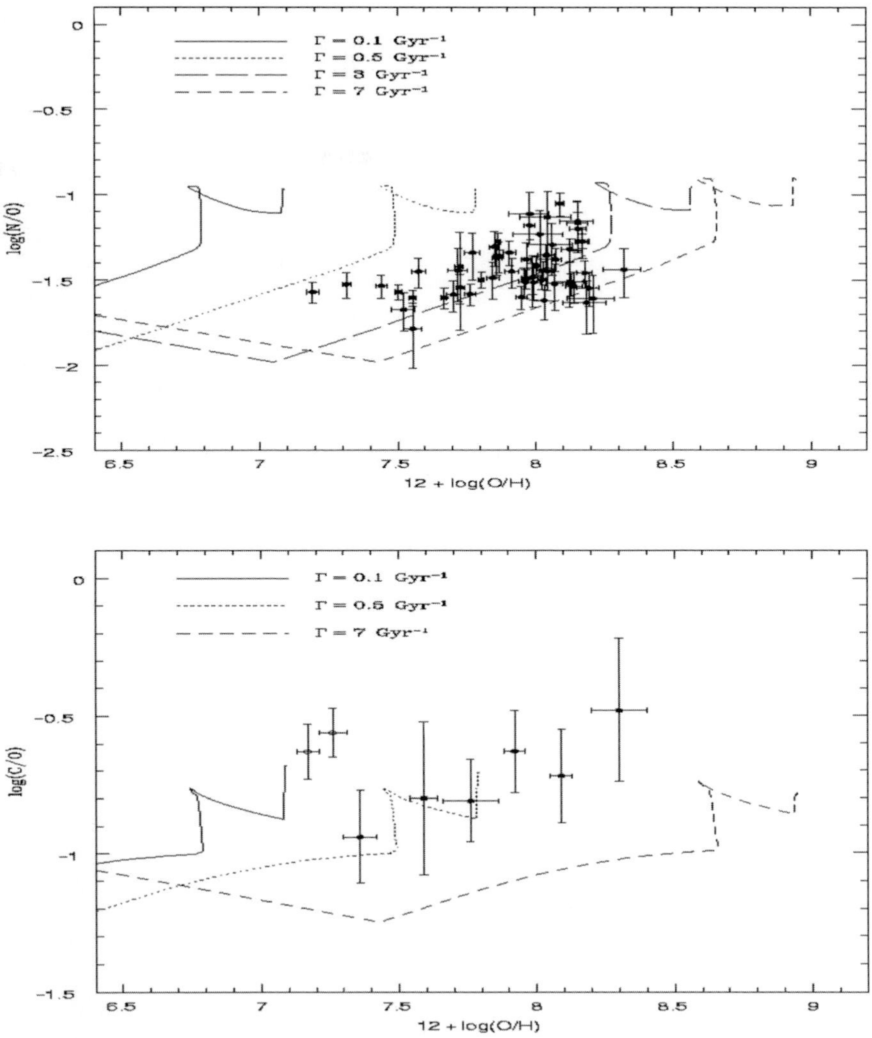

FIGURE 6.13. Upper panel: predicted log(N/O) versus 12 + log(O/H) for a model with three bursts of SF separated by quiescent periods and different SF efficiencies here indicated by Γ. Lower panel: predicted log(C/O) versus $12 + \log(\mathrm{O/H})$. The data in both panels are from Kobulnicky and Skillman (1996). The models assume a dark-matter halo ten times larger than the luminous mass and $S = 0.3$ (Bradamante *et al.* 1998), see the text.

In Figure 6.13 we show some of the results of Bradamante *et al.* (1998) compared with data on BCGs: it is evident that the spread in the chemical properties can be simply reproduced by different SF efficiencies, which translate into different wind efficiencies.

In Figure 6.14 we show the results of the chemical-evolution models of Henry *et al.* (2000). These models take into account exponential infall but not outflow. They suggested that the SF efficiency in extragalactic H II regions must have been low and that this effect coupled with the primary N production from intermediate mass stars can explain the plateau in log(N/O) observed at low $12 + \log(\mathrm{O/H})$. Henry *et al.* (2000) also concluded that ^{12}C is produced mainly in massive stars (yields published by Maeder (1992)) whereas ^{14}N is produced mainly in intermediate-mass stars (yields published by HG97). This conclusion, however, should be tested also on the abundances of stars in the Milky Way,

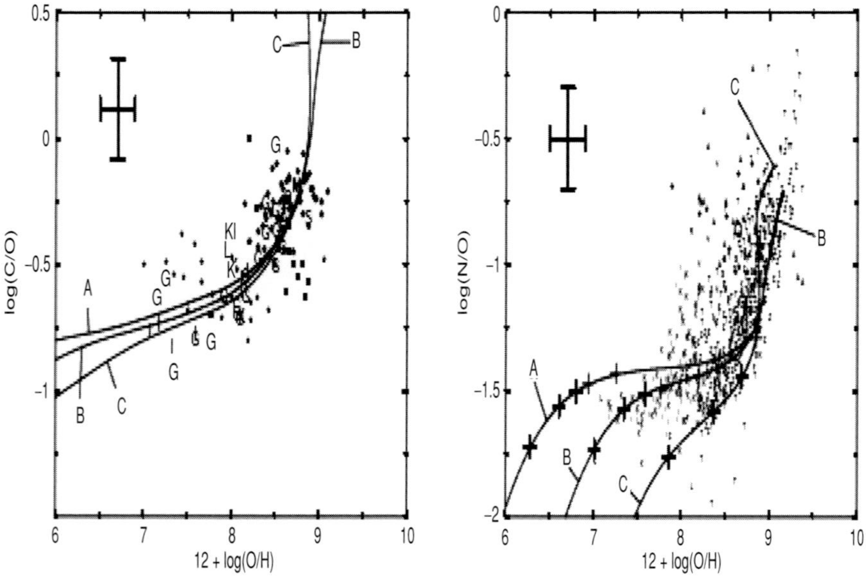

FIGURE 6.14. A comparison between numerical models and data for extragalactic H II regions and stars (filled circles, filled boxes, and filled diamonds); M and S mark the positions of the Galactic H II regions and the Sun, respectively. Their best model is model B with an efficiency of SF of $\nu = 0.03$. From Henry *et al.* (2000).

where the flat behavior of [C/Fe] versus [Fe/H] from $[\mathrm{Fe/H}] = -2.2$ up to $[\mathrm{Fe/H}] = 0$ suggests a similar origin for the two elements, namely partly from massive stars and mainly from low- and intermediate-mass ones (Chiappini *et al.* 2003b).

Concerning the [O/Fe] ratios, we show results from Thuan *et al.* (1995) in Figure 6.15, where it is evident that generally BCGs have overabundant [O/Fe] ratios.

Very recently, an extensive study from the SDSS of chemical abundances from emission lines in a sample of 310 metal-poor emission-line galaxies appeared (Izotov *et al.* 2006). The global metallicity in these galaxies ranges from $\sim 7.1(Z_\odot/30)$ to $\sim 8.5(0.7\,Z_\odot)$. The SDSS sample is merged with 109 BCGs containing objects of extremely low metallicity. These data, shown in the lower panel of Figure 6.15, substantially confirm previous ones showing that abundances of α-elements do not depend on the O abundance, suggesting a common origin for these elements in stars with $M > 10 M_\odot$, except for a slight increase of Ne/O with metallicity, which is interpreted as due to a moderate dust depletion of O in metal-rich galaxies. An important finding is that all the galaxies studied are found to have $\log(N/O) > -1.6$, which indicates that none of these galaxies is a truly young object, unlike the DLA systems at high redshift which have a $\log(N/O) \sim -2.3$.

6.3.4 *Results from chemo-dynamical models: IZw18*

IZw18 is the metal-poorest local galaxy, thus resembling a primordial object. Probably it experienced no more than two bursts of star formation including the present one. The age of the oldest stars in this galaxy is still unknown, although recently Tosi *et al.* (2006) suggested an age possibly >2 Gyr. The oxygen abundance in IZW18 is $12 + \log(\mathrm{O/H}) = 7.17$–$7.26$, ∼15–20 times lower than the Solar abundance of oxygen ($12 + \log(\mathrm{O/H}) = 8.39$ (Asplund *et al.* 2005) and $\log(\mathrm{N/O} = -1.54)$ to -1.60 (Garnett *et al.* 1997).

Recently, FUSE provided abundances also for H I in IZw18: the evidence is that the abundances in H I are lower than those in H II (Aloisi *et al.* 2003; Lecavelier des Etangs

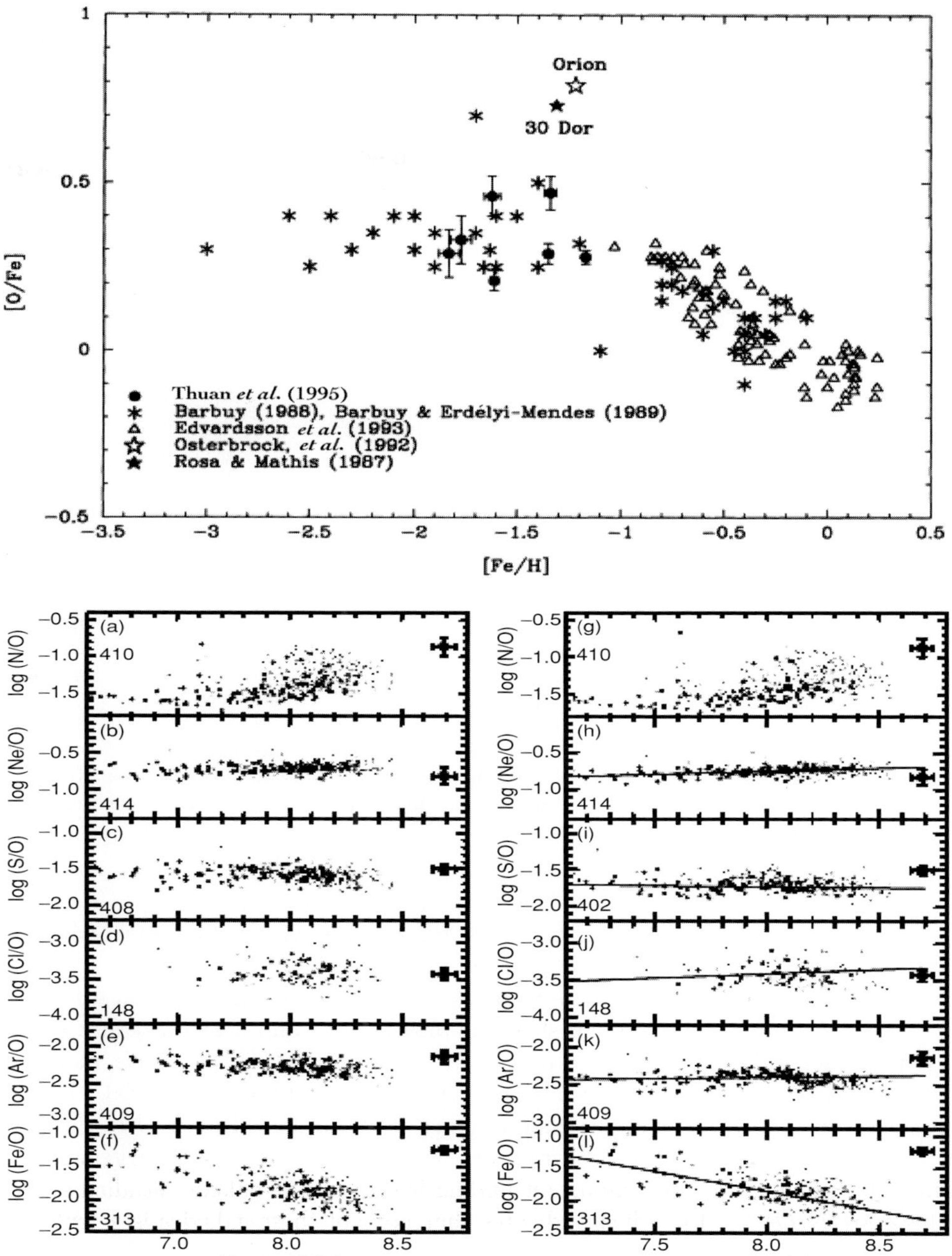

FIGURE 6.15. Upper panel: [O/Fe] versus [Fe/H] observed in a sample of BCGs by Thuan *et al.* (1995) (filled circles); open triangles and asterisks are data for disk and halo stars shown for comparison. Adapted from Thuan *et al.* (1995). Lower panel: new data from Izotov *et al.* (2006). The large filled circles represent the BCGs whereas the dots are for the SDSS galaxies. Abundances in the left panel are calculated as in Thuan *et al.* (1995) whereas those in the right panel are calculated as in Izotov *et al.* (2006) (see the original papers for details). Adapted from Izotov *et al.* (2006).

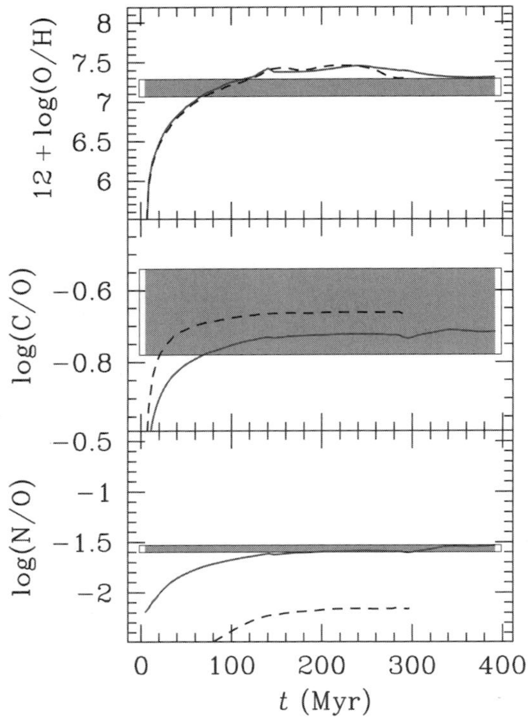

FIGURE 6.16. Predicted abundances for the H II region in IZw18 (dashed lines represent a model adopting the yields of Meynet & Maeder (2002) for $Z = 10^{-5}$, whereas the continuous line refers to a higher metallicity ($Z = 0.004$). Observational data are represented by the shaded areas. From Recchi *et al.* (2004).

et al. 2003). In particular, Aloisi *et al.* (2003) found the largest difference relative to the H II data.

Recchi *et al.* (2001), using chemo-dynamical (two-dimensional) models, studied first the case of IZw18 with only one burst at the present time and concluded that the starburst triggers a galactic outflow. In particular, the metals leave the galaxy more easily than does the unprocessed gas and, among the types of enriched material, the SN Ia ejecta leave the galaxy more easily than do other ejecta. In fact, Recchi *et al.* (2001) had reasonably assumed that Type Ia SNe can transfer almost all of their energy to the gas, since they explode in an already hot and rarefied medium after the SN II explosions. As a consequence of this, they predicted that the [α/Fe] ratios in the gas inside the galaxy should be larger than the [α/Fe] ratios in the gas outside the galaxy. They found, at variance with results of previous studies, that most of the metals are already in the cold gas phase after 8–10 Myr since the superbubble does not break immediately and thermal conduction can act efficiently. Recchi *et al.* (2004) extended the model to a two-burst case, still with the aim of reproducing the characteristics of IZw18. The model well reproduces the chemical properties of IZw18 with a relatively long episode of SF lasting 270 Myr plus a recent burst of SF that is still going on. In Figure 6.16 we show the predictions of Recchi *et al.* (2004) for the abundances in the H II regions of IZW18 and in Figure 6.17 those for the H I region, showing that there is little difference between the H II and H I abundances, which is more in agreement with the data of Lecavelier des Etangs *et al.* (2004).

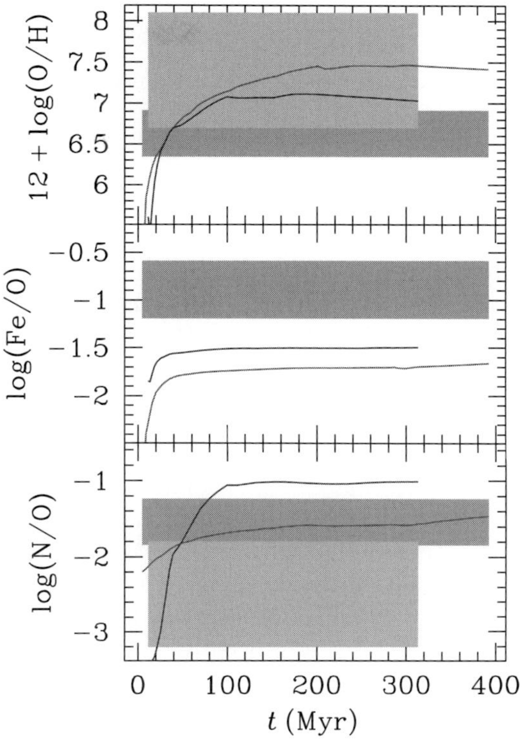

FIGURE 6.17. Predicted abundances for the H I region. The models are the same as in Figure 6.16. Observational data are represented by the shaded areas. The upper shaded area in the panel for oxygen and the lower shaded area in the panel for N/O represent the data of Lecavelier des Etangs *et al.* (2003). From Recchi *et al.* (2004).

6.4. Elliptical galaxies–quasars–ICM enrichment

6.4.1 *Ellipticals*

We recall here some of the most important properties of ellipticals or early-type galaxies (ETG), which are systems made of old stars with no gas and no ongoing SF. The metallicity of ellipticals is measured only by means of metallicity indices obtained from their integrated spectra, which are very similar to those of K giants. In order to pass from metallicity indices to [Fe/H] one needs then to adopt a suitable calibration, which is often based on population-synthesis models (Worthey 1994). We also summarize the most common scenarios for the formation of ellipticals.

6.4.2 *Chemical properties*

The main properties of the stellar populations in ellipticals are as follows.

- There exist the well-known color–magnitude and color–σ_o (velocity dispersion) relations indicating that the integrated colors become redder with increasing luminosity and mass (Faber 1977; Bower *et al.* 1992). These relations are interpreted as a metallicity effect, although there exists a well-known degeneracy between metallicity and age of the stellar populations in the integrated colors (Worthey 1994).
- The index Mg_2 is normally used as a metallicity indicator since it does not depend much upon the age of stellar populations. There exists for ellipticals a well-defined Mg_2–σ_o relation, equivalent to the already-discussed mass–metallicity relation for star-forming galaxies (Bender *et al.* 1993; Bernardi *et al.* 1998; Colless *et al.* 1999).

- Abundance gradients in the stellar populations inside ellipticals are found (Carollo *et al.* 1993; Davies *et al.* 1993). Kobayashi & Arimoto (1999) derived the average gradient for ETGs from a large compilation of data, for which it is $\Delta[\mathrm{Fe/H}]/\Delta r \sim -0.3$, with the average metallicity in ETGs of $\langle[\mathrm{Fe/H}]\rangle_* \sim -0.3$ dex (from -0.8 to $+0.3$ dex).
- A very important characteristic of ellipticals is that their central dominant stellar population (dominant in the visual light) exhibits an overabundance, relative to the Sun, of the Mg/Fe ratio, $\langle[\mathrm{Mg/Fe}]\rangle_* > 0$ (from 0.05 to +0.3 dex) (Peletier 1989; Worthey *et al.* 1992; Weiss *et al.* 1995; Kuntschner *et al.* 2001).
- In addition, the overabundance increases with increasing galactic mass and luminosity, $\langle[\mathrm{Mg/Fe}]\rangle_*$ versus σ_o, (Worthey *et al.* 1992; Matteucci 1994; Jorgensen 1999; Kuntschner *et al.* 2001).

6.4.3 *Scenarios for galaxy formation*

The most common ideas on the formation and evolution of ellipticals can be summarized as follows.

- They formed by an early monolithic collapse of a gas cloud or early merging of lumps of gas where dissipation plays a fundamental role (Larson 1974; Arimoto & Yoshii 1987; Matteucci & Tornambè 1987). In this model SF proceeds very intensively until a galactic wind is developed and SF stops after that. The galactic wind is voiding the galaxy of all its residual gas.
- They formed by means of intense bursts of star formation in merging subsystems made of gas (Tinsley & Larson 1979). In this picture SF stops after the last burst and gas is lost via ram pressure stripping or galactic wind.
- They formed by early merging of lumps containing gas and stars in which some dissipation is present (Bender *et al.* 1993).
- They formed and continue to form in a wide redshift range and preferentially at late epochs by merging of stellar systems that formed early (e.g. Kauffmann *et al.* 1993, 1996).

Pipino & Matteucci (2004), by means of recent revised monolithic models taking into account the development of a galactic wind (see Section 6.3), computed the relation [Mg/Fe] versus mass (velocity dispersion) and compared it with the data published by Thomas *et al.* (2002). Thomas (1990) already showed how hierarchical semi-analytical models cannot reproduce the observed [Mg/Fe] versus mass trend, since in this scenario massive ellipticals have longer periods of star formation than do smaller ones. In Figure 6.18, the original figure from Thomas *et al.* (2002) is shown, on which we have also plotted our predictions. In the Pipino & Matteucci (2004) model it is assumed that the most-massive galaxies assemble faster and form stars faster than do less-massive ones. The IMF adopted is the Salpeter one. In other words, more-massive ellipticals seem to be older than less-massive ones, in agreement with what has been found for spirals (Boissier *et al.* 2001). In particular, in order to explain the observed $\langle[\mathrm{Mg/Fe}]\rangle_* > 0$ in giant ellipticals, the dominant stellar population should have formed on a timescale no longer than (3–5) $\times 10^8$ yr (Weiss *et al.* 1995; Pipino & Matteucci 2004).

6.4.4 *The ellipticals–quasars connection*

We know now that most if not all massive ETGs host an AGN for some time during their life. Therefore, there is a strict link between quasar activity and the evolution of ellipticals.

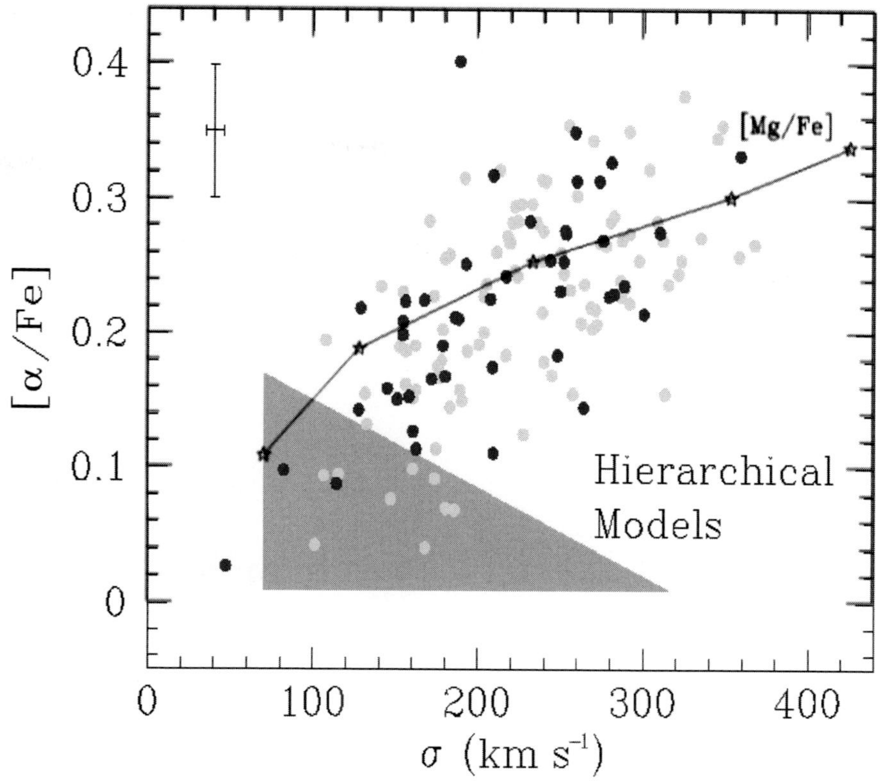

FIGURE 6.18. The relation [α/Fe] versus velocity dispersion (mass) for ETGs. The continuous line represents the prediction of the Pipino & Matteucci (2004) model. The shaded area represents the prediction of hierarchical models for the formation of ellipticals. The symbols are the observational data. Adapted from Thomas *et al.* (2002).

6.4.5 *The chemical evolution of QSOs*

It is very interesting to study the chemical evolution of QSOs by means of the broad emission lines in the QSO region. In the first studies Wills *et al.* (1985) and Collin-Souffrin *et al.* (1986) found that the abundance of Fe in QSOs, as measured from broad emission lines, turned out to be about a factor of ten more than the Solar one, which constituted a challenge for chemical-evolution model makers. Hamman & Ferland (1992) from N V/C IV line ratios for QSOs derived the N/C abundance ratios and inferred the QSO metallicities. They suggested that N is overabundant by factors of 2–9 in the high-redshift sources ($z > 2$). Metallicities 3–14 times the Solar one were also suggested in order to produce such a high N abundance, under the assumption of a mainly secondary production of N. To interpret their data they built a chemical-evolution model, a Milky Way-like model, and suggested that these high metallicities are reached in only 0.5 Gyr, implying that QSOs are associated with vigorous star formation. At the same time, Padovani & Matteucci (1993) and Matteucci & Padovani (1993) proposed a model for QSOs in which QSOs are hosted by massive ellipticals. They assumed that after the occurrence of a galactic wind the galaxy evolves passively and that for massess $> 10^{11}\ M_\odot$ the gas restored by the dying stars is not lost but feeds the central black hole. They showed that in this context the stellar mass-loss rate can explain the observed AGN luminosities. They also found that Solar abundances are reached in the gas in no more

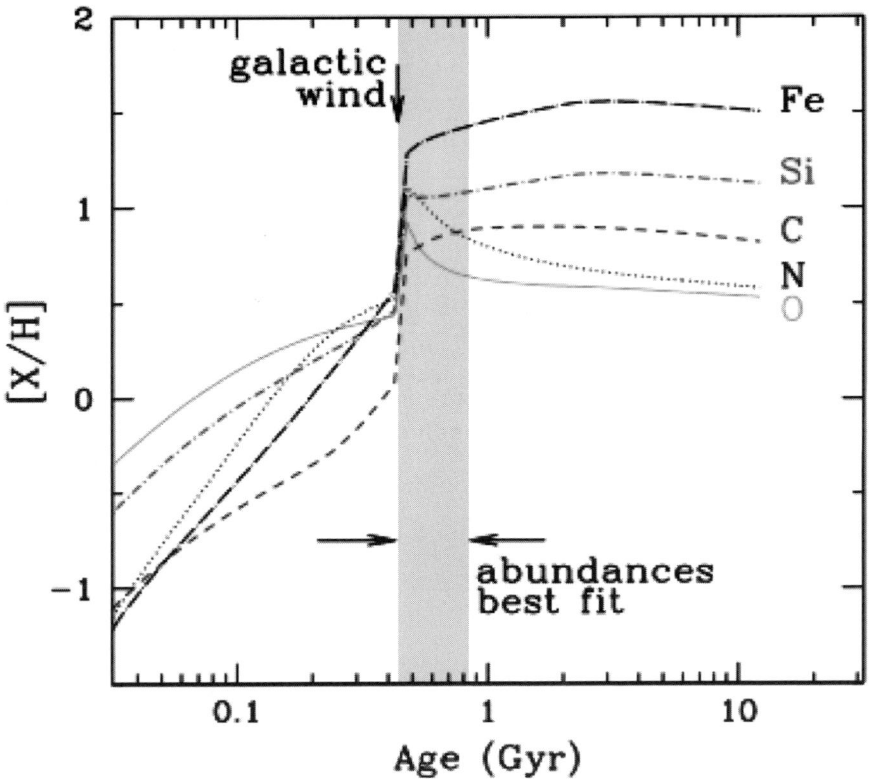

FIGURE 6.19. The temporal evolution of the abundances of several chemical elements in the gas of an elliptical galaxy with luminous mass of $10^{11} M_{\odot}$. Feedback effects are taken into account in the model (Pipino & Matteucci 2004), as described in Section 6.3. The arrow pointing down indicates the time for the occurrence of the galactic wind. After this time, the SF stops and the elliptical evolves passively. All the abundances after the time for the occurrence of the wind are those that we observe in the broad-emission-line region. The shaded area indicates the abundance sets which best fit the line ratios observed in the QSO spectra. From Maiolino *et al.* 2006.

than 10^8 yr, explaining in a natural way the standard emission lines observed in high-z QSOs. The predicted abundances could explain the data available at that time and solve the problem of the quasi-similarity of QSO spectra at different redshifts. Finally, they suggested also a criterion for establishing the ages of QSOs on the basis of the $[\alpha/\text{Fe}]$ ratios observed from broad emission lines; see also Hamman & Ferland (1993).

Much more recently, Maiolino *et al.* (2005, 2006) used more than 5000 QSO spectra from SDSS data to investigate the metallicity of the broad-emission-line region in the redshift range $2 < z < 4.5$ and over the luminosity range $-24.5 < M_B < -29.5$. They found substantial chemical enrichment in QSOs already at $z = 6$. Models for ellipticals by Pipino & Matteucci (2004) were used as a comparison with the data and they well reproduce the data, as one can see in Figure 6.19. In this figure the evolutions of the abundances of several chemical elements in the gas of a typical elliptical are shown. The elliptical suffers a galactic wind at around 0.4 Gyr since the beginning of star formation. This wind voids the galaxy of all the gas present at that time. After this time, the SF stops and the galaxy evolves passively. All the gas restored after the galactic-wind event by dying stars can in principle feed the central black hole, thus the abundances shown in Figure 6.19, after the time of the wind, can be compared with the abundances measured in the broad-emission-line region. As one can see, the predicted Fe abundance after the galactic wind is always higher than that of O, owing to the Type Ia SNe which continue to produce Fe even after SF has stopped. On the other hand, O and α-elements cease to be produced when the SF halts. The predicted abundances and those derived from the QSO spectra are in very good agreement and indicate ages for these objects of between 0.5 and 1 Gyr.

Finally, in the context of the joint formation of QSOs and ellipticals we recall the work of Granato *et al.* (2001), who include the energy feedback from the central AGN in ellipticals. This feedback produces outflows and stops the SF in a downsizing fashion, in agreement with the chemical properties of ETGs indicating a shorter period of SF for the more-massive objects.

6.4.6 *The chemical enrichment of the ICM*

The X-ray emission from galaxy clusters is generally interpreted as thermal bremsstrahblung in a hot gas (10^7–10^8 K). There are several emission lines (O, Mg, Si, S) including the strong Fe K-line at around 7 keV which was discovered by Mitchell *et al.* (1976). Iron is the best-studied element in clusters. For $kT \geq 3$ keV the intracluster medium (ICM) Fe abundance is constant and $\sim 0.3 X_{\mathrm{Fe}_\odot}$ in the central cluster regions; the existence of metallicity gradients seems evident only in some clusters (Renzini 2004). At lower temperatures, the situation is not so simple and the Fe abundance seems to increase. The first works on chemical enrichment of the ICM even preceded the discovery of the Fe line (Gunn & Gott 1972; Larson & Dinerstein 1975). In the following years other works appeared, such as those of Vigroux (1977), Himmes & Biermann (1988), and Matteucci & Vettolani (1988). In particular, Matteucci & Vettolani (1988) started a more detailed approach to the problem that was followed by David *et al.* (1991), Arnaud (1992), Renzini *et al.* (1993), Elbaz *et al.* (1995), Matteucci & Gibson (1995), Gibson & Matteucci (1997), Lowenstein & Mushotzky (1996), Martinelli *et al.* (2000), Chiosi (2000), and Moretti *et al.* (2003). In the majority of these papers it was assumed that galactic winds (mainly from ellipticals and S0 galaxies) are responsible for the chemical enrichment of the ICM. In fact, ETGs are the dominant type of galaxy in clusters and Arnaud (1992) found a clear correlation between the mass of Fe in clusters and the total luminosity of ellipticals. No such correlation was found for spirals in clusters. Alternatively, the abundances in the ICM are due to ram pressure stripping (Himmes & Biermann 1988) or derive from a chemical enrichment from pre-galactic Population III stars (White & Rees 1978).

In Matteucci & Vettolani (1988) the Fe abundance in the ICM relative to the Sun, $X_{\mathrm{Fe}}/X_{\mathrm{Fe}_\odot}$, was calculated as $(M_{\mathrm{Fe}})_{\mathrm{pred}}/(M_{\mathrm{gas}})_{\mathrm{obs}}$ to be compared with the observed ratio $(X_{\mathrm{Fe}}/X_{\mathrm{Fe}_\odot})_{\mathrm{obs}} = 0.3$–$0.5$ (Rothenflug & Arnaud 1985). They found a good agreement with the observed Fe abundance in clusters if all the Fe produced by ellipticals and S0, after SF has stopped, is eventually restored into the ICM and if the majority of gas in clusters has a primordial origin. Low values for [Mg/Fe] and [Si/Fe] at the present time were predicted, due to the short period of SF in ETGs and to the Fe produced by Type Ia SNe. With the Salpeter IMF they found that the Type Ia SNe contribute $\geq$50% of the total Fe in clusters. This leads to a bimodality of the $[\alpha/\mathrm{Fe}]$ ratios in the stars and in the gas in the ICM, since the stars have overabundances of $[\alpha/\mathrm{Fe}] > 0$ whereas the ICM should have $[\alpha/\mathrm{Fe}] \leq 0$. The same conclusion was drawn and given more emphasis later by Renzini *et al.* (1993). More recently, Pipino *et al.* (2002) computed the chemical enrichment of the ICM as a function of redshift by considering the evolution of the cluster luminosity function and an updated treatment of the SN feedback. They adopted Woosley & Weaver (1995) yields for Type II SNe and the Nomoto *et al.* (1997) W7 model for Type Ia SNe and a Salpeter IMF. They also predicted Solar or undersolar

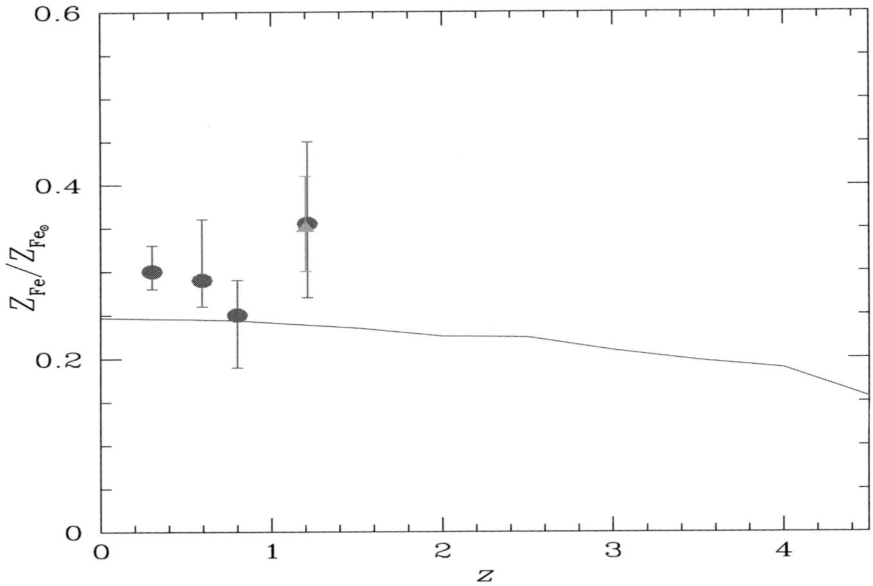

FIGURE 6.20. Observed Fe abundance and predicted Fe abundance in the ICM as a function of redshift: data from Tozzi *et al.* (2003)(dark circles), with a model (continuous line) from Pipino *et al.* (2002), in which the formation of ETGs was assumed to occur at $z = 8$.

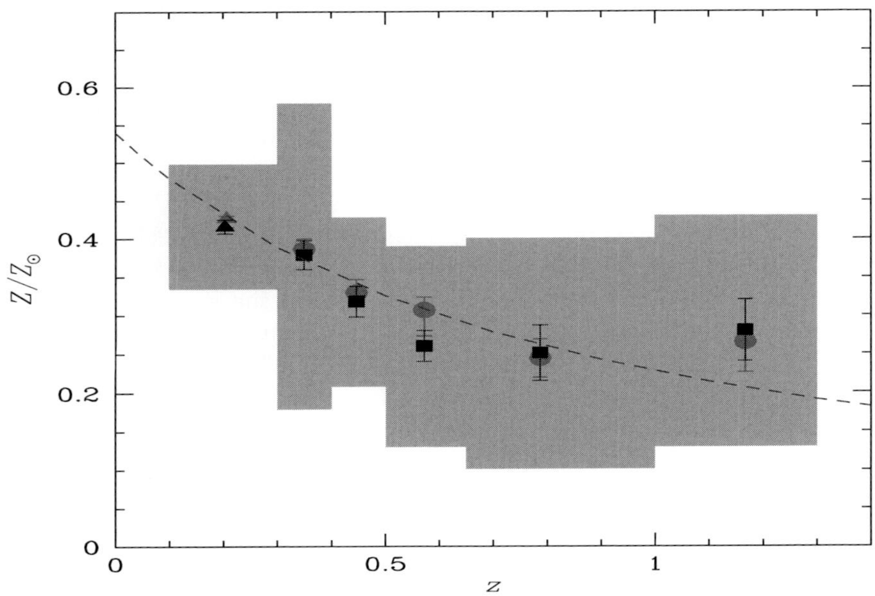

FIGURE 6.21. New data (always relative to Fe) from Balestra *et al.* (2006) showing an increase of the Fe abundance in the ICM on going from $z = 1$ to $z = 0$. Error bars refer to the 1σ confidence level. The big shaded area represents the rms dispersion. From Balestra *et al.* (2006).

[α/Fe] ratios in the ICM. The observational data on abundance ratios in clusters are still uncertain and vary from cluster centers, where they tend to be Solar or undersolar, to the outer regions, where they tend to be oversolar (e.g. Tamura *et al.* 2004). So, no firm conclusions can be drawn on this point. Concerning the evolution of the Fe abundance in the ICM as a function of redshift, most of the above-mentioned models predict very little or no evolution of the Fe abundance from $z = 1$ to $z = 0$ (Pipino *et al.* 2002). This prediction seemed to be in good agreement with data from Tozzi *et al.* (2003) as shown in Figure 6.20. However, more recently, more data on Fe abundance for high-redshift clusters have appeared, showing a different behavior.

In Figure 6.21 we show the data of Balestra *et al.* (2006), who claim to find an increase, by at least a factor of two, of the Fe abundance in the ICM on going from $z = 1$ to $z = 0$. Clearly, if we assume that only ellipticals have contributed to the Fe abundance in the ICM, this effect is difficult to explain unless we assume that recent SF has occurred in ellipticals. Another possible explanation could be that spiral galaxies contribute Fe when they become S0 as a consequence of ram pressure stripping, and this morphological transformation might have started just at $z = 1$.

6.4.7 *Conclusions on the enrichment of the ICM*

From what has been discussed before, we can draw the following conclusions.

- Elliptical galaxies are the dominant contributors to the abundances and energetic content of the ICM. A constant Fe abundance of $\sim 0.3\, Z_{\mathrm{Fe}_\odot}$ is found in the central regions of clusters hotter than 3 keV (Renzini 2004).
- Good models for the chemical enrichment of the ICM should reproduce the iron mass measured in clusters plus the [α/Fe] ratios inside galaxies and in the ICM as well as the Fe mass-to-light ratio (IMLR $= M_{\mathrm{Fe}_{\mathrm{ICM}}}/L_{\mathrm{B}}$, with L_{B} being the total blue luminosity of member galaxies, as defined by Renzini *et al.* (1993). Abundance ratios are very powerful tools for imposing constraints on the evolution of ellipticals and of the ICM.
- Models in which a top-heavy IMF for the galaxies in clusters is not assumed (a Salpeter IMF can reproduce best the properties of local ellipticals) predict $[\alpha/Fe] > 0$ inside ellipticals and $[\alpha/\mathrm{Fe}] \leq 0$ in the ICM. Observed values are still too uncertain to allow one to draw firm conclusions on this point.

Acknowledgments

This research has been supported by INAF (the Italian National Institute for Astrophysics), Project PRIN-INAF-2005-1.06.08.16

REFERENCES

Alibès, A., Labay, J., Canal, R. 2001, A&A, 370, 1103

Aloisi, A., Savaglio, S., Heckman, T. M., Hoopes, C. G., Leitherer, C., Sembach, K. R. 2003, ApJ, 595, 760

Argast, D., Samland, M., Gerhard, O. E., Thielemann, F.-K. 2000, A&A, 356, 873

Arimoto, N., Yoshii, Y. 1987, A&A 173, 23

Arnaud, M., Rothenflug, R., Boulade, O., Vigroux, L., Vangioni-Flam, E. 1992, A&A, 254, 49

Asplund, M., Grevesse, N., Sauval, A. J. 2005, ASP (Astronomical Society of the Pacific) Conf. Series, 336, p. 55

Balestra, I., Tozzi, P., Ettori, S., Rosati, P., Borgani, S., Mainieri, V., Norman, C., Viola, M. 2006, A&A in press, astro-ph/0609664

Barbuy, B., Grenon, M. 1990, in *Bulges of Galaxies*, eds. B. J. Jarvis & D. M. Terndrup, ESO/CTO Workshop, p. 83

Barbuy, B., Ortolani, S., Bica, E. 1998, A&AS, 132, 333

Bender, R., Burstein, D., Faber, S. M. 1993, ApJ, 411, 153

Berman, B. C., Suchov, A. A., 1991, Astrophys. Space Sci. 184, 169

Bernardi, M., Renzini, A., da Costa, L. N., Wegner, G. *et al.* 1998, ApJ, 508, L143

Boissier, S., Prantzos, N. 1999, MNRAS, 307, 857

Boissier, S., Boselli, A., Prantzos, N., Gavazzi, G. 2001, MNRAS, 321, 733

Bradamante, F., Matteucci, F., D'Ercole, A. 1998, A&A, 337, 338

Calura, F. 2004, PhD Thesis, Trieste University

Calura, F., Matteucci, F., Vladilo, G. 2003, MNRAS, 340, 59

Carollo, C. M., Danziger, I. J., Buson, L. 1993, MNRAS, 265, 553

Cayrel, R., Depagne, E., Spite, M., Hill, V., Spite, F., Franois, P., Plez, B., Beers, T. *et al.* 2004, A&A, 416, 117

Chabrier, G. 2003, PASP, 115, 763

Chang, R. X., Hou, J. L., Shu, C. G., Fu, C. Q. 1999, A&A, 350, 38

Chiappini, C., Hirschi, R., Meynet, G., Ekström, S., Maeder, A., Matteucci, F. 2006, A&A, 449, L27

Chiappini, C., Matteucci F., Gratton R. 1997, ApJ, 477, 765

Chiappini, C., Matteucci, F., Meynet, G. 2003b, A&A, 410, 257

Chiappini, C., Matteucci, F., Padoan, P. 2000, ApJ, 528, 711

Chiappini, C., Matteucci, F., Romano, D. 2001, ApJ, 554, 1044

Chiappini, C., Romano, D., Matteucci, F. 2003a, MNRAS, 339, 63

Chiosi, C. 1980, A&A, 83, 206

Chiosi, C. 2000, A&A, 364, 423

Colless, M., Burstein, D., Davies, R. L., McMahan, R. K., Saglia, R. P., Wegner, G. 1999, MNRAS, 303, 813

Collin-Souffrin, S., Joly, M., Pequignot, D., Dumont, S. 1986, A&A, 166, 27

David, L. P., Forman, W., Jones, C. 1991, ApJ, 376, 380

Davies, R. L., Sadler, E. M., Peletier, R. F. 1993, MNRAS, 262, 650

Dopita, M. A., Ryder, S. D. 1994, ApJ, 430, 163

Eggen, O. J., Lynden-Bell, D., Sandage, A. R. 1962, ApJ, 136, 748

Elbaz, D., Cesarsky, C. J., Fadda, D., Aussel, H. *et al.* 1999, A&A, 351, 37

Ellison, S. L., Songaila, A., Schaye, J., Pettini, M. 2000, AJ, 120, 1175

Erb, D. K., Shapley, A. E., Pettini, M., Steidel, C. C., Reddy, N. A., Adelberger, K. L. 2006, ApJ, 644, 813

François, P., Matteucci, F., Cayrel, R., Spite, M., Spite, F., Chiappini, C. 2004, A&A, 421, 613

Garnett, D. R., Shields, G. A. 1987, ApJ, 317, 82

Garnett, D. R., Skillman, E. D., Dufour, R. J., Shields, G. A. 1997, ApJ, 481, 174

Gibson, B. K., Matteucci, F. 1997, ApJ, 475, 47

Goswami, A., Prantzos, N. 2000, A&A, 359, 191

Granato, G. L., Silva, L., Monaco, P., Panuzzo, P., Salucci, P., De Zotti, G., Danese, L. 2001, MNRAS, 324, 757

Greggio, L., Renzini, A. 1983, A&A, 118, 217

Grevesse, N., Sauval, A. J. 1998, Space Sci. Rev., 85, 161

Gunn, J. E., Gott, J. R. III 1972, ApJ, 176, 1

Hachisu, I., Kato, M., Nomoto, K. 1996, ApJ, 470, L97

Hachisu, I., Kato, M., Nomoto, K. 1999, ApJ, 522, 487

Hamman, F., Ferland, G. 1993, ApJ, 418, 11

Henry, R. B. C., Edmunds, M. G., Koeppen, J. 2000, ApJ, 541, 660

Hill, V., François, P., Spite, M., Primas, F., Spite, F. 2000, A&A, 364, L19

Himmes, A., Biermann, P. 1988, A&A, 86, 11

Holweger, H. 2001, Joint SOHO/ACE workshop *Solar and Galactic Composition*, ed. R. F. Wimmer-Schweingruber (New York: American Institute of Physics), p. 23

Iben, I. Jr., Tutukov, A. V. 1984a, ApJS, 54, 335

Iben, I. Jr., Tutukov, A. 1984b, ApJ, 284, 719

Ishimaru, Y., Arimoto, N. 1997, PASJ, 49, 1

Iwamoto, K., Brachwitz, F., Nomoto, K., Kishimoto, N., Umeda, H., Hix, W. R., Thielemann, F.-K. 1999, ApJS, 125, 439

Izotov, Y. I., Stasińska, G., Meynet, G., Guseva, N. G., Thuan, T. X. 2006, A&A, 448, 955

Jimenez, R., Padoan, P., Matteucci, F., Heavens, A. F. 1998, MNRAS 299, 123

Jorgensen, I. 1999, MNRAS, 306, 607

Josey, S. A., Arimoto, N. 1992, A&A, 255, 105

Kauffmann, G., Charlot, S., White, S. D. M. 1996, MNRAS, 283, L117

Kauffmann, G., White, S. D. M., Guiderdoni, B. 1993, MNRAS, 264, 201

Kennicutt, R. C. Jr. 1989, ApJ, 344, 685

Kennicutt, R. C. Jr. 1998, ARAA, 36, 189

Kobayashi, C., Arimoto, N. 1999, ApJ, 527, 573

Kobulnicky, H. A., Skillman, E. D. 1996, ApJ, 471, 211

Kodama, T., Yamada, T., Akiyama, M., Aoki, K., Doi, M., Furusawa, H., Fuse, T., Imanishi, M. *et al.* 2004, ApJ, 492, 461

Kroupa, P., Tout, C. A., Gilmore, G. 1993, MNRAS, 262, 545

Kuntschner, H., Lucey, J. R., Smith, R. J., Hudson, M. J., Davies, R. L. 2001, MNRAS, 323, 625

Lacey, C. G., Fall, S. M. 1985, ApJ, 290, 154

Lanfranchi, G., Matteucci, F. 2003, MNRAS, 345, 71

Lanfranchi, G., Matteucci, F. 2004, MNRAS, 351, 1338

Larson, R. B. 1972, Nature, 236, 21

Larson, R. B. 1974, MNRAS, 169, 229

Larson, R. B. 1976, MNRAS, 176, 31

Larson, R. B. 1998, MNRAS, 301, 569

Larson, R. B., Dinerstein, H. L. 1975, PASP, 87, 911

Lecavelier des Etangs, A., Desert, J.-M., Kunth, D. 2003, A&A, 413, 131

Lequeux, J., Kunth, D., Mas-Hesse, J. M., Sargent, W. L. W. 1995, A&A, 301, 18

Lequeux, J., Peimbert, M., Rayo, J. F., Serrano, A., Torres-Peimbert, S. 1979, A&A, 80, 155

Loewenstein, M., Mushotzky, F. 1996, ApJ, 466, 695

Maeder, A. 1992, A&A, 264, 105

Maiolino, R., Cox, P., Caselli, P., Beelen, A., Bertoldi, F., Carilli, C. L., Kaufman, M. J., Menten, K. M. *et al.* 2005, A&A, 440, L51

Maiolino, R., Nagao, T., Marconi, A., Schneider, R., Pedani, M., Pipino, A, Matteucci, F. *et al.* 2006, Mem. Soc. Astron. It., 77, 643

Mannucci, F., Della Valle, M., Panagia, N., Cappellaro, E., Cresci, G., Maiolino, R., Petrosian, A., Turatto, M. 2005, A&A, 433, 807

Mannucci, F., Della Valle, M., Panagia, N. 2006, MNRAS, 370, 773

Marconi, G., Matteucci, F., Tosi, M. 1994, MNRAS, 270, 35

Martin, C. L. 1996, ApJ, 465, 680

Martin, C. L. 1998, ApJ, 506, 222

Martin, C. L. 1999, ApJ, 513, 156

Martinelli, A., Matteucci, F., Colafrancesco, S. 2000, A&A, 354, 387

Matteucci, F. 1994, A&A, 288, 57

Matteucci, F. 2001, *The Chemical Evolution of the Galaxy* (Dordrecht: Kluwer)

Matteucci, F., Chiosi, C. 1983, A&A, 123, 121

Matteucci, F., François, P. 1989, MNRAS, 239, 885

Matteucci, F., Gibson, B. K. 1995, A&A, 304, 11

Matteucci, F., Greggio, L. 1986, A&A, 154, 279

Matteucci, F., Molaro, P., Vladilo, G. 1997, A&A, 321, 45

Matteucci, F., Padovani, P. 1993, ApJ, 419, 485

Matteucci, F., Raiteri, C. M., Busso, M., Gallino, R., Gratton, R. 1993, A&A, 272, 421

Matteucci, F., Recchi, S. 2001, ApJ, 558, 351

Matteucci, F., Tornambé, A. 1987, A&A, 185, 51

Matteucci, F., Vettolani, G. 1988, A&A, 202, 21

McWilliam, A., Rich, R. M. 1994, ApJS, 91, 749

Menanteau, F., Jimenez, R., Matteucci, F. 2001, ApJ, 562, L23

Meynet, G., Maeder, A. 2002, A&A, 390, 561

Moretti, A., Portinari, L., Chiosi, C. 2003, A&A, 408, 431

Nomoto, K., Hashimoto, M., Tsujimoto, T., Thielemann, F.-K. *et al.* 1997, Nucl. Phys. A, 616, 79

Oey, M. S. 2000, ApJ, 542, L25

Padovani, P., Matteucci, F. 1993, ApJ, 416, 26

Papaderos, P., Fricke, K. J., Thuan, T. X., Loose, H.-H. 1994, A&A, 291, L13

Pardi, M. C., Ferrini, F., Matteucci, F. 1994, ApJ, 444, 207

Peletier, R. 1989, PhD Thesis, University of Groningen

Pilyugin, I. S. 1993, A&A, 277, 42

Pipino, A., Matteucci, F., Borgani, S., Biviano, A. 2002, New Astron., 7, 227

Pipino, A., Matteucci, F. 2004, MNRAS, 347, 968

Pipino, A., Matteucci, F. 2006, MNRAS, 365, 1114

Portinari, L., Chiosi, C. 2000, A&A, 355, 929

Prantzos, N. 2003, A&A, 404, 211

Prantzos, N., Boissier, S. 2000, MNRAS, 313, 338

Recchi, S., Matteucci, F.,D'Ercole, A. 2001, MNRAS, 322, 800

Recchi, S., Matteucci, F., D'Ercole, A., Tosi, M. 2004, A&A, 426, 37

Renzini, A. 2004, in *Clusters of Galaxies: Probes of Cosmological Structure and Galaxy Evolution*, ed. J. S. Mulchay, A. Dressler, & A. Oemler (Cambridge: Cambridge University Press), p. 260

Renzini, A., Ciotti, L. 1993, ApJ, 416, L49

Renzini, A., Ciotti, L., D'Ercole, A., Pellegrini, S. 1993, ApJ, 416, L49

Rothenflug, R., Arnaud, M. 1985, A&A, 144, 431

Salpeter, E. E. 1955, ApJ, 121, 161

Sandage, A. 1986, A&A, 161, 89

Scalo, J. M. 1986, Fund. Cosmic Phys. 11, 1

Scalo, J. M. 1998, in *The Stellar Initial Mass Function* (San Francisco, CA: Astronomical Society of the Pacific), p. 201

Schechter, P. 1976, ApJ, 203, 297

Schmidt, M. 1959, ApJ, 129, 243

Schmidt, M. 1963, ApJ, 137, 758

Schneider, R., Salvaterra, R., Ferrara, A., Ciardi, B. 2006, MNRAS, 369, 825

Searle, L., Zinn, R. 1978, ApJ, 225, 357

Skillman, E. D., Terlevich, R., Melnick, J. 1989, MNRAS, 240, 563

Springel, V., Hernquist, L. 2003, MNRAS, 339, 312

Tamura, T., Kaastra, J. S., den Herder, J. W. A., Bleeeker, J. A. M., Peterson, J. R. 2004, A&A, 420, 135

Thielemann, F. K., Nomoto, K., Hashimoto, M. 1996, ApJ, 460, 408

Thomas, D., Greggio, L., Bender, R. 1999, MNRAS, 302, 537

Thomas, D., Maraston, C., Bender, R., Mensez de Oliveira, C. 2005, ApJ, 621, 673

Thomas, D., Maraston, C., Bender, R. 2002, in *Reviews in Modern Astronomy*, ed. R. E. Schielicke, Vol. 15, p. 219

Thuan, T. X., Izotov, Y. I., Lipovetsky, V. A. 1995, ApJ, 445, 108

Tinsley, B. M. 1980, Fund. Cosmic Phys., 5, 287

Tinsley, B. M., Larson, R. B. 1979, MNRAS, 186, 503

Tornambé, A. 1989, MNRAS, 239, 771

Tosi, M. 1988, A&A, 197, 33

Tosi, M., Aloisi, A., Annibali, F. 2006, IAU Symp. 35, p. 19

Tozzi, P., Rosati, P., Ettori, S., Borgani, S., Mainieri, V., Norman, C. 2003, ApJ, 593, 705

Tremonti, C. A., Heckman, T. M., Kauffmann, G., Brinchmann, J., Charlot, S., White, S. D. M., Seibert, M., Peng, E. W. *et al.* 2004, ApJ, 613, 898

Tsujimoto, T., Shigeyama, T., Yoshii, Y. 1999, ApJ, 519, 63

van den Hoek, L. B., Groenewegen, M. A. T. 1997, A&AS, 123, 305

Vigroux, L. 1977, A&A, 56, 473

Vladilo, G. 2002, A&A, 391, 407

Weiss, A., Peletier, R. F., Matteucci, F. 1995, A&A, 296, 73

Whelan, J., Iben, I. Jr. 1973, ApJ, 186, 1007

White, S. D. M., Rees, M. J. 1978, MNRAS, 183, 341

Wills, B. J., Netzer, H., Wills, D. 1985, ApJ, 288, 94

Woosley, S. E., Weaver, T. A. 1995, ApJS, 101, 181 (WW95)

Worthey, G. 1994, ApJS, 95, 107

Worthey, G., Faber, S. M., Gonzalez, J. J. 1992, ApJ, 398, 69

Worthey, G., Trager, S. C., Faber, S. M. 1995, ASP Conf. Ser., 86, 203

Wyse, R. F. G., Gilmore, G. 1992, AJ, 104, 144

Wyse, R. F. G., Silk, J. 1989, ApJ, 339, 700

Zaritsky, D., Kennicutt, R. C., Huchra, J. P. 1994, ApJ, 420, 87

7. Galactic sources of emission lines

STEPHEN S. EIKENBERRY

7.1. Introduction

Ultimately, the overwhelming majority of emission-line sources in the Universe are "galactic sources" – meaning discrete objects located within a particular galaxy (rather than some global property of a galaxy or some source not located in a galaxy). However, the most common of these, H II regions, are so ubiquitous that they are being covered elsewhere in this volume as the "baseline" source of emission lines. In addition, most of the other chapters are devoted to line emission either integrated over entire galaxies (or significant portions thereof) or from active galactic nuclei.

Given that coverage, I will focus this chapter primarily on "stellar" sources of line emission in the Milky Way other than H II regions – including young stellar objects, massive and/or evolved stars, and stellar remnants (planetary nebulae, supernova remnants, and accreting compact objects in binary systems). I will also put considerable emphasis on emission lines with rest wavelengths in the near-infrared waveband, due to the importance of this waveband for probing the dusty planar regions of the Milky Way where most of these sources are to be found.

In the sections below, I will begin with a review of important diagnostic optical emission lines and a more-detailed overview of key (rest-wavelength) infrared emission lines. I will then move on to "nebular" sources of emission lines (omitting H II regions, but including planetary nebulae and supernova remnants). Next, I will cover "stellar" sources (including young stellar objects, "main-sequence" massive stars, and evolved Wolf–Rayet stars). After that, I will move on to emission lines arising from accreting stellar remnants (cataclysmic variables, neutron-star X-ray binaries, and black-hole X-ray binaries).

7.2. Overview of infrared emission lines

As noted above, other chapters in this volume address some of the primary diagnostic optical emission lines associated with H II regions. These include the Balmer series of hydrogen emission lines, as well as O II, N II, and S II. However, rest-wavelength infrared (IR) lines, particularly those in the wavelength range of $1.0 - 2.5\,\mu\text{m}$, while less widely known to most of the astronomical community, are of critical importance for studying Galactic sources of line emission. In the following subsections, I will review why IR emission is important for Galactic objects, and observational aspects for IR line emission (detectors/instruments, IR atmospheric windows, IR atmospheric emission). I will then move on to review key emission-line features from hydrogen (particularly Paschen and Brackett series), heavier-element emission, and particularly molecular line emission.

7.2.1 *Why infrared emission lines?*

Many readers may ask the question, "Infrared – why bother?" (In fact, many professional astronomers feel that way even now.) From a simple atomic-physics perspective, the IR looks less than promising. Hydrogen, the most-abundant element in the Universe, has its strongest transitions – the Lyman series – in the ultraviolet (UV) waveband, rather far from the IR. Of course, then, one could ask "Why the optical?," given that the Balmer

The Emission-Line Universe, ed. J. Cepa. Published by Cambridge University Press.

series is intrinsically weaker than Lyman lines. The primary answer to that question is that the Universe in general (and the Earth's atmosphere in particular) is not transparent to Lyman radiation. Since most cosmic hydrogen is in its ground state, it will absorb Lyman photons. Interstellar dust also wreaks havoc on these short wavelengths, both for Lyman series and for other non-hydrogen transitions. As a result, optical lines such as the Balmer series are often more powerful probes than the Lyman series (at least in the local Universe) despite their weaker atomic strengths.

While the Earth's atmosphere is less "friendly" to the IR waveband than it is to the optical (see Section 7.2.2), "cosmic transparency" is even better for the IR – to the point that for many Galactic situations it is the *only* waveband of the UV, optical, and IR set which is useful for observing emission-line sources. Again, the primary reason for this is the simple physics of radiative processes. With typical grain sizes of <1.0 μm, interstellar dust grains simply do not have much cross-section for scattering or absorbing photons at wavelengths $\lambda > 1.0\,\mu\mathrm{m}$. These dust grains are of course strongly concentrated in the plane of the Milky Way, as are most star-forming regions, massive stars, planetary nebulae, supernovae remnants, etc., making the IR waveband invaluable for studies of these objects.

Figure 7.1 shows a view of our Galactic Center as seen in the "optical" B, R, and I bands, as compared with the near-IR J (1.1–1.4 μm), H (1.5–1.8 μm) and K_s (1.95–2.4 μm) bands. While the optical pictures are actually deeper than the IR in terms of magnitudes, the vastly superior penetrating power of the IR ($A_K \sim 0.1\ A_V$ – a factor of ten in magnitudes!!) reveals many more sources of emission. For the Galactic Center, $A_K \sim 3$ mag, meaning that >6% of the emitted IR photons make it through the dust, while $A_V \sim 30$ mag, meaning that only one optical photon of 10^{12} emitted makes it through.

7.2.2 *Observing in the infrared waveband*

While we have just seen that "longer is better" for observing wavelengths, at least in terms of Galactic dust extinction, very important non-cosmic effects usually limit sensitive IR observations to $\lambda < 2.5\,\mu\mathrm{m}$ (from both space and the ground, for now). There are several factors that contribute to this approximate limit, which I will now review.

7.2.2.1 Infrared observing technology

In recent years, IR detector and instrument technology has seen tremendous gains, and now closely resembles optical detection technology in many ways. However, there are important – even critical – differences between the two, which tend to make IR instrumentation much more challenging than optical. These differences all spring from the fundamental fact that in this waveband – 1.0–2.5 μm – the outstanding optical detection technology of the silicon-based charge-coupled device (CCD) is no longer functional. The energies of individual IR photons are simply too low to excite valence-band electrons across the silicon bandgap into the conduction band where they can be collected and "detected," or measured. This requires erstwhile IR observers to seek out and use alternative semiconductor devices for their instrumentation.

The IR detectors of choice for this waveband currently come in two primary flavors – HgCdTe and InSb arrays. Owing to the trillions of dollars invested in silicon technology over the past few decades, neither of these can match the cosmetic quality, large format, and relatively low cost of CCD detectors. HgCdTe is a ternary alloy and its tunable composition allows one to vary its bandgap energy and thus cutoff wavelength for detection. The most well-developed HgCdTe mixture, in terms of cosmetic quality, array format, quantum efficiency, noise properties, and simple number of IR arrays in

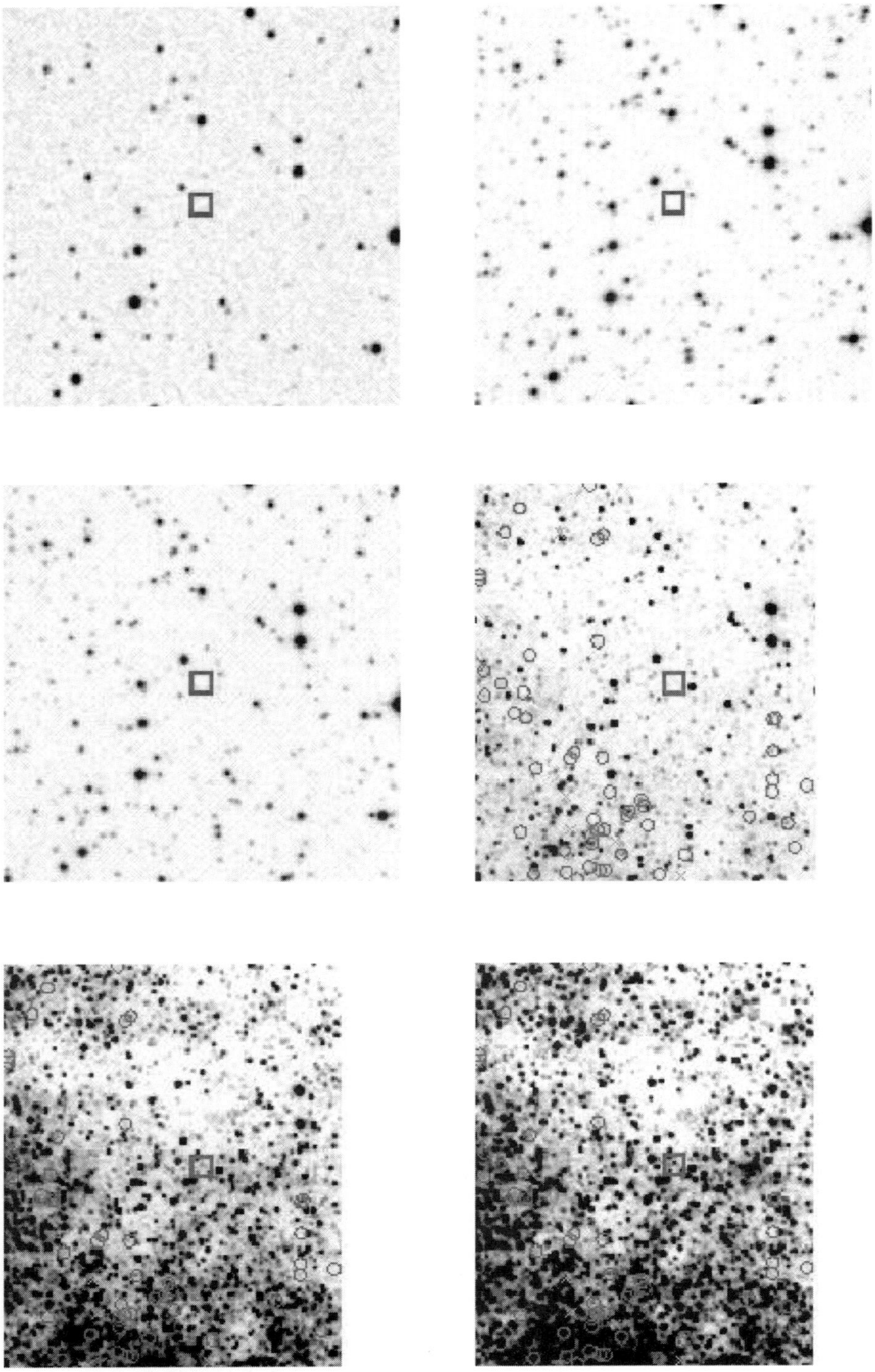

FIGURE 7.1. Multi-wavelength views of the Galactic Center region in the DPOSS B band (top left); DPOSS R band (top right); DPOSS IR band (middle left); 2MASS J band (middle right); 2MASS H band (bottom left); and 2MASS K_s band (bottom right).

use, has a cutoff wavelength of 2.5 μm. The current state-of-the-art HgCdTe arrays have 2048 × 2048-pixel formats, quantum efficiency of ∼70%, and reading noise of about ten electrons. InSb technology also produces high-quality arrays, with a bandpass cutoff of ∼5.5 μm. InSb arrays have similar formats to HgCdTe arrays, and higher quantum efficiency (>90% – nearly perfect). However, the advantage of the latter can be offset by the higher reading noise (about 25 electrons) and the disadvantage (for ground-based applications) of the longer cutoff wavelength.

In addition to the relatively poorer quality of IR detectors compared with CCD technology, their longer-wavelength cutoff also enforces a crucial global design requirement – cryogenic cooling. If the detectors have temperatures such that $kT \sim E_{\mathrm{bandgap}}$, then the thermal distribution of electron energies will excite electrons over the semiconductor bandgap to the conduction band, making them appear identical to photoelectrons – a primary source of "dark current" in detectors. Since $E_{\mathrm{bandgap}} = hc/\lambda_{\mathrm{c}}$, where λ_{c} is the cutoff wavelength of the detector material, to avoid overwhelming dark current the detectors must be cooled to $T \ll hc/k\lambda_{\mathrm{c}}$. For HgCdTe with a 2.5-μm cutoff, this corresponds to $T \sim 65$ K, whereas for InSb with a 5.5-μm cutoff, it corresponds to $T \sim 30$ K. Similarly, any optical materials (lenses/mirrors) and their support structures must also be cooled enough to prevent the Wien tail of their blackbody glow from swamping the detectors (albeit, this requirement usually allows slightly higher temperatures than for the detectors themselves). These temperatures require large and cumbersome vacuum/cryogenic systems for IR instruments, in addition to unusual, difficult, and expensive optical and structural materials.

7.2.2.2 The infrared atmosphere

In addition to the differences in instrument-technology requirements, another major issue for IR observations (at least from the ground) is the Earth's atmosphere. While the Earth's atmosphere is largely transparent and dark in the optical waveband, it is both partially opaque AND largely bright in the IR (!). Figure 7.2 shows the atmospheric transmission curve for the Earth over the UV–IR bandpass.

In the waveband range 1.0–2.5 μm, the atmosphere transmits well only in certain atmospheric "windows." These constitute the near-IR bandpasses of the J(1.1–1.4 μm), H(1.5–1.8 μm), and K(1.95–2.45 μm) bands. Note that there are also windows at longer wavelengths than this, which are even less susceptible to dust extinction than JHK bands. However, at these wavelengths, thermal emission at temperatures commonly found in the Earth's atmosphere dominates the cosmic "sky background," making sensitive observations at these wavelengths very difficult from the ground (with the possible exception of from very cold locations such as the South Pole).

While thermal atmospheric emission is typically small in the JHK bandpass (except on warm nights at the long end of the K band – which has led to the common use of the shorter-cutoff K_{s} filter), the atmosphere is still much brighter than in the optical waveband. In the J and H bands, as well as the short part of the K band, this is primarily due to a veritable forest of telluric OH emission lines (see Figure 7.3), while thermal emission begins to become non-negligible in the long-wavelength end of the K band.

7.2.3 *Important infrared emission lines*

While Galactic objects certainly have optical emission, and much of the observational work to date revolves around them, those lines will be presented and reviewed elsewhere in this volume. There are several important sets of IR emission lines that are key for studies of Galactic sources that have become extinct, which will not be reviewed elsewhere, so I will go over them here.

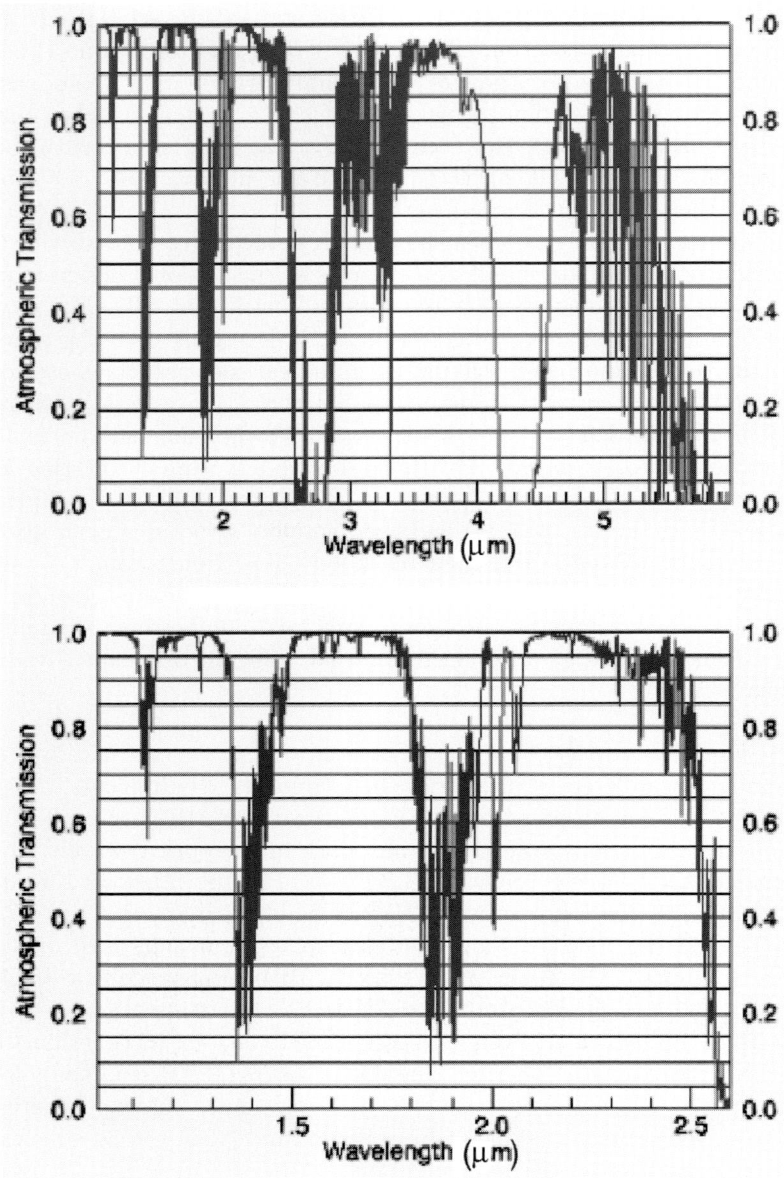

FIGURE 7.2. An atmospheric-transmission curve for Mauna Kea (adapted from Gemini Observatory website).

7.2.3.1 Hydrogen emission in the infrared

The primary "infrared" hydrogen emission series are the Paschen and Brackett series, with their lowest energy levels being 3 and 4, respectively. This is to be compared with the "ultraviolet" Lyman series (lowest energy level 1) and the "optical" Balmer series (lowest energy level 2). Table 7.1 summarizes the key transitions for each.

An important factor to note from the above is that none of the strongest "α" transitions for these series lies in a wavelength range easily observable from the ground. The strongest Paschen-series line observable from the ground is the Paβ transition, but for Brackett series the strongest is Brγ. Similarly, all but the transitions near the continuum limit

TABLE 7.1. Hydrogen emission lines in the infrared

Line	Transition	Wavelength (μm)
Paα	4–3	1.875
Paβ	4–3	1.281
Paγ	4–3	1.094
Paδ	4–3	1.004
Pa limit	inf–3	0.820
Brα	5–4	4.053
Brβ	6–4	2.626
Brγ	7–4	2.166
Brδ	8–4	1.945
Br limit	inf–3	1.818
Pfα	6–5	7.460
Pfβ	7–5	4.650
Pfγ	8–5	3.720
Pfδ	9–5	3.300
Pf limit	inf–5	2.280

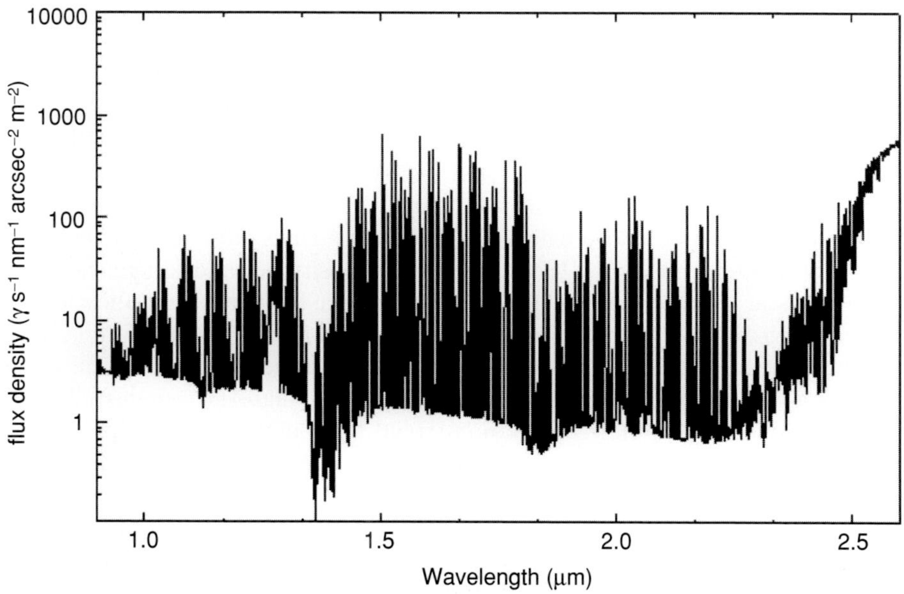

FIGURE 7.3. An IR atmospheric emission spectrum for $H_2O = 1.6$ mm and air mass = 1.5 (adapted from Gemini Observatory website).

of the Pfund series are either blocked by the atmosphere or lie within the thermally dominated portion of the IR waveband.

7.2.3.2 Metal lines

Here, I use "metal" in the astronomical sense (which is to say, the least physically justified) to indicate all elements other than hydrogen. Table 7.2 gives a list of a few important observable lines.

TABLE 7.2. Some metal emission lines in the infrared

Line	Wavelength (μm)
He I	1.083
He I	2.058
He I	2.112
He II	2.189
Fe II	2.089
[Fe II]	1.643

7.2.3.3 Molecular lines

Another important difference between the IR and optical wavebands is that in the IR we can see important molecular features in emission. Some features, particularly hydrides and H_2O "steam bands," are important in *absorption* in the atmospheres of late-type stars and brown dwarfs in the sub-micrometer region. However, these are seldom seen in emission. Therefore, the most-prominent *emission* features are typically from molecular hydrogen (H_2) and carbon monoxide (CO).

7.3. Galactic nebular sources of emission lines

As noted in the introduction to this chapter, the "fundamental unit" for emission-line studies is typically the H II region. Thus we also begin the discussion of Galactic sources with H II regions (which in any case are fundamentally Galactic in nature). However, due to the emphasis of other chapters on these, I will simply emphasize how Galactic studies of distant H II regions can differ from "standard" H II-region studies. After that, I will move on to planetary nebulae – which can be simply envisioned as H II regions where the affected interstellar medium (ISM) has the compositional and kinematic imprint of the progenitor star. Note of course that this imprint has non-trivial implications for the emission lines arising from planetary nebulae as opposed to H II regions. Similar analogies apply to supernova remnants – though here the "kinematic imprint" of the progenitor star includes a shock wave moving as fast as at $\sim 0.1c$ through the circumstellar medium and the ISM, which is of course highly non-trivial for the emission-line properties, and typically dominates line excitation from the hot (or, for old remnants, already-cooled) interior.

In the subsections below, I address each of these source classes in turn.

7.3.1 H II *regions*

Once again, I assume that the detailed properties of H II regions (both Galactic and extragalactic) are covered by other chapters in this volume. However, I also assume that few/none of these will emphasize the *infrared* properties of H II regions. If they are available, observations of optical (or better yet, UV) emission lines such as the Balmer (Lyman) series provide superior physical probes of H II regions. However, since much of our Galaxy's star-forming mass lies in regions with $A_V > 10$ mag, IR observations are often the only practical choice available.

As their name implies, hydrogen lines often provide the most diagnostic insights into H II regions in the optical, and that truism extends to the IR region as well. Furthermore, even for relatively nearby H II regions, differential extinction (both along the line of sight

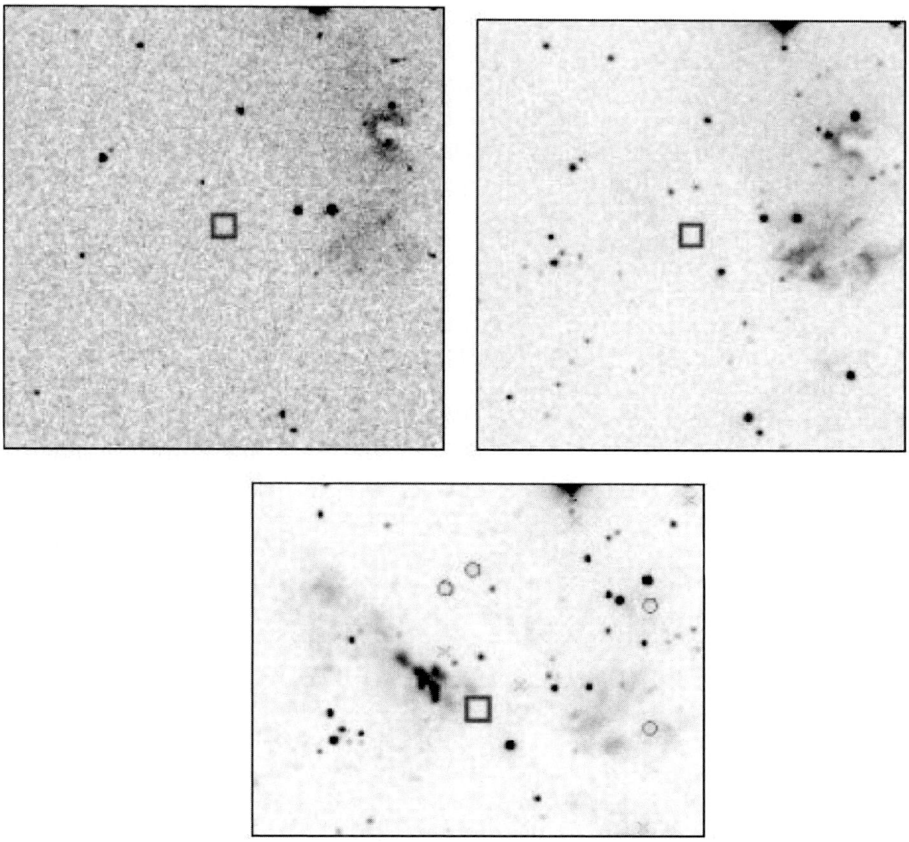

FIGURE 7.4. Multi-wavelength views of the Cepheus A star-forming H II region: POSS B band (top left); POSS "IR" band (top right); and 2MASS J band (bottom).

and within the immediate vicinity of the H II region itself) may make IR observations even more powerful than optical ones in some aspects (see Figure 7.4).

However, the fact that the "α" transitions for the Paschen and Brackett series are not available for high-sensitivity observations from the ground presents a non-trivial drawback. For instance, one of the most-powerful basic diagnostics for H II regions is measurement of the "Balmer decrement" (from Hα to Hβ). One could try to measure a "Paschen decrement," but Paα lies in the middle of a telluric "band of avoidance" between the atmospheric H and K windows, generally requiring space-based observations for this measurement (though some novel approaches being developed may alleviate, if not necessarily eliminate, this problem). Similarly, a "Brackett decrement" using Brα and Brγ **is** feasible from the ground – but Brα is well into the thermal-background-dominated regime, and lies at the ragged edge of an atmospheric window (thus requiring space-based observations, and a cooled telescope even then, for truly high sensitivity).

One *can* use an alternative approach combining Brγ and Paβ – both located well within "good" atmospheric windows – to create an "IR hydrogen decrement" analogous to the Balmer decrement in the optical. However, this line pair has no common energy levels (Brγ is a 7–4 transition, and Paβ is a 5–3 transition – see Table 1.1), requiring additional assumptions about the observed system and thus introducing significant uncertainties into the physical interpretation of the measurement.

7.3.2 *Planetary nebulae*

Planetary nebulae (PNe) are a second class of "Galactic" nebular emission lines. They are interesting as the (near-)final evolutionary phase of the majority of stars in the Universe. They are also responsible for the return of significant amounts of chemically enriched material to the ISM. In addition, they are truly beautiful objects to behold, as demonstrated wonderfully by the Hubble Space Telescope.

7.3.2.1 PNe emission-line basics

The emission lines from PNe share with lines from H II regions a similar basic idea, in that they arise from a hot stellar central source of ionizing radiation illuminating the "ISM." In this case, however, the "ISM" is dominated by the ejected circumstellar envelope, which has a different composition, density structure, and kinematic structure than the "standard" ISM. Nevertheless, detailed modeling of the PNe can provide similar diagnostics to those of H II regions with respect to electron temperature, electron density, and ionic abundances. In addition, IR line features such as H_2 and Fe provide insight into shocks and diagnostics to distinguish them from radiative excitation. The emission lines also show the kinematics of outflows and morphological features in the planetary nebula.

For electron-density diagnostics, key line pairs include [O II] and [S II] and, at slightly higher densities, [Cl III] and [Ar IV] – similar to H II regions. Analogous diagnostics for electron temperature also parallel H II-region diagnostics. In addition, combining these with models for the PNe can provide insights into ionic abundances. As we will see below, though, the fact that the structure of PNe is more complex than that of many H II regions can complicate these models significantly. Major additional sources of uncertainty for PNe include their distance and internal differential extinction due to dust formation.

Figure 7.5 shows typical near-IR spectra of PNe. As one might expect, in the K-band window, dominant features tend to be Brγ (2.165 μm) and He I (2.058 μm). However, in some of these spectra, molecular H_2 emission features approach the strength of, and even dominate over, Brγ and He I. Other notable features include He II as well as higher-ionization forbidden lines of Fe and Kr.

The H_2 lines, as noted above, can provide diagnostics of shock excitation versus radiative excitation of emission lines. This is due to the fact that H_2 can be excited by both fluorescence and thermal (collisional/shock) mechanisms. Thus, at low densities, we find that UV excitation of cool ($T \sim 100$ K) molecular gas causes the ratio of the 2.12-μm and 2.25-μm features to be fairly steady at 1.7. These are 1–0 S(1) and 2–1 S(1) transitions, respectively. At higher densities ($> 10^4$ cm^{-3}), this ratio increases and becomes a useful diagnostic of temperature (up to $T \sim 1000$ K).

Another useful diagnostic IR feature in PNe is [Fe II]. This is because Fe in the ISM is usually depleted onto dust grains, producing very low abundances in the gas phase of this species. However, shocks (particularly slow shocks such as those commonly found in PNe) break up these dust grains as they pass through the ISM. Thus, shocks will greatly increase the local abundance of gas-phase Fe (at least temporarily), and their occurrence is revealed by the presence of [Fe II] lines. In the case of PNe, the shocks are so slow (in comparison with, for instance, supernova remnants), that only the fastest-moving PNe shocks seem to lead to Fe emission lines.

7.3.2.2 Morphology and outflow kinematics

Contrary to simple expectations, spherical stars tend NOT to produce spherical planetary nebulae when they die. Rather, most PNe are highly spherically asymmetric.

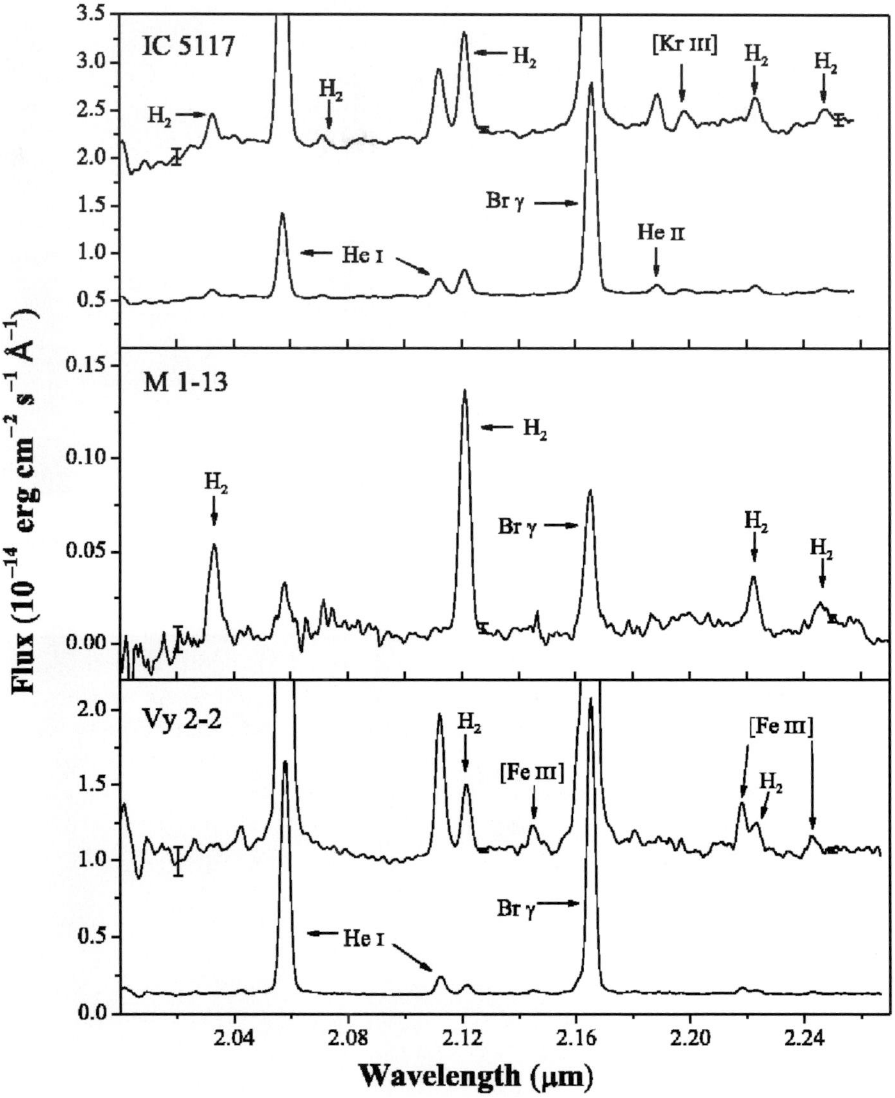

FIGURE 7.5. The near-IR spectrum of a planetary nebula. Reproduced from Likkel *et al.* (2006).

However, PNe also tend not to be irregular – rather they are very symmetric in an aspheric way. In addition, in some cases there is eye-catching evidence for collimated outflows (see Figure 7.6).

The collimation in these PNe ranges from relatively mild, via "medium," to a high degree of collimation (see Figure 7.7). Furthermore, point symmetry seems to be pervasive in these objects (see Figure 7.7). This point symmetry is usually associated with bipolarity, a progressive variation in the direction of the outflows, and episodic events of (collimated) mass loss. Thus, point symmetry in PNe indicates the presence of a bipolar rotating episodic jet or collimated outflow (BRET). In a true BRET, the morphology is also reflected in the kinematics, and we see that this is the case in many PNe (Figure 7.8). Of course, emission lines are critical for making these diagnoses.

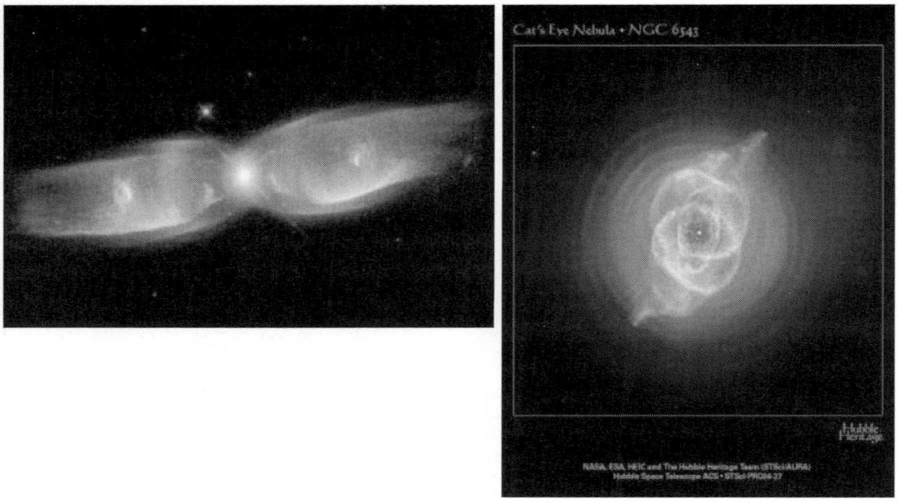

FIGURE 7.6. Morphologically evident outflows in PNe (figures reproduced from the Hubble Space Telescope website).

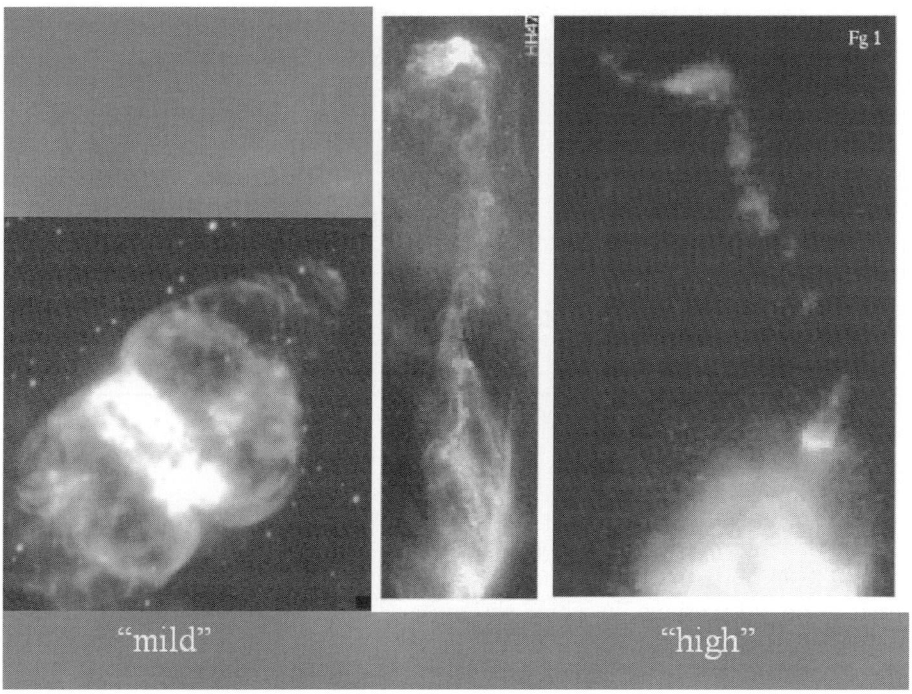

FIGURE 7.7. Examples of different degrees of collimation in PNe (figure courtesy of J. A. Lopez).

7.3.3 *Supernova remnants*

Supernovae are among the most-interesting phenomena in the astrophysics of recent decades, and their remnants are also critical for our understanding of the cosmos. Supernovae are the final events in the lives of the massive stars in the Universe. Their remnants are critical for chemical enrichment of the Universe as well as the return of kinetic energy

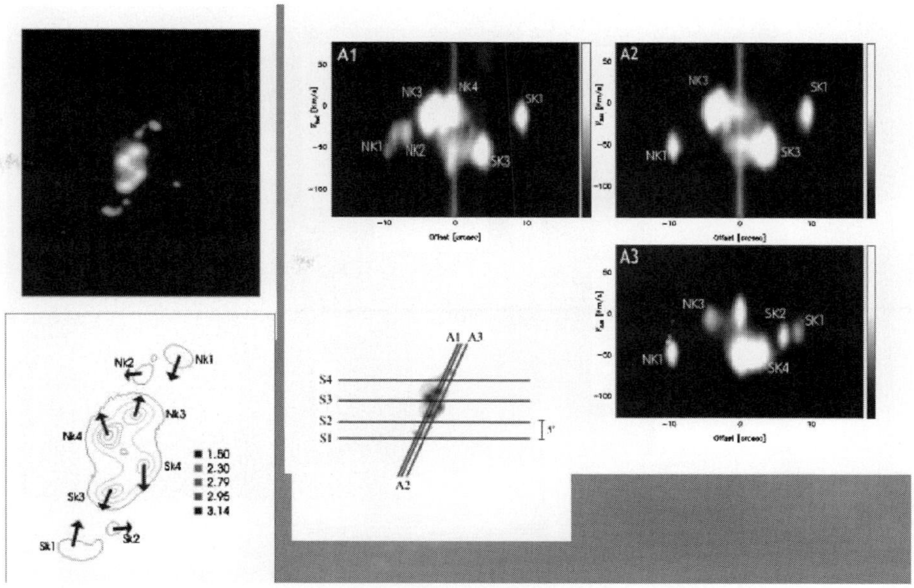

FIGURE 7.8. Morphological/kinematic connection in BRETs (figure courtesy of J. A. Lopez).

to the ISM – they are the delivery phase for "feedback" of stellar populations on galaxy evolution. They are also the birthplaces of neutron stars and black holes, and can provide important insights into the physics of these exotic objects.

There are two basic types of supernova remnants (SNRs): shell-like and center-filled. The young (<5000 yr) shell-like SNRs are typically dominated by material from the progenitor star, whereas older shell SNRs (5000–20 000 yr) are typically dominated by ISM material entrained by the SNR expansion. As their name implies, they typically appear to be hollow "shells" in their structural properties (see Figure 7.9).

Center-filled SNRs, on the other hand, have (as their name also indicates) a "filled" appearance, rather than being hollow. These SNRs, also known as "plerions," are typically powered by pulsars and/or pulsar-wind nebulae. The archetypal example is the Crab Nebula (Figure 7.10), where we can clearly see that the relativistic particle wind from the pulsar is filling the SNR center with synchrotron-emitting particles. Plerions are almost uniformly young (<5000 yr).

A third class of SNR is constituted by the "composite" remnants, such as the Vela SNR, which have both a pronounced shell-like structure and some (typically small/faint) center-filling from a pulsar. Owing to their morphology and pulsar power, these are often considered to be old plerions, in which the pulsar's center-filling is fading away and the expanding SNR shell is clearly separated from the center.

The optical emission spectrum of most SNRs contains very little continuum emission – most of the light comes out in the form of emission lines (a partial exception of course being the blue synchrotron glow from the center of the Crab Nebula, for instance). The spectrum of CTB 1, for instance, has strong lines of H, [O II], [O III], [S II], and [N II] (Figure 7.11). We can also see fainter lines of many other species, including He I, He II, [O I], [N I], [Ne III], [Fe II], [Fe III], [Ca II], and [Ar III].

Figure 7.12 shows a near-IR spectrum of a fast-moving knot (FMK) in Cas A. We can see very similar species to the optical spectrum, indicating a close link between these wavebands. However, we can also see high-ionization "coronal" emission lines of [Si VI] and [Si X].

FIGURE 7.9. An example of a shell SNR – Cas A (figure reproduced from the Hubble Heritage archive).

FIGURE 7.10. The Crab Nebula – an example of a plerion SNR (figure reproduced from the Chandra X-ray Observatory website).

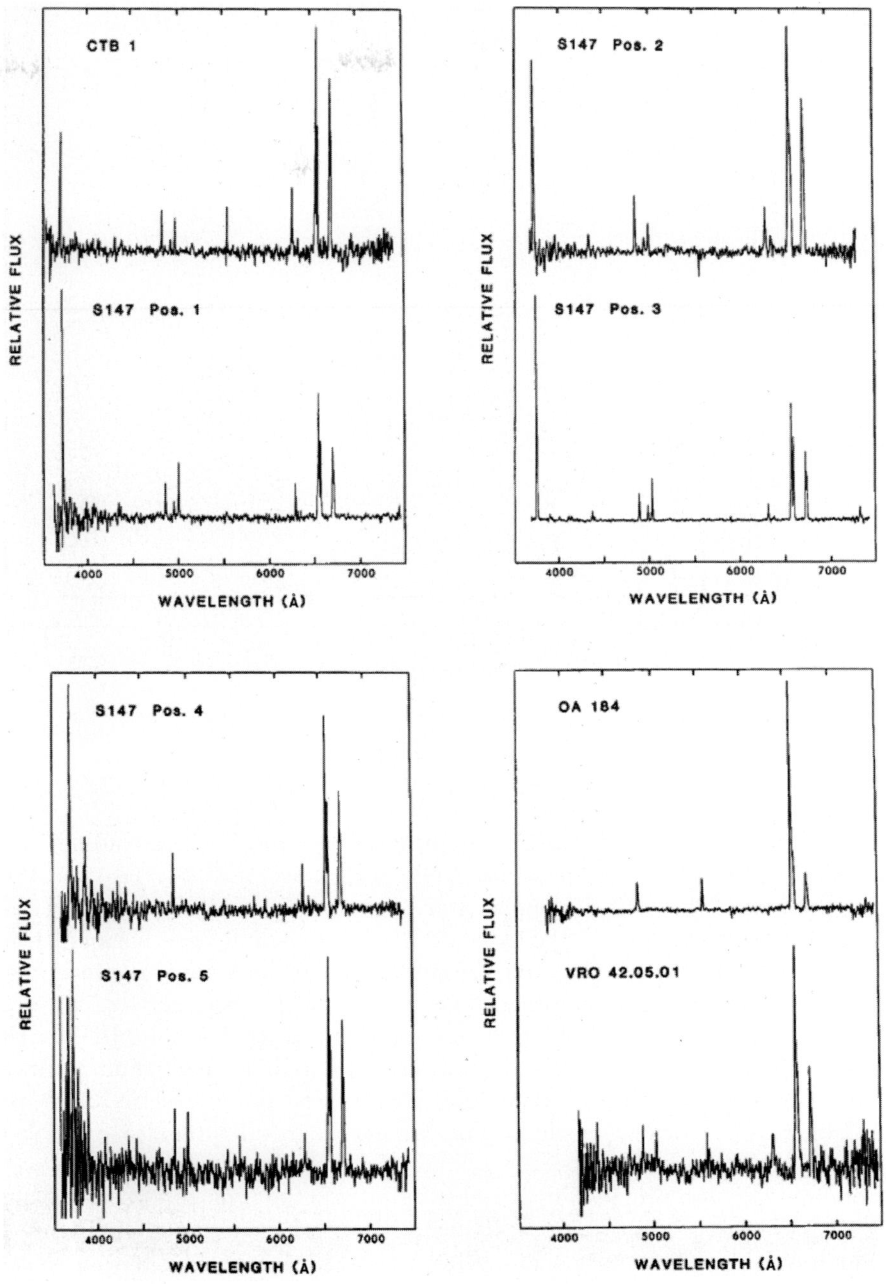

FIGURE 7.11. The optical spectrum of the CTB1 SNR, reproduced from Fesen *et al.* (1985).

If we then turn to another feature in Cas A – a "quasi-stellar floccule" – we also see similar species to the optical (Figure 7.13). However, here we see that the He I and [Fe II] features are very strong – in fact, much stronger than even the hydrogen lines! If we then turn to a quasi-stellar floccule feature in the Kelper SNR, we see very similar spectral features. Thus, similar features in different SNR can resemble one another even more closely than different features from the same SNR!

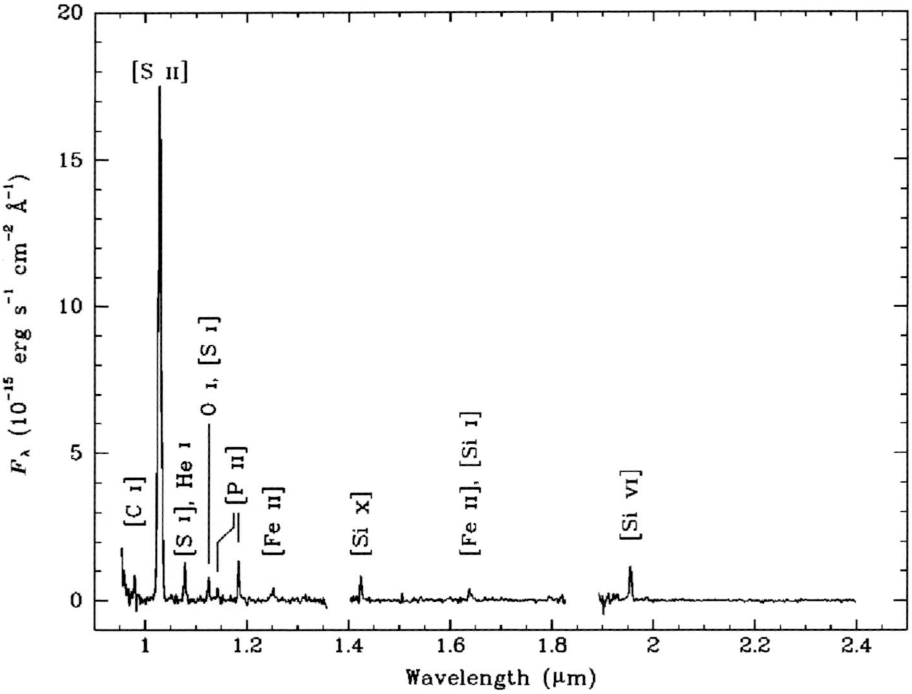

FIGURE 7.12. The IR spectrum of a fast-moving knot in Cas A. Adapted from Gerardy & Fesen (2001).

These emission lines hold a wealth of information on the SNR, including its identity/nature, electron density/temperature, ionic abundance, shocks, and even the late evolutionary stages of the SNR progenitor star.

Owing to the high extinction in the Galactic Plane, most SNR have been identified by their shell-lock morphology in the radio. However, H II regions are a significant source of confusion, insofar as their "sculpting" by massive-star-forming regions can produce similar morphologies. Fortunately, SNR emission lines in the optical (and IR) can resolve this problem. Figure 7.14 shows a plot of various SNR and H II regions in a parameter space defined by line ratios of [O I], [O II], and Hβ. As we can see, the SNRs are cleanly separated from the H II regions on such a plot. Similar diagnostics are also available using [S II].

Emission lines can also provide important insights into the progenitor star's evolution, particularly in the case of very young (<100 yr) SNRs, such as SNR 1987A (Figure 7.15). As the supernova blast wave propagates through the ISM, it excites a wide variety of emission lines. For massive stars such as these, of course, the local ISM is in fact dominated by ejecta from the progenitor star (prior to the supernova event). This allows us to probe the composition, density, and kinematics of these ejecta, and thus tie them to the progenitor star's evolution.

I will wrap up SNRs with a quick note regarding [Fe II] emission. As noted above, Fe in the ISM is usually depleted onto dust grains, producing very low abundances in the gas phase of this species. However, shocks break up these dust grains as they pass through the ISM and greatly increase the local abundance of gas-phase Fe, as revealed by the presence of [Fe II] lines. Thus the [Fe II] lines in SNR provide excellent shock diagnostics, and can be used effectively to identify SNRs in contrast to confusing H II

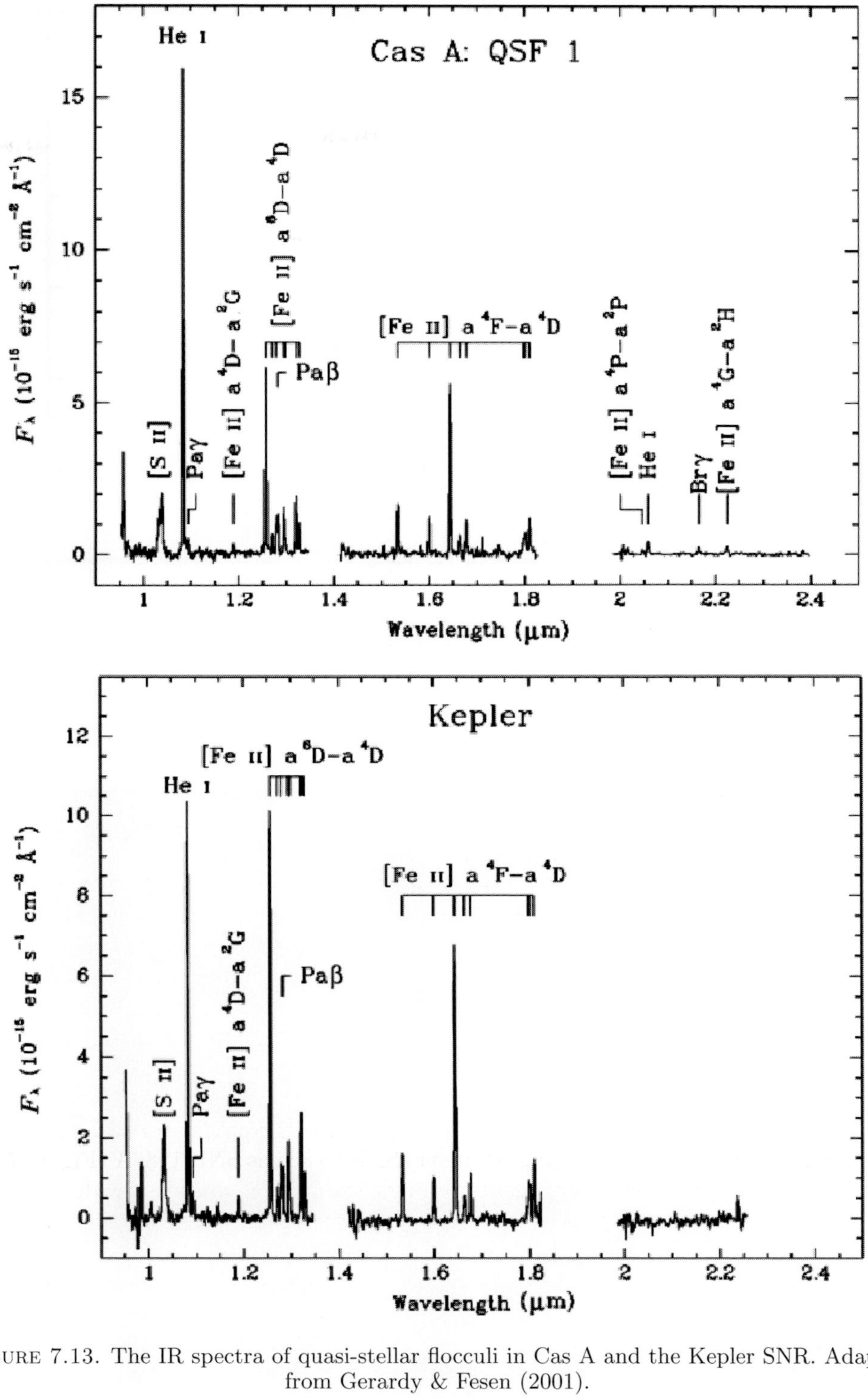

FIGURE 7.13. The IR spectra of quasi-stellar flocculi in Cas A and the Kepler SNR. Adapted from Gerardy & Fesen (2001).

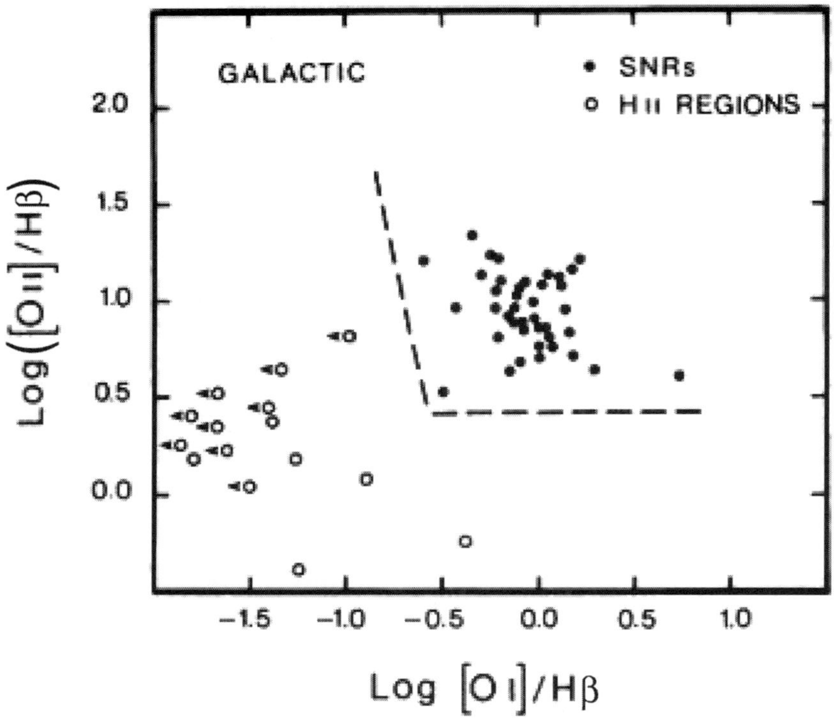

FIGURE 7.14. Separation of SNRs from H II regions using emission-line ratios. Adapted from Fesen *et al.* (1985).

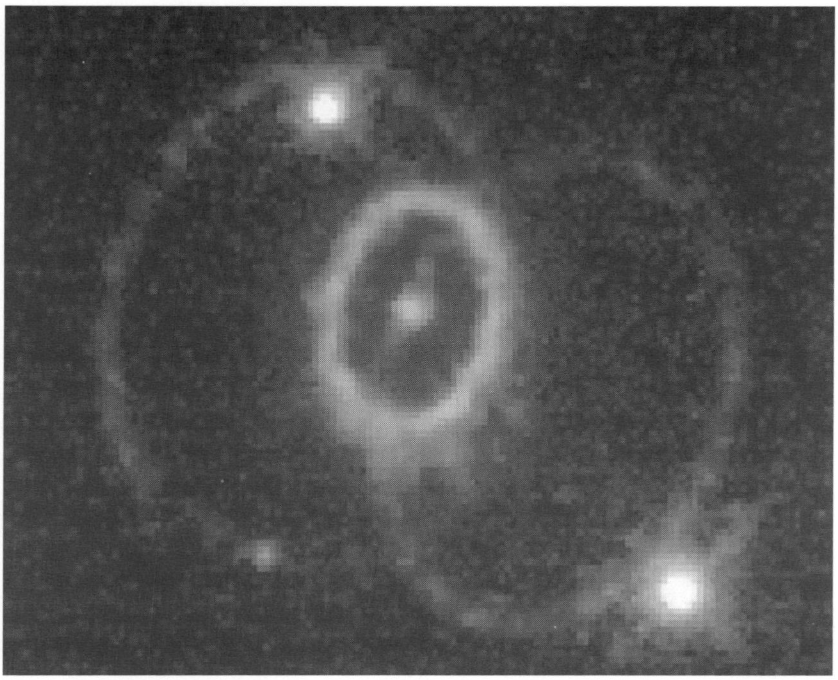

FIGURE 7.15. Remnant of SN 1987A (figure reproduced from the Hubble Space Telescope website).

regions. This is particularly important given that the optical diagnostics noted above are often observationally unavailable for most SNRs due to extinction.

7.4. Galactic stellar sources of emission lines

Now that we have covered the "nebular" Galactic sources of emission lines, I will turn to "stellar" sources of the same. By "stellar," I mean relatively compact (i.e. nearly point-like at typical angular resolution) objects with emission lines driven by the photospheric output of a star.

A recurring theme for stellar emission-line sources is the idea that "emission equals extension." That is, stellar emission-line activity essentially requires some sort of extended atmosphere beyond a "normal" photosphere – chromosphere arrangement. This is a fundamental feature of radiative transfer in stellar atmospheres. In a "standard" stellar atmosphere, any "piece" of the atmosphere is roughly in local thermodynamic equilibrium (LTE). This means that (at least approximately) thermal collisional excitation and de-excitation of atomic energy levels occur in equilibrium with the local blackbody radiation field. Hence the number of "line" photons emitted due to a "downward" shift in electron energy levels is exactly balanced by the number of line photons absorbed due to "upward" shifts in energy levels. Thus, no lines can be formed. As any basic text on stellar atmospheres explains, the ubiquitous absorption features found in most stellar spectra are due to non-LTE variations in the atmosphere.

Emission lines also require non-LTE conditions to hold for their formation. More specifically, they require that some process produce a population inversion of electron energy levels, so that an excess of electrons at high energy levels (relative to the number produced by absorption of incident "line" photons) produces a corresponding excess of emitted "line" photons via downward energy-level shifts. Exactly such an inversion is produced when a portion of the stellar atmosphere "sees" a radiation field that is "hard" (i.e. of higher energy or "hotter") relative to the local temperature – we adopt that as our working concept for an "extended" atmosphere. A standard atmosphere avoids such a situation because temperature changes are small over distance scales for which the optical depth has values $\tau \sim 1$.

In the following subsections, I will review prominent stellar sources of emission lines in the Galaxy, including young stellar objects, massive stars, magnetically active stars, and compact-object binary systems.

7.4.1 *Young stellar objects*

By "young stellar objects" here, I primarily refer to objects making the transition from accreting protostars to hydrogen-burning main-sequence stars. I include in this class T Tauri stars, Herbig–Haro objects, and massive protostars. Note that these are not necessarily exclusive classifications – it may be possible for individual objects to belong to more than one of these classes simultaneously. However, as we shall see below, each class has its own defining characteristics.

7.4.1.1 T Tauri stars

T Tauri (or "TT") stars represent the final formation phase for low-mass stars – the most numerous stars in the Universe. As such, they are a critical part of the star-formation process. TT stars are generally thought to be low-mass protostars near the end of their accretion phase (in fact, some are only weakly accreting or perhaps even non-accreting). They are thought to consist of a central object (the protostar, or perhaps even a zero-age main-sequence object) surrounded by an accretion disk. (In the non-accreting case, this

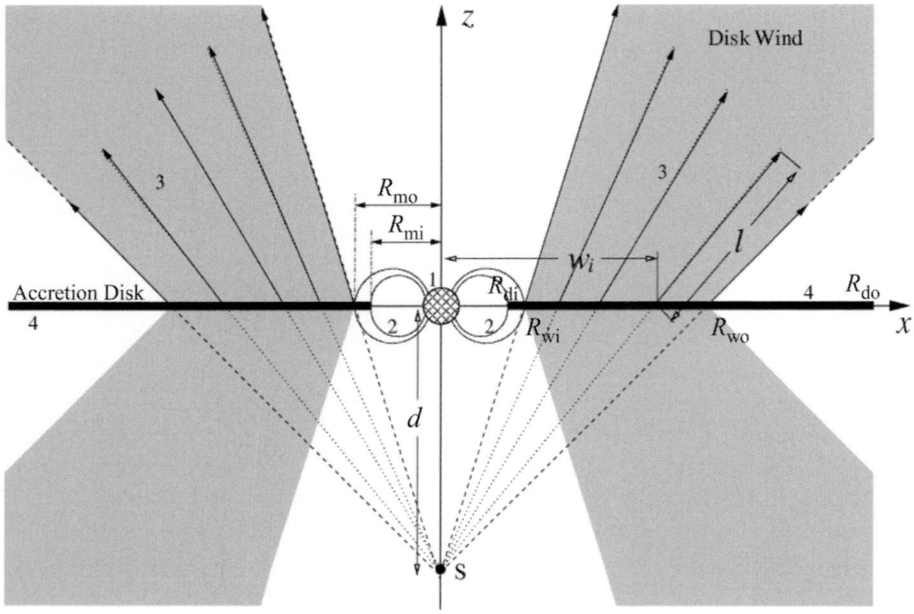

FIGURE 7.16. A schematic diagram of a T Tauri star system, including the central protostar, accretion disk, disk central hole with protostar magnetosphere, wind region, and possible central jet outflow. Adapted from Kurosawa *et al.* (2006).

may be a "remnant disk" – the last vestiges of material left in the disk after accretion has halted.) TT objects typically have significant outflows as well. The accretion through the inner disk is typically thought to be slowed (or halted) by the magnetic field of the central object, and magnetospheric/disk interactions are likely to power parts of the outflow as well (particularly jets). This can be seen schematically in Figure 7.16.

TT stars have a continuum spectrum usually accompanied by strong emission lines – typically very strong hydrogen series. They can also exhibit strong He I (1.083-μm) emission, which has been tied to windy outflows. Somewhat surprisingly, TT stars can also exhibit strong C III and O IV emission lines. These high ionization states are typical for temperatures $T > 15\,000$ K, which is significantly higher than the main-sequence photospheric temperatures for such low-mass objects. This is probably indicative of accretion-powered "hot spots" in the system, indicating that accretion is a significant energy source in these systems.

Much can be deduced about TT stars by investigating the profiles of the strong H emission lines. The complex line profiles indicate highly non-trivial disk geometries (see Figure 7.17). These profiles also exhibit significant time variability, sometimes on timescales as fast as a few days. This indicates significant variations in the accretion flow in the TT-star system – in geometry, in accretion rate $\mathrm{d}M/\mathrm{d}t$, or in both. The He I (1.083-μm) line profiles (Figure 7.18) are also good diagnostics for the TT outflow. They are known to be very sensitive tracers of wind activity. More recently, they have been shown to trace mass infall (accretion) as well. Thus, we see that TT-star emission lines can provide good insights into accretion and outflow in the system.

7.4.1.2 Herbig–Haro Objects

Herbig–Haro ("HH") objects are another important formation phase for some low-mass stars, and are not necessarily unrelated to T Tauri stars. The hallmark of HH

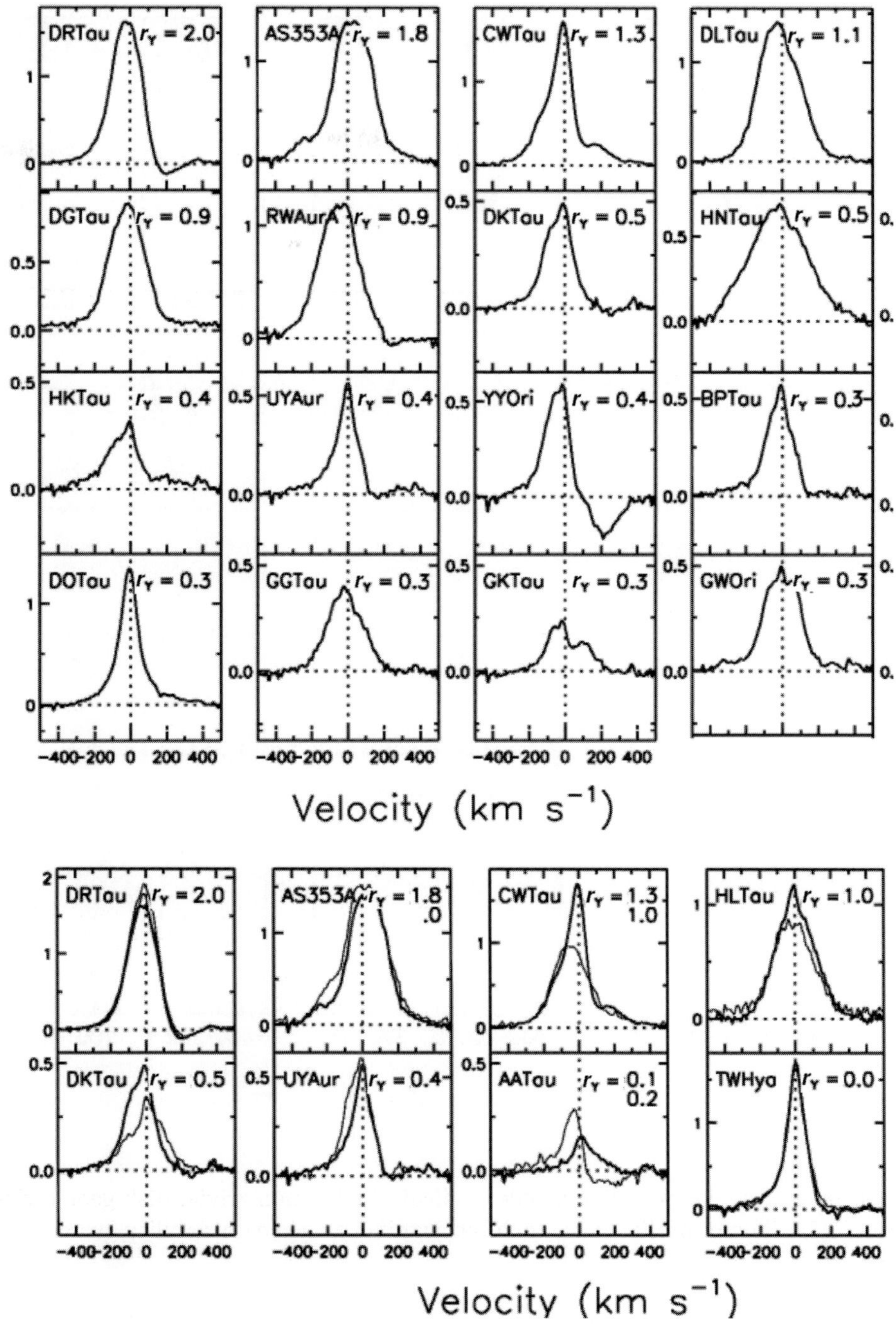

FIGURE 7.17. T Tauri-star Hα line profiles, showing complex geometry and time variability. Adapted from Edwards *et al.* (2006).

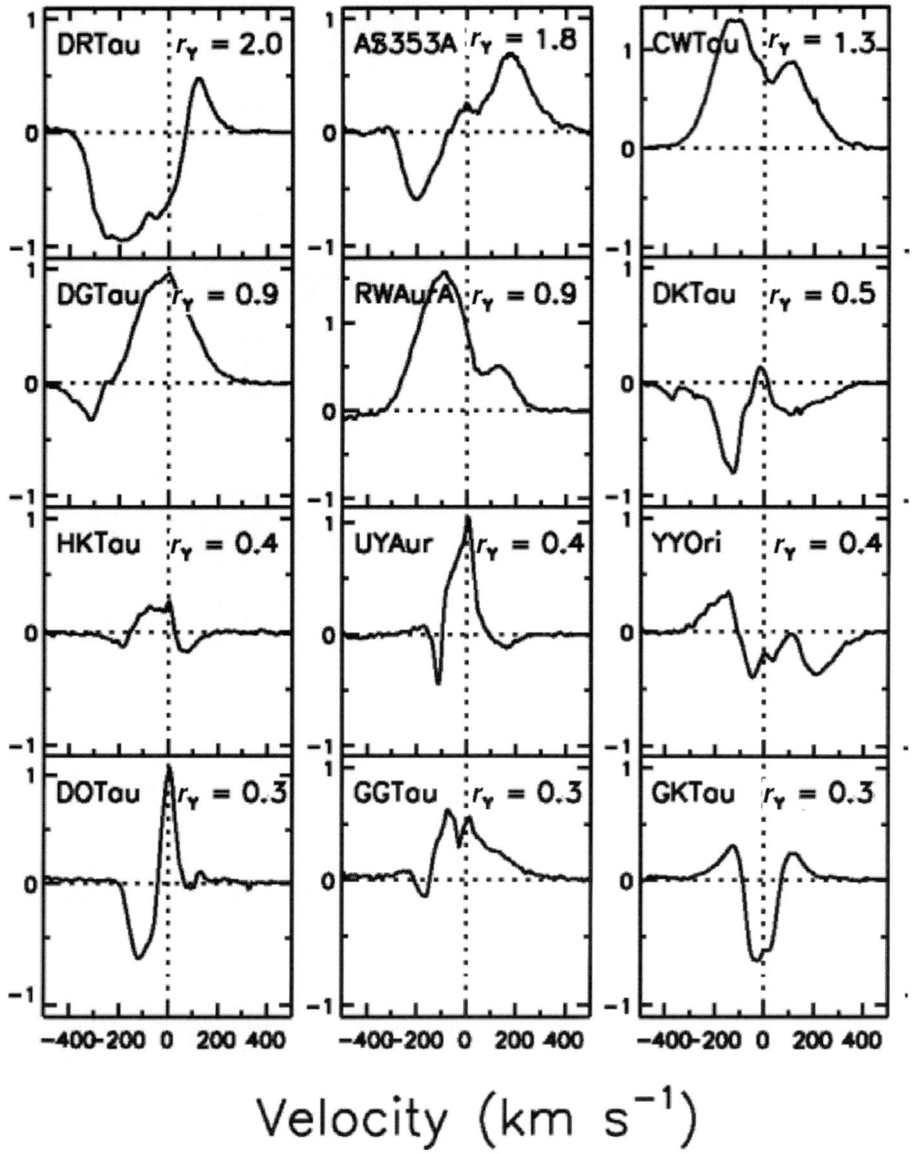

FIGURE 7.18. T Tauri-star He I 1.083-μm line profiles, showing complex geometry in the disk wind. Adapted from Edwards *et al.* (2006).

objects is collimated outflows. Such outflows are known to carry away significant angular momentum, which is of course the critical activity for enabling accretion. Figure 7.19 shows pictures of the prototypical objects HH1 and HH2, where the outflows can be clearly seen. As can be seen in Figure 7.19, HH objects are also quite common in many star-forming regions.

The outflows from HH objects can have very-well-collimated jets, with opening angles of a few degrees or even smaller. The example shown in Figure 7.20, for instance, illustrates a projected opening half-angle of $\theta_{1/2} < 0.5°$. In addition, the jet outflows often have complicated structures of bright spots or knots along the flow. These are typically

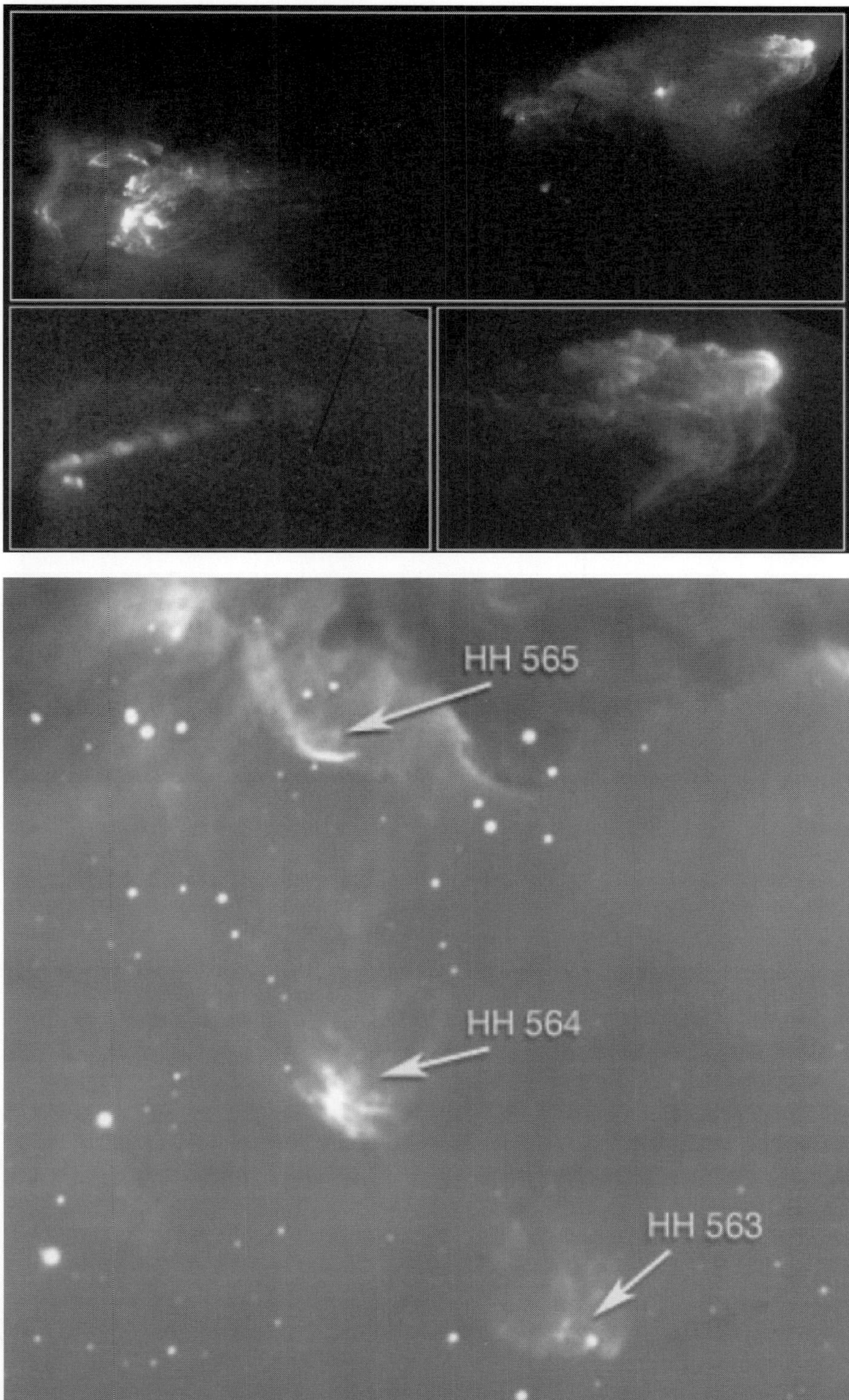

FIGURE 7.19. Top: close-up of two Herbig–Haro objects (figure reproduced from Hubble Space Telescope website). Bottom: three Herbig–Haro objects together in a single star-forming field (figure reproduced from Reipurth & Bally (2001)).

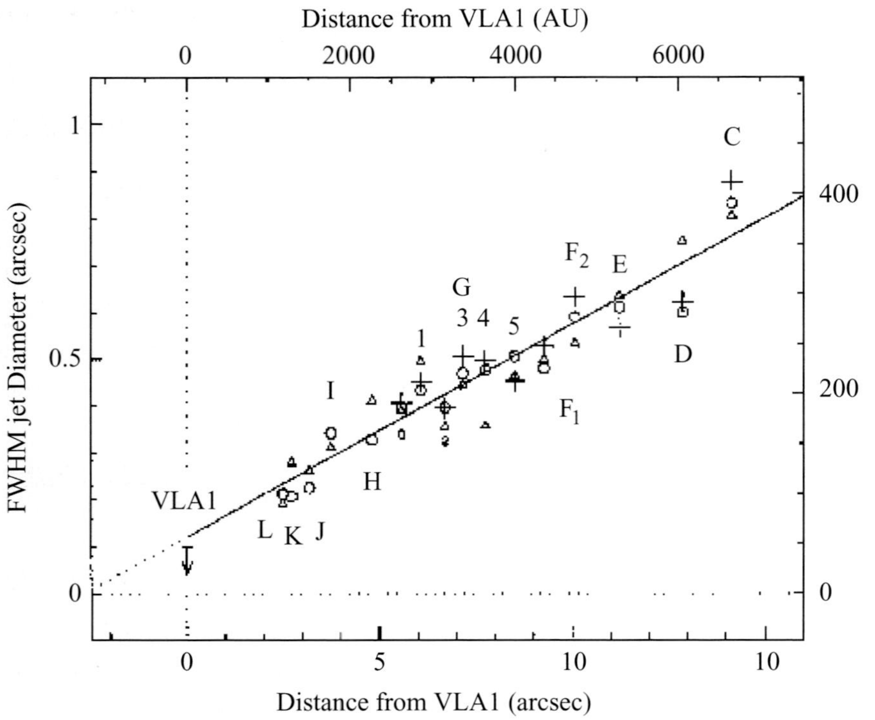

FIGURE 7.20. Jet diameter versus distance, illustrating that the opening half-angle is <0.5° Adapted from Reipurth & Bally (2001).

thought to be shocks – both internal shocks and collisions with the ISM. The optical spectra of HH objects are thus, not surprisingly, dominated by the forbidden transitions that typically accompany collisional-shock excitation (Figure 7.21). These are typically much brighter than the hydrogen lines which are more dominant in radiatively excited environments. In the IR, it is common to see strong H_2 emission lines, indicating that significant molecular material is also present in the outflow region. The IR spectra also reveal strong [Fe II] emission, again indicating grain disruption typically associated with shock activity.

7.4.1.3 Young massive stars

Massive protostars are another related set of objects, which are again an important phase for the formation of many stars. In particular, there is some belief that massive-star formation may follow a different process than that which applies for low-mass stars. The details of the accretion process of course have critical implications for the final stellar mass. The mass distribution of high-mass stars in turn has a dramatic impact on galaxy evolution (chemical enrichment, ISM kinetic energetics, etc.) and the production of compact objects (neutron stars and black holes).

Despite the importance of this subject, relatively little work has been done to date. This is in part due to the relative rarity of high-mass stars compared with their numerous low-mass counterparts, and also due to the faster evolutionary track they take – this phase is both numerically rarer and temporally shorter than low-mass star-formation. Furthermore, work in the IR is essential to penetrate the high extinction typically associated with these objects.

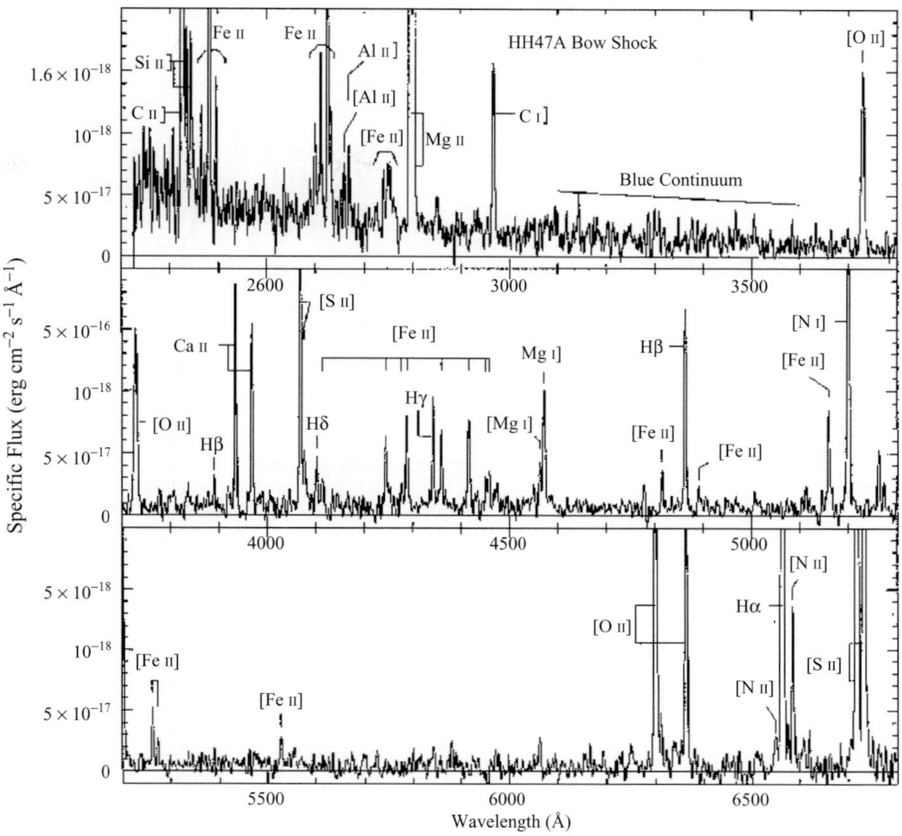

FIGURE 7.21. Spectra of a Herbig–Haro object in the optical. Adapted from Reipurth & Bally (2001).

Blum *et al.* recently published some important high-spectral-resolution line profiles for massive protostars. Figure 7.22 shows the molecular CO bandhead *in emission* for one of these objects. The double-peaked profile of the individual line transitions clearly indicates the presence of an accretion disk (similar to the case of lower-mass TT stars). Likewise, the Brγ line profile also has a double-peaked "disk" profile, as does Brα (Figure 7.22), albeit with a more-asymmetric profile. From these results, one may be tempted to conclude that all stars in fact share the same formation process via disk accretion. However, Blum *et al.* point out that these signatures are NOT seen for stars earlier than O8 or O9. Thus, either the disks of these stars dissipate *much* more quickly than do the slightly lower-mass objects seen above, or else they were never there – potentially implying the existence of a different formation process.

7.4.2 *Massive stars*

Massive stars are in many ways the most important stars in the Universe. While low-mass stars are vastly more numerous and contain the dominant fraction of the stellar mass in most galaxies, the high-mass stars dominate in most other ways. In particular, due to the rough proportionality of $L \propto M^3$, massive stars dominate the luminosity output of most galaxies that contain them. Owing to the strong winds and outflows they generate during their lives – and especially their spectacular deaths in supernova explosions – massive stars also dominate the kinetic-energy budget for the ISM in most galaxies. Finally, their

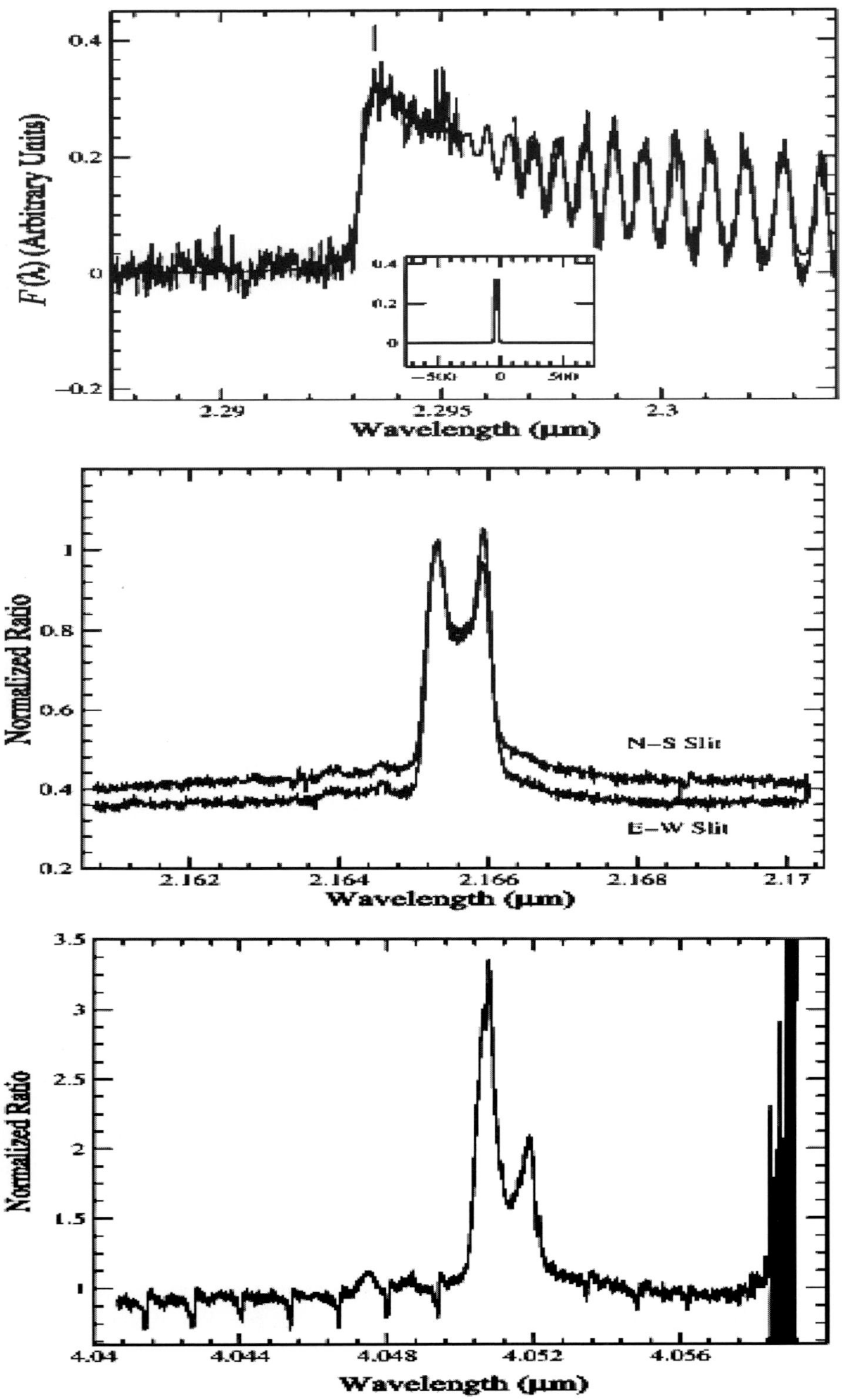

FIGURE 7.22. Spectra of massive protostars in the IR in (top) the CO bandhead; (middle) Brackett γ; and (bottom) Brackett α. Adapted from Blum *et al.* (2004).

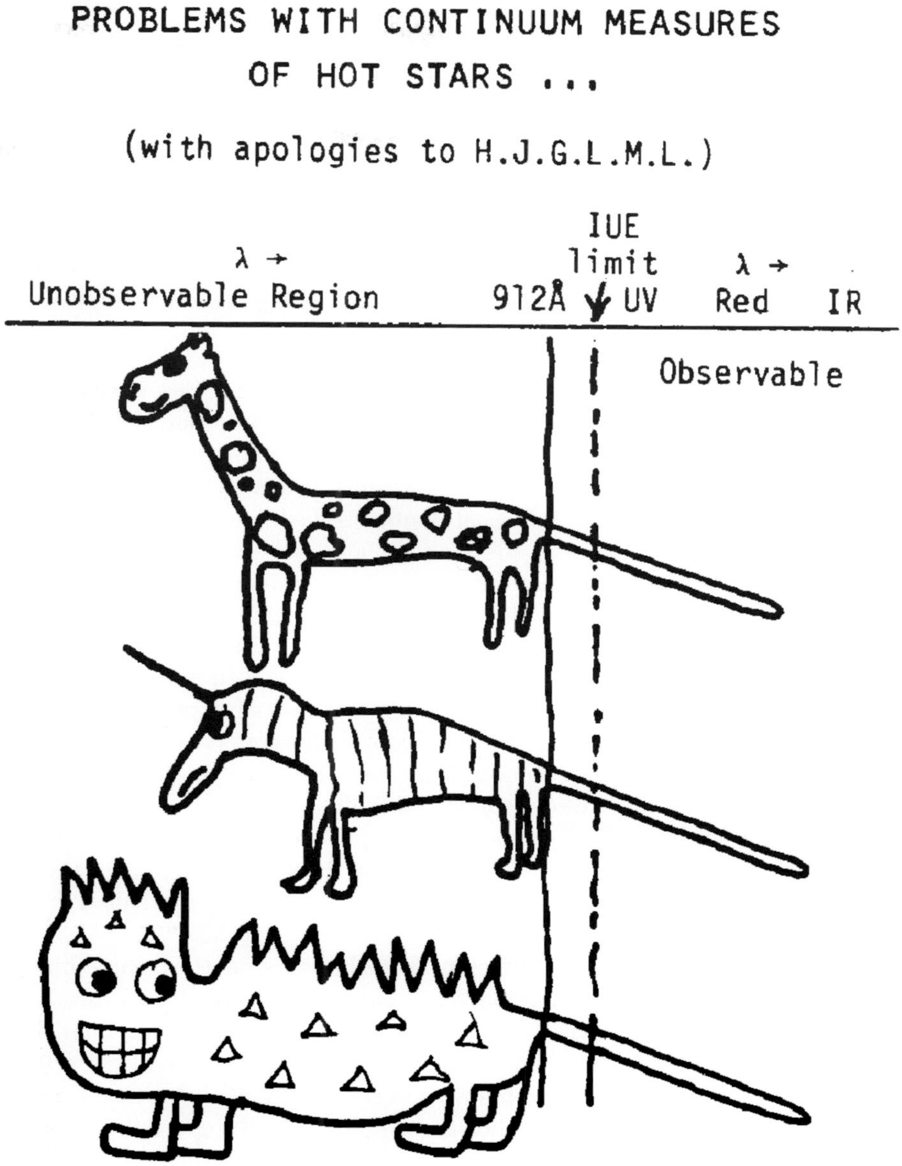

FIGURE 7.23. An illustration of the dangers of judging massive stars by their optical/IR spectral properties. Reproduced from Conti (1986).

deaths are also prime contributors to the chemical enrichment of galaxies as a function of age.

The most-massive stars are also notoriously unstable objects. Their high temperatures combined with significant outflows are a natural source of strong emission lines, making them among the most prominent sources of Galactic emission-lines. We typically classify these stars by their emission-line properties in the optical and IR, but this is a dangerous undertaking given the fact that the overwhelming majority of their energy output is in the unobservable UV range – as illustrated in a quasi-famous illustration by Peter Conti (Figure 7.23). Furthermore, in an excellent review of the IR spectra of massive stars,

Morris *et al.* (1996) show that these stars can even *change* their classification from one class to another and back again on timescales of just a few years!

In the subsections below, we consider in turn the emission-line properties of Be and B[e] stars, Of stars, luminous blue variables, and Wolf–Rayet stars.

7.4.2.1 Be and B[e] stars

Be stars are the most numerous class of massive stars exhibiting significant line emission. As their name implies, they are basically "normal" B stars with a strong emission line – typically hydrogen lines. Hα is the primary means for identifying these stars, though Brγ is the only such diagnostic for deeply embedded stars or those located at large distances in the disk of the Milky Way.

The obvious question one would next ask is "What makes some B stars have emission lines?" The most-prominent theory for the last 50 years in Be stars is that, like some young stellar objects (YSOs), Be stars have a circumstellar disk of material. The relatively hard radiation field from the B-star photosphere impinges on the cooler disk material and thus radiatively pumps the observed line emission – primarily from neutral hydrogen. The disk-like geometry produces a classical double-peaked emission-line profile like the one shown in Figure 7.24.

However, unlike YSOs, Be-star disks are very likely "excretion disks" as opposed to accretion disks – that is, they result from outflows rather than inflows. It is now fairly well established that for Be stars there is a clear correlation between stellar surface rotation (v sin i) and Hα FWHM (e.g. Hanuschik 1988). This is taken to indicate that the disks are rotationally supported, probably by coupling of the stellar photosphere (possibly magnetically?) to the disk itself. This basic concept is illustrated in Figure 7.25.

Tycner *et al.* (2006) have recently obtained interferometry in the Hα line for two very-nearby Be stars, in which they resolve the emission-line region. They find that the disk, rather than having a standard "thin-disk" profile (roughly equivalent to a square step function in latitude centered on the stellar equator as shown in Figure 7.25), actually has a broad Gaussian-like profile – more closely resembling an equatorial enhancement to a quasi-spherical outflow than a classical disk.

B[e] stars, are both similar to and different from Be stars, as their name hints. They are fundamentally B stars with emission lines. However, for B[e] stars many "metal" lines are also present, together with, as the brackets indicate, forbidden lines. Be stars, on the other hand, are dominated by hydrogen Balmer transitions and little or no metal forbidden-line transitions. Figure 7.26 shows a spectrum of the B[e] star CI Cam. Note the large number of metal transitions – Jaschek and Andillat (2000) report 450 identified line transitions in this single star spectrum! Optical emission lines of the B[e] star MWC 349 have "flat-topped" profiles typical of an optically thick, spherically symmetric geometry. Furthermore, for many B[e] stars there are IR excesses consistent with warm dust in their circumstellar disks. This combination of features, as well as the typical lack of clearly identifiable normal "photospheric" absorption lines, seems to indicate that the B[e] stars are highly evolved stars, each of which is deeply enshrouded in its own circumstellar envelope.

7.4.2.2 Luminous blue variables

Luminous blue variables (LBVs) are the most-luminous known individual stars. With luminosities several *million* times that of the Sun and apparent masses approaching (or exceeding) the theoretical limits of star formation, they are not surprisingly highly unstable as well. The most-famous LBVs include η Car and P Cygni. P Cygni is well known as the prototypical star for the line profile that bears its name. A typical "P

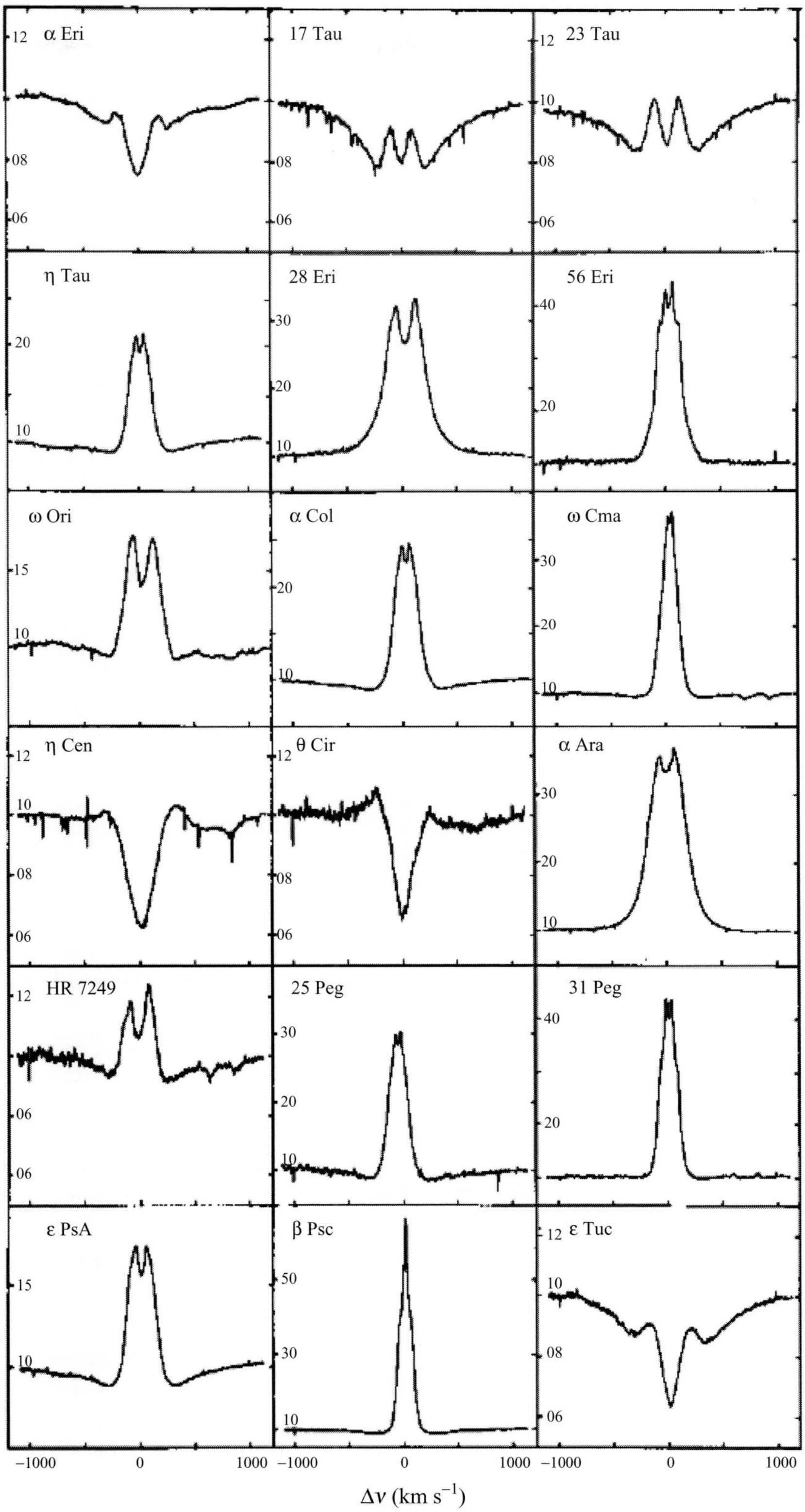

FIGURE 7.24. Hα line profiles from Be stars. From Hanuschik *et al.* (1988).

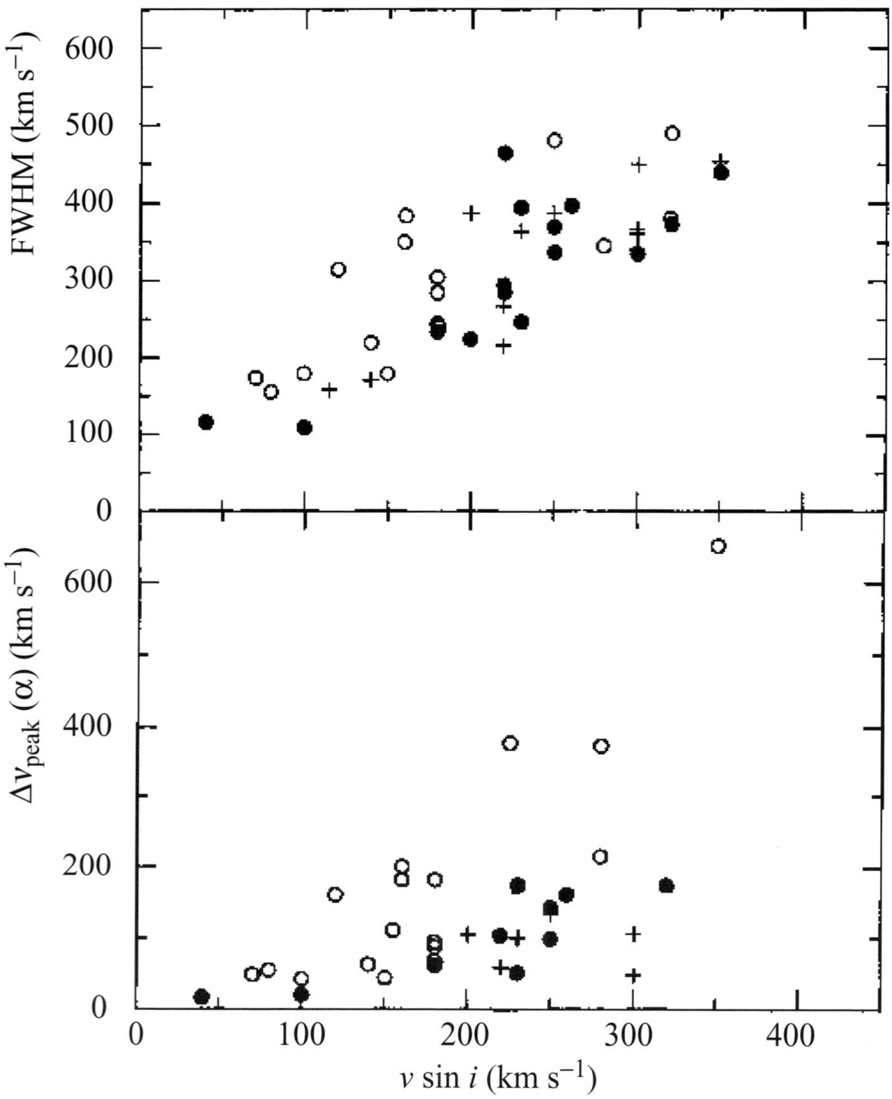

FIGURE 7.25. Rotational support correlations for Be stars. From Hanuschik *et al.* (1988).

Cygni-profile" line is shown in Figure 7.27. The broad blue absorption edge coupled with a redder emission edge is indicative of a strongly cooling wind outflow from a star. The wind regions closest to the stellar photosphere are exposed to the hard radiation field from the star, producing uniform line emission. However, the bluest features arise from the portion of the wind moving directly towards the observer – which of course means that those photons must traverse even-cooler material moving at the same projected velocity as the emitting material, creating strong line absorption. The line of sight to line-emitting material moving significantly away from the "head-on" direction also traverses cooler material, but with different projected velocities, which make the material effectively transparent to the line emission.

LBV stars are somewhat famously "eruptive" in their nature – during much of the 1800 s, η Car went from relative obscurity at $m_V \sim 7$ mag to being the brightest star

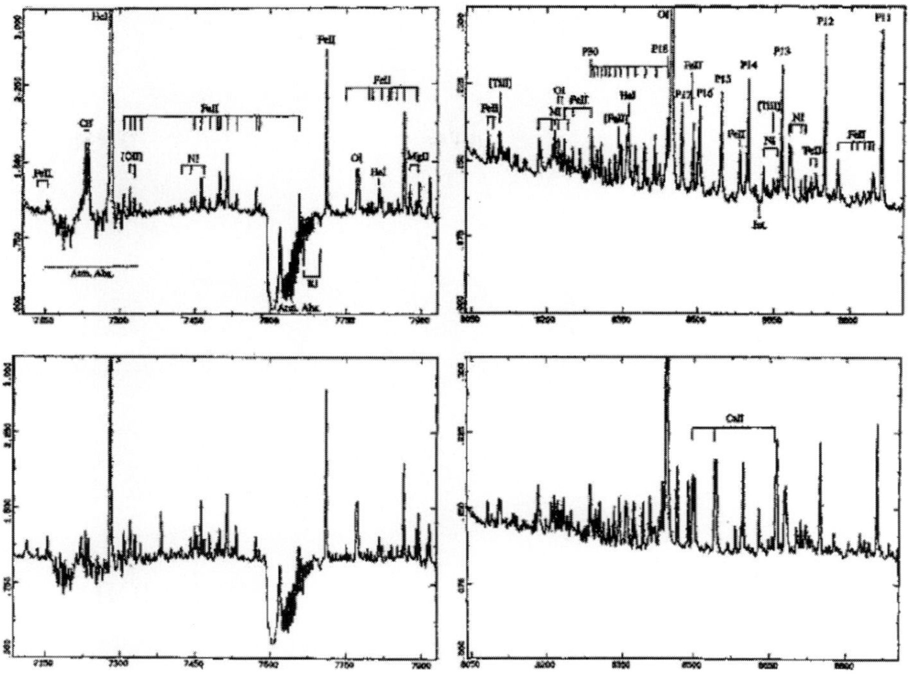

FIGURE 7.26. The spectrum of the B[e] star CI Cam (MWC 84). From Jaschek & Andrillat (2000). Unfortunately the labeling of this figure is illegible, but it's nonetheless worth reproducing to illustrate the large number of metal transitions of B[e] stars.

in the sky, before fading away again a few decades later. This eruption resulted in the formation of the famous "Homunculus" cloud of material surrounding η Car now (Figure 7.28).

The optical spectra of LBVs are typically dominated by Balmer lines, He I emission, and Fe and [Fe] lines, most with strong P Cygni profiles. Their IR spectra show the same features, though they also often include Mg II and Na I emission as well.

Another interesting set of LBVs is the IR-discovered stars the "Pistol Star" (Figer *et al.* 1997) and LBV 1806-20 (van Kerkwijk *et al.* 1995; Eikenberry *et al.* 2004). Both of these stars have very high luminosities of $\sim(3-5)\times 10^6 L_\odot$ (Figer *et al.* 1997; Eikenberry *et al.* 2004), and there is evidence of nebular remnants and/or radio shells indicating LBV-type past eruptions. Their IR spectra are remarkably similar to each other, and closely resemble those of other LBVs – particularly AG Car and η Car (see Figure 7.29). Together with η Car, they three constitute what appear to be the three most-luminous stars known in the Milky Way to date. It is important to note that two of the three most-luminous stars in the Galaxy were discovered serendipitously in the past decade. One would think that their instrinsic luminosity would make such stars obvious targets. However, neither of these two is easily visible in the optical due to $A_V \sim 25-30$ mag of extinction. This again emphasizes the importance of more IR observational work – particularly spectroscopy – for probing the massive-stellar content of our own Galaxy.

7.4.2.3 Wolf–Rayet stars

Wolf–Rayet (WR) stars are massive objects that tend to show strong He emission, but relatively weak H emission. This is thought to indicate that WR stars have shed

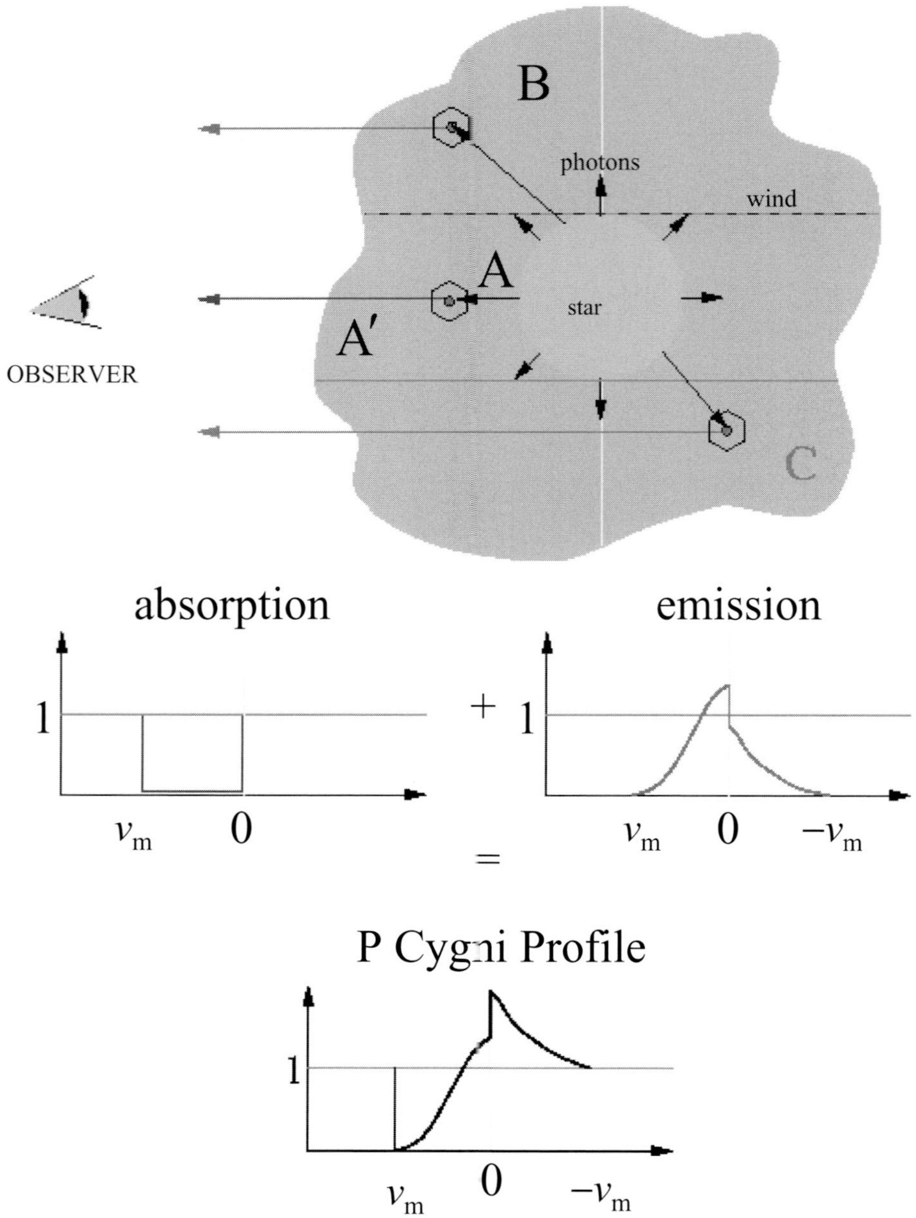

FIGURE 7.27. Typical P Cygni line profile and schematic illustration. Adapted reproduced from Kudritzki (2000).

most/all of their hydrogen envelope, and are thus highly evolved massive stars. They may be the final evolutionary phase for massive stars before their life-ending supernova explosions. There are two primary subtypes of WR stars – the "WN" stars, which typically have strong N II and N III emission lines (see Figure 7.30); and those of "WC" subtype, which have the same N lines as well as C III and C IV emission features. The WC9d stars also have thermal excess in the IR, indicating the presence of dusty envelopes (see Figure 7.31).

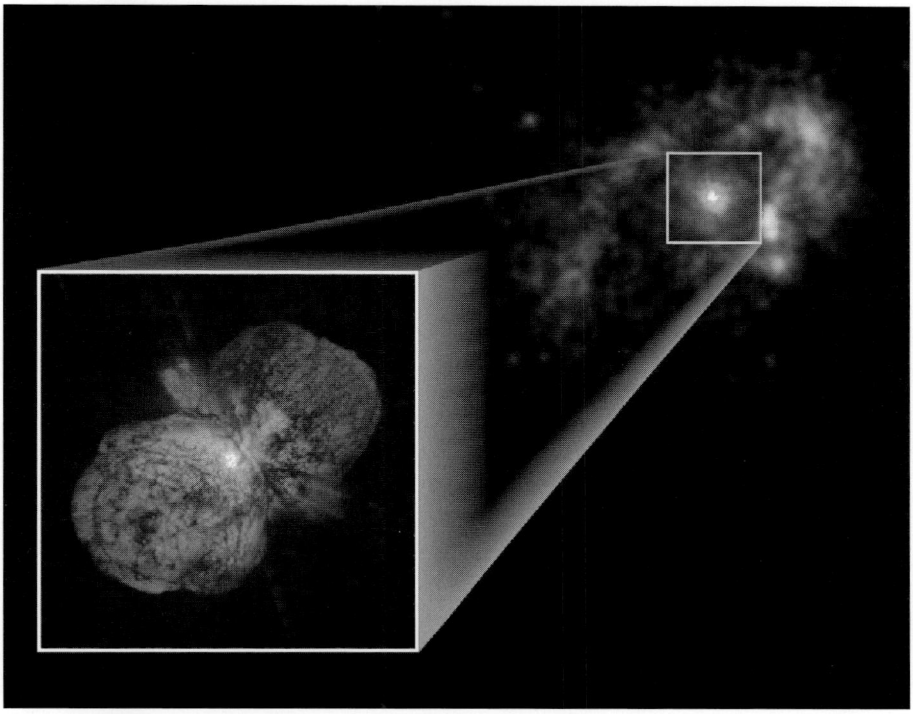

FIGURE 7.28. The Homunculus structure surrounding Eta Carina (figure reproduced from the Hubble Space Telescope website).

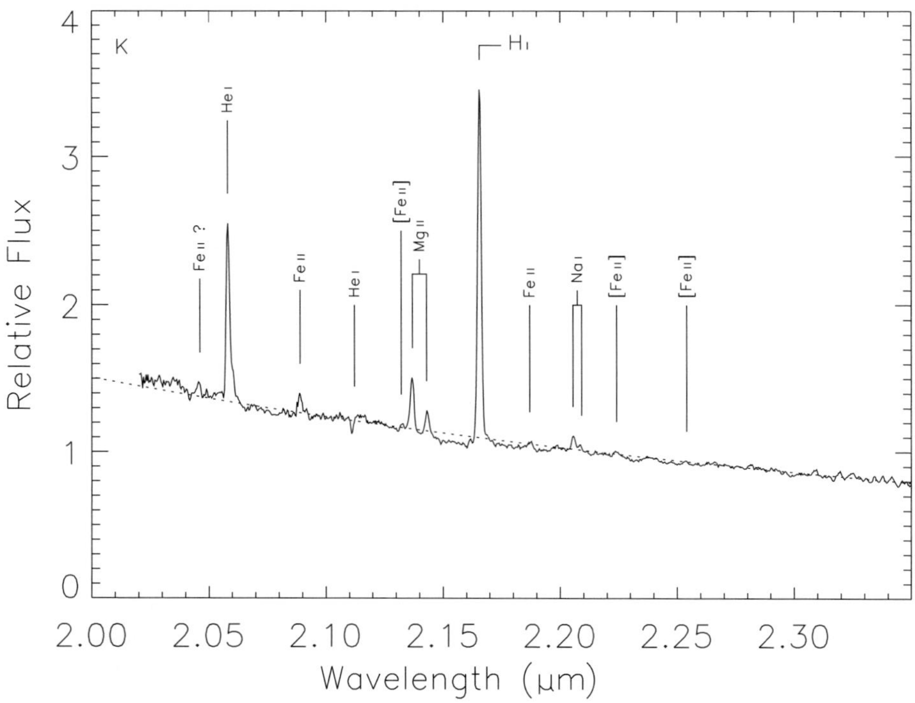

FIGURE 7.29. The near-IR spectrum of LBV 1806–20. Adapted from Eikenberry *et al.* (2004).

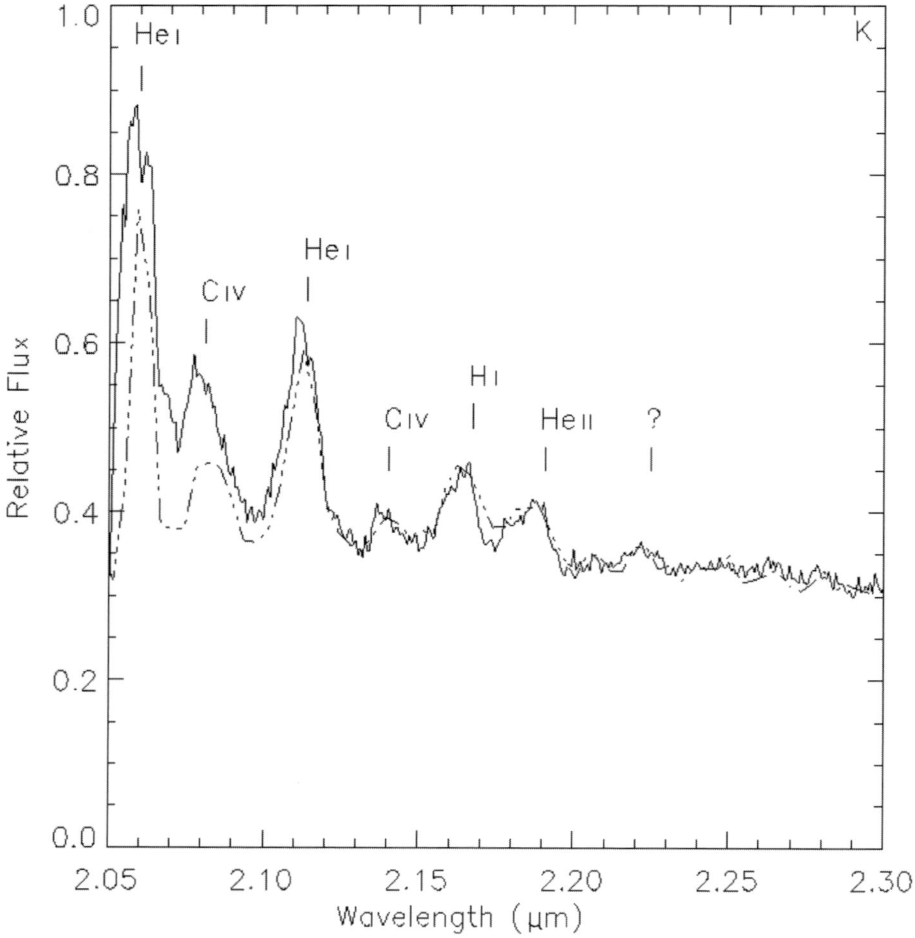

FIGURE 7.30. The near-IR spectrum of the WN Wolf–Rayet star near LBV 1806–20.

7.4.3 *Magnetospherically active stars*

In some sense, the Sun itself is a Galactic source of emission lines. The Sun has an active magnetosphere, which produces detectable line emission. These lines are in fact crucial diagnostics for many aspects of Solar physics. As far as we can tell, probably *all* stars have emission lines from their magnetospheres. However, "normal" stars like the Sun are seldom considered "emission-line objects," largely because the line luminosity is very low – both in absolute terms and as a fraction of the total system luminosity.

The dM stars also have low-luminosity line emission, but as a fraction of total luminosity it can be quite significant, resulting in moderate-to-high equivalent widths. These are magnetospherically active objects, and other magnetospherically active stars also possess emission lines. However, this class (like stars in general) is numerically dominated by the low-mass M stars. While the luminosities for these objects are low, they are so numerous that they are often found as foreground contaminants for surveys of more-distant/luminous objects. Figure 7.32 shows some dM star spectra, including a "normal" stellar continuum with Hα in emission. Even magnetospherically active brown dwarfs with Hα emission have recently been discovered.

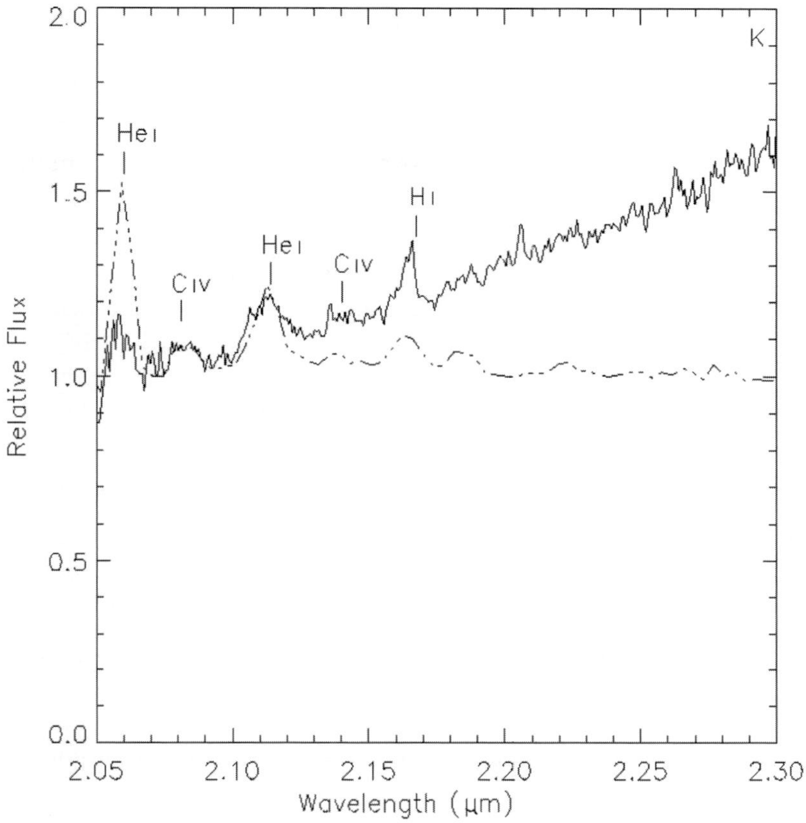

FIGURE 7.31. The near-IR spectrum of the WC9d Wolf–Rayet star near LBV 1806–20.

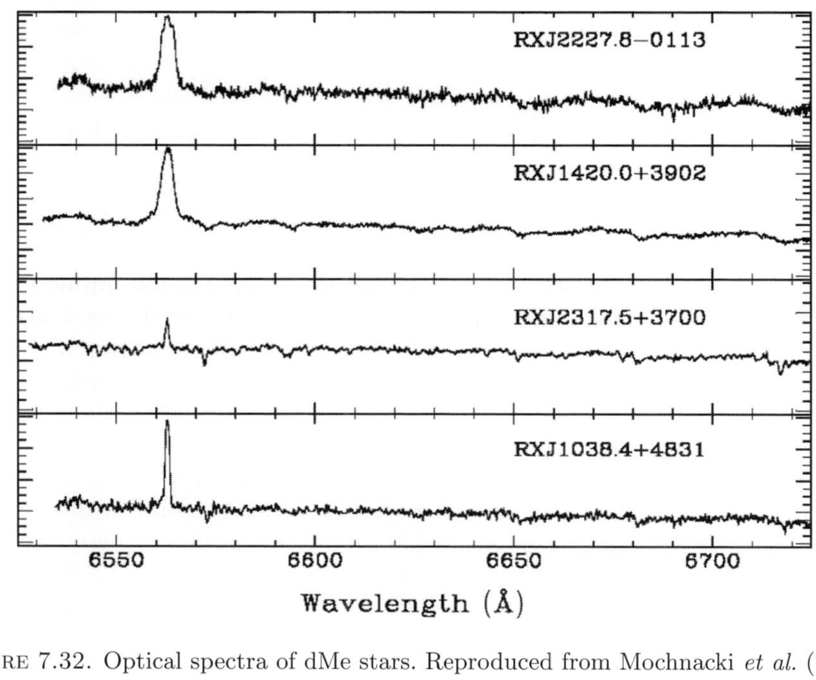

FIGURE 7.32. Optical spectra of dMe stars. Reproduced from Mochnacki *et al.* (2002).

7.4.4 *Galactic compact-object sources of emission lines*

As the final section of this work, we turn to Galactic compact-object systems as sources of emission lines. These include cataclysmic variables (CVs) as well as X-ray binaries – both neutron-star and black-hole binaries.

7.4.4.1 Cataclysmic variables

Cataclysmic variables (CVs) are accretion-powered binary systems with a white dwarf as the compact object. This general class includes a wide variety of systems, such as novae, dwarf novae, polars, and intermediate polars – a veritable bestiary of CVs. As was seen with the T Tauri stars above, accretion can produce much higher temperatures in a system than one would expect from "standard" photospheres. In the case of CVs, these temperatures are in the range of $\sim 10^5$ K and higher. Thus, photoionization often dominates the emission-line spectrum in these systems, up to and including very highly-ionized states of metals. The emission lines arise from all surfaces irradiated by "hard" radiation fields – the accretion disk, the donor-star face, the disk "hot spot" at its outer edge, and the accretion column where matter is funneled onto the white dwarf's surface by its magnetic field (if there is one).

The emission lines from CVs can provide important probes of physical conditions over a broad swath of the system's emitting regions. In particular, we expect different lines/ionization states to arise from different regions, so that we can separate them (though typically this is quite model-dependent, of course). By measuring the line strengths, we can diagnose the characteristics of the radiation field. Furthermore, by measuring line wavelengths and widths, we can constrain the motions of material in the system.

Perhaps most interestingly, for CVs we can carry out "Doppler tomography" to construct detailed pictures of the system's geometry and physics. This is because line wavelengths map line velocities, which in turn can be mapped to positions via a model of the system (Figure 7.33). Thus, line profiles can give some idea of the system's structure. Furthermore, as the CV system rotates orbitally, our line of sight changes, shifting the mapping and providing complementary information. By following through a complete revolution of the system orbit, we can use the line profiles to create a "tomogram" of the system. This can provide detailed insight into the structure of a system that is spatially completely unresolved!

7.4.4.2 X-ray binaries

X-ray binaries (XRBs) are analogous to CVs in that they are also accretion-powered binary systems with a compact object. However, rather than the white dwarfs in CVs, XRBs typically contain either a neutron star or a black hole as their compact object. Again, the accretion process produces unusually high temperatures in these systems, and the deeper potential wells provided by these compact objects mean that the temperatures can and do reach higher levels ($\sim 10^7$ K or higher) – resulting in the significant X-ray emission that earns their name. As with CVs, XRBs often display high ionization states of some metals, up to and including He-like Fe ions! Emission lines again arise from essentially all surfaces irradiated by the hard field produced here: the accretion disk, donor-star face, hot spot, and accretion column. In the case of XRBs, jet outflows are also frequently seen with emission lines. As with CVs, then, XRB emission lines provide diagnostics of all of these regions, including properties of the irradiating field and

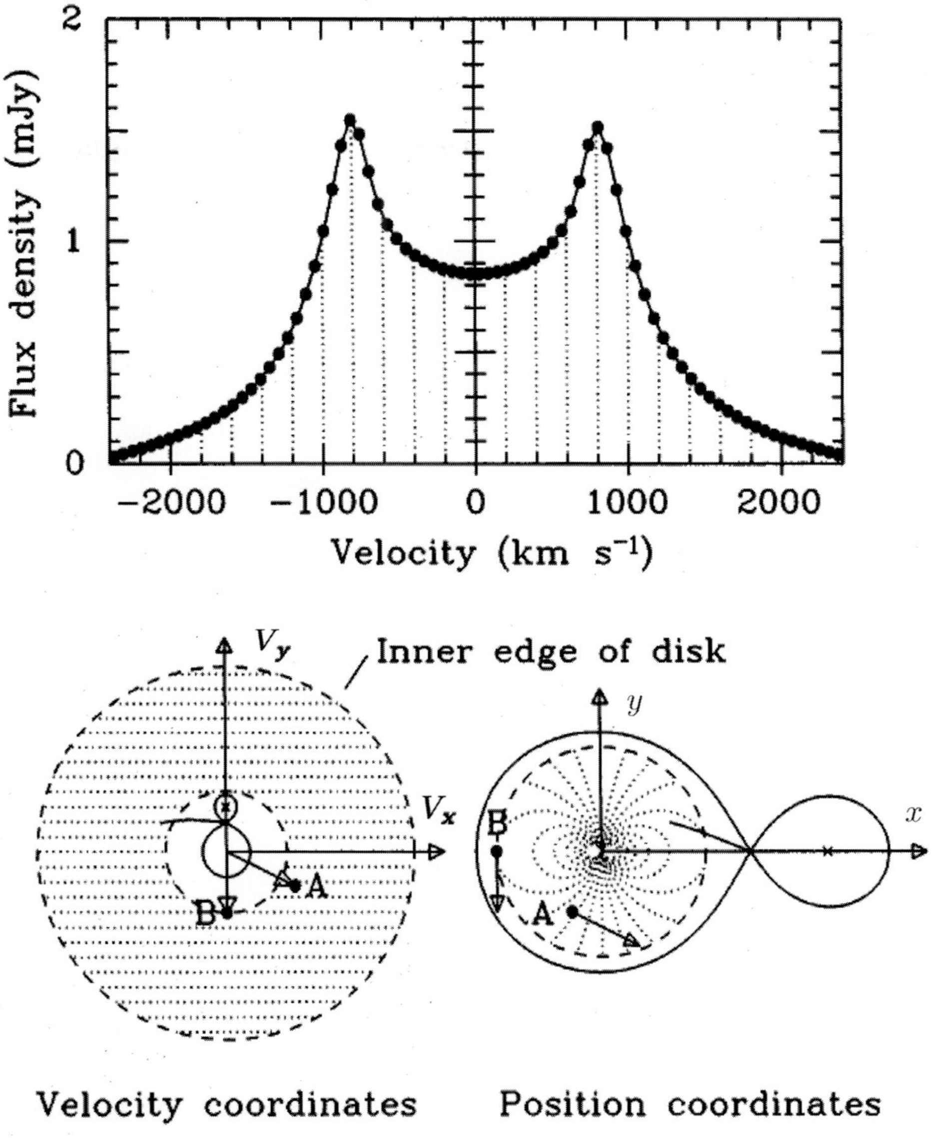

FIGURE 7.33. An illustration of the concept of Doppler tomography. Adapted from Horne & Marsh (1988).

motions within the system. In the case of XRBs, knowledge of these motions can allow measurement of the mass function, which thus provides a lower limit on the mass of the compact object, M_x.

The fundamental division in XRBs is between black-hole binaries and neutron-star binaries. However, it is observationally non-trivial to distinguish between these types. In the absence of a mass measurement, there is no currently accepted reliable diagnostic for this. Even for the small-ish subset with mass functions, the lower mass limit may be inconclusive as a diagnostic between neutron stars and black holes. Thus, we

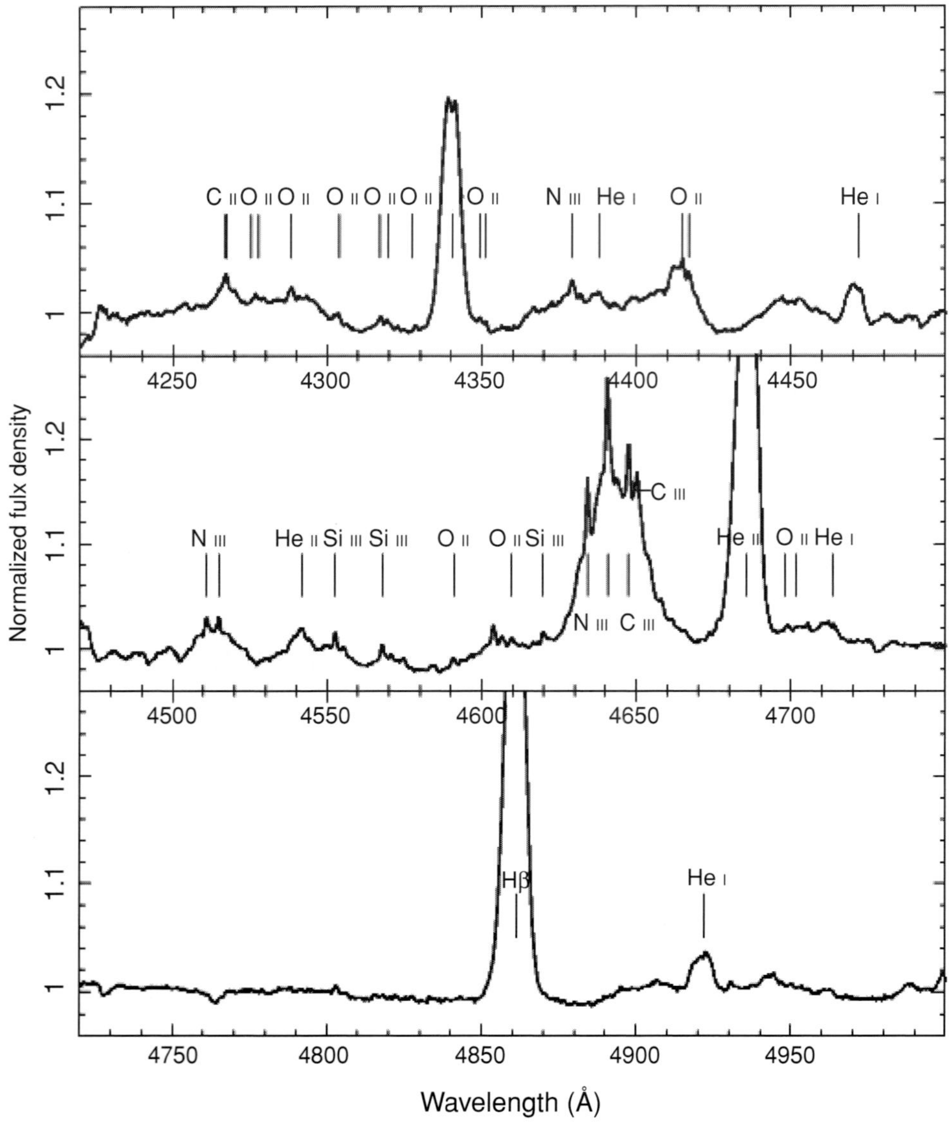

FIGURE 7.34. The optical spectrum of the LMXB Sco X-1. Adapted from Steeghs & Casares (2002).

typically adopt an observationally easier (if somewhat less informative) division between "low-mass" and "high-mass" X-ray binaries. It is important to note that the mass referred to here is NOT that of the compact object – rather it is the mass (in practice, spectral type or luminosity class) of the donor star.

For low-mass XRBs (LMXBs), the spectrum is typically dominated by the accretion disk itself, rather than by the low-mass (and thus typically low-luminosity) donor star. Typical features are broad emission lines of H, He I, and He II. We also often see high ionization states of metals (Figure 7.34). Radial velocity measurements can give mass-function measurements, and line profiles can even yield Doppler tomograms like those obtained for CVs. The IR spectra also show similar features. These features are often

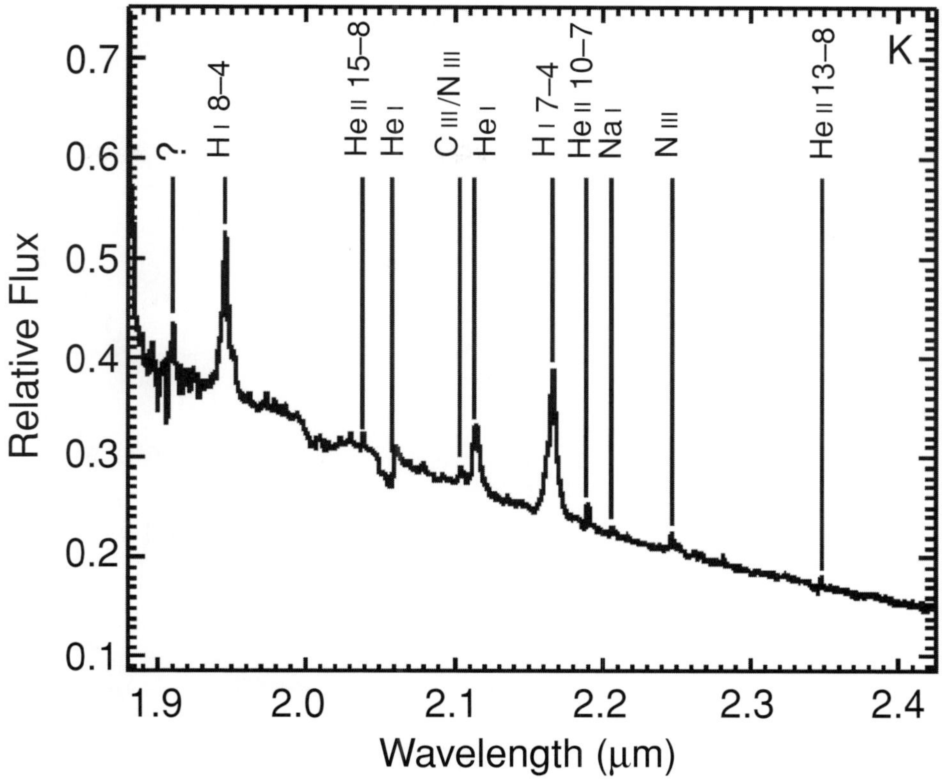

FIGURE 7.35. The infrared spectrum of the HMXB Edd-1. Adapted from Mikles *et al.* (2006).

highly time-variable, as is the typical accretion flow in LMXBs in general. In addition, we can see rapid correlated line variability, such as the continuum/Brγ correlation seen during jet-flaring events from the microquasar LMXB GRS 1915+105 (Eikenberry *et al.*, 1998).

High-mass XRBs (HMXBs) can show similar accretion-disk features, but they can also be masked, obscured, or confused by similar features from the massive donor star itself! Figure 7.35 shows the IR spectrum of the Galactic Center HMXB Edd-1. In addition, the B[e] system shown above – CI Cam – is itself thought to be a HMXB system (though often in quiescence).

Finally, we conclude with a look at the power of emission lines for diagnosing some of these system – in this case SS433. As we can see in Figure 7.36, SS433 has the "standard" XRB emission lines – broadened H, He, etc. However, we can also see what appear to be red/blueshifted features. In fact, with sufficient sensitivity, we find that essentially every "rest-wavelength" emission feature has these counterparts. Furthermore, these line measurements reveal velocities as high as $v \sim 0.26c$ (!!). In the radio, we see collimated outflows, and on combining the two we see that SS433 has emission lines emanating from two oppositely directed relativistic jets! A kinematic model of these jets shows that they are precessing with a period of $P \sim 164d$, and recent work shows that this period has been essentially stable for > 20 yr. Detailed models and observations of these jet emission lines are providing important probes into the formation of relativistic jets in this extremely interesting system.

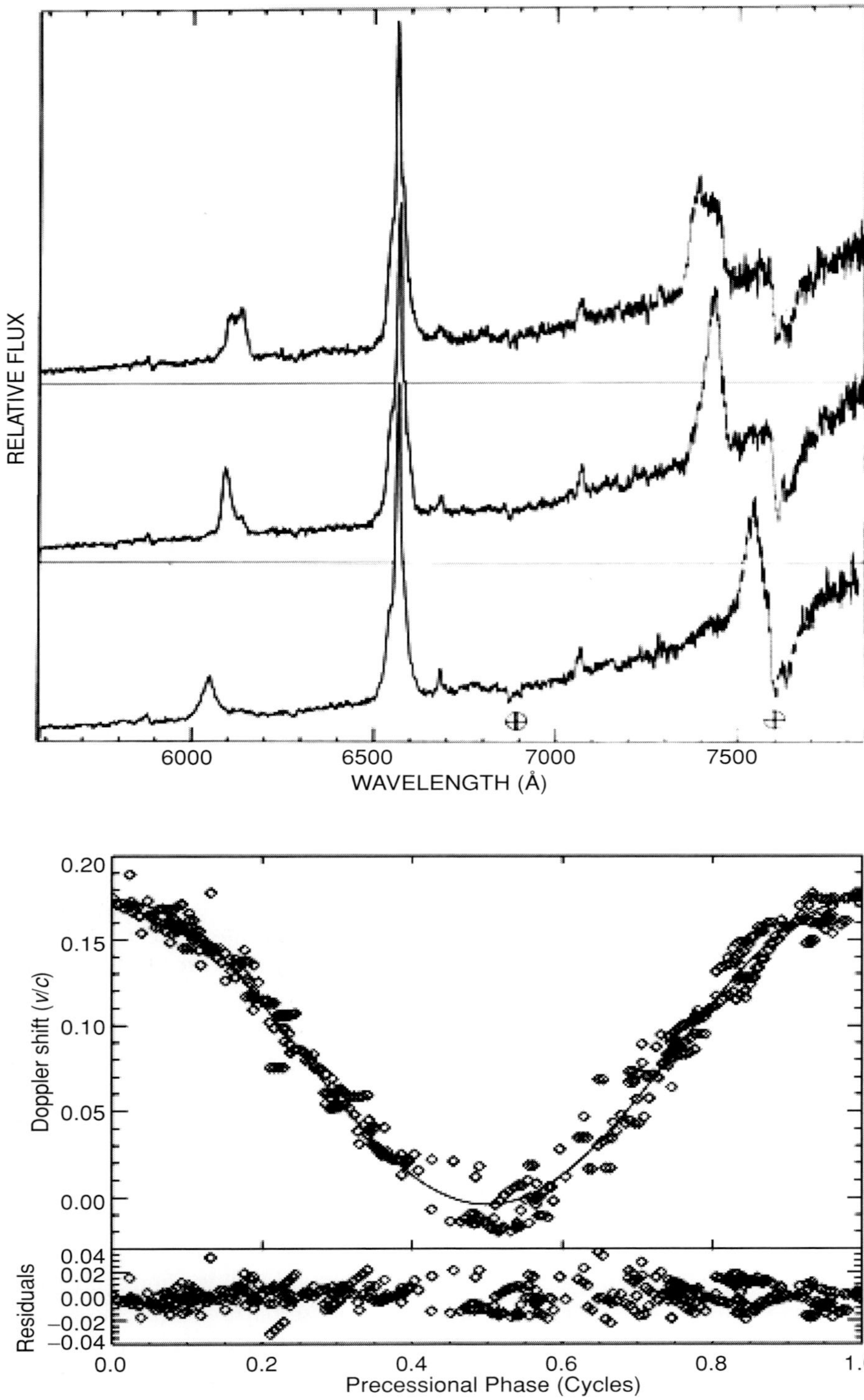

FIGURE 7.36. The optical spectrum of SS433, adapted from Margon *et al.* (1984), and inferred disk precession, adapted from Eikenberry *et al.* (2001).

REFERENCES

Blum, R., Barbosa, C. L., Daminelli, A., Conti, P. S. Ridgway, S. 2004, ApJ, 617, 1167

Conti, P. S. 1986, in *Luminous Stars and Associations in Galaxies*, ed. C. W. H. de Loore, A. J. Willis, & P. Laskarides (Dordrecht: Reidel), p. 199

Edwards, S., Fischer, W., Hillenbrand, L., Kwan, J. 2006, ApJ, in press

Eikenberry, S. S. *et al.* 2004, ApJ, 616, 506

Eikenberry, S. S., Cameron, P. B., Fierce, B. W., Kull, D. M., Dror, D. H., Houck, J. R., Margon, B. 2001, ApJ, 561, 1027

Fese, R., Blair, W., Kirshner, R. P. 1985, ApJ 292, 29

Hanuschik, R. W., Kozok, J. R., Kaiser, D. 1988, A&A, 189, 147

Jaschek, R., Andrillat, Y. 2000, in *The Be Phenomenon in Early-Type Stars* (San Francisco, CA: Astronomical Society of the Pacific), pp. 83–86.

Kurosawa, R., Harries, T., Symington, N. 2006, MNRAS, in press

Likkel, L., Dinerstein, H. L., Lester, D., Kindt, A., Batrig, K. 2006, ApJ, 131, 1515

Margon, B. 1984, ARAA, 22, 507

Marsh, T., Horne, K. 1988, MNRAS, 235, 269

Mikles, V., Eikenberry, S. S., Muno, M., Bandyopadhyay, R., Patel, S. G. 2006, ApJ, 651, 408

Mochnacki, S. W. *et al.*, 2002, AJ, 124, 2868

Reipurth, B., Bally, J. 2001, ARAA, 39, 403

Steeghs, D., Casares, J. 2002, ApJ, 568, 273

8. Narrow-band imaging

SERGIO PASCUAL AND BERNABÉ CEDRÉS

8.1. Introduction

The imaging method which uses narrow-band filters is the first and most direct way to obtain information about the emission lines. In contrast to broad-band photometry, in which several emission lines are integrated within one filter, the narrow-band filters are designed to transmit only one emission line (in ideal cases).

This method has considerable advantages over spectroscopic methods: we are able to obtain spatial resolution, i.e. several objects fit in only one exposure frame; the required time for an observation run is shortened; and lastly, the observation procedures and reduction are less complicated than those of a spectroscopic survey. Usually during a run information for only a few lines (three or four at best) can be obtained, in contrast to the full spectrum obtained in spectroscopy: see, for example, the works of Belley & Roy (1992), where hundreds of H II regions are presented with fluxes in three emission lines using direct imaging, and van Zee *et al.* (1998), with only about a dozen H II regions in ten different emission lines employing optical spectroscopy.

Narrow-band imaging is a powerful tool for characterizing the physical properties of the star-forming regions in nearby galaxies. With a few lines, we are able to determine, for example, the abundance, temperature, or initial mass function of the ionizing stars, thus obtaining clues to the physical processes that occur in the core of the H II regions; see, for example, Cedrés & Cepa (2002). This can be seen in Figure 8.1, where the composite broad-band image (with the B, R, and K bands) and an Hα image of the galaxy UCM 2325+2318 can be compared. The star-forming regions are clearly defined in the narrow-band image.

Narrow-band imaging has successfully been used to build samples of diverse types of emission-line objects: see for example Rauch (1999) for galactic planetary nebulae, and Okamura *et al.* (2002), Castro-Rodríguez *et al.* (2003), and Arnaboldi *et al.* (2003) for extragalactic PNe.

Emission-line galaxies (hereafter ELGs) are an invaluable resource for our understanding of the evolution of galaxies in the Universe. Faint galaxies can be difficult to confirm spectroscopically, whereas the ELGs are generally easy to identify. Furthermore, the emission lines are produced within regions related either to star formation or to the phenomenon of active galactic nuclei (AGNs).

Narrow-band imaging reduces significantly the contribution of the sky brightness, since it is admitted in a small range of wavelengths. The sky background, which is the most-significant limitation in the detection of objects in broad-band images, is greatly reduced. A small wavelength range also increases the contrast (Thompson *et al.* 1995) between the emission line and the continuum. In order to use the narrow-passband filters with maximum efficiency, regions of the night-sky spectrum with a minimum background are selected. In the optical wavelengths, the windows in the Meinel OH bands around 8200 Å and 9200 Å have been used to center narrow-band filters.

Narrow-band imaging produces volume-limited samples, since the narrow observed bands correspond to small windows in redshift space, with approximately constant luminosity limits. In the case of detecting lines used as star-formation tracers (such as Hα),

The Emission-Line Universe, ed. J. Cepa. Published by Cambridge University Press.

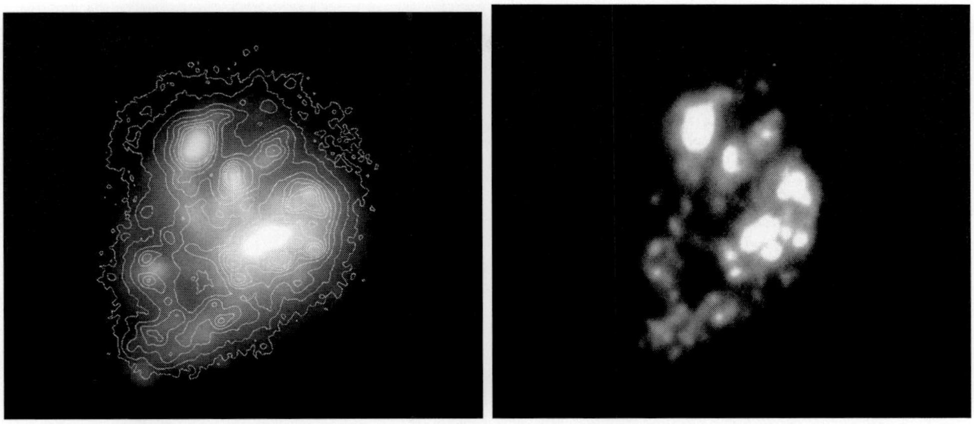

FIGURE 8.1. Images of galaxy UCM 2325+2318. On the left is shown a composite B, R, and K image with Hα contours. On the right is shown a narrow-band image of the galaxy.

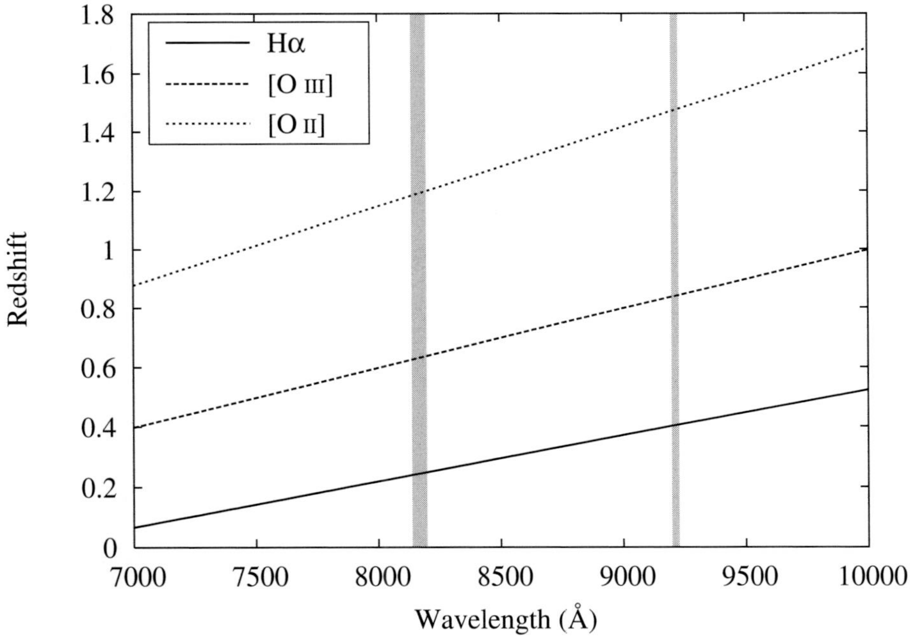

FIGURE 8.2. Wavelength of the emission lines as a function of the redshift of the galaxy. The emission lines are represented by lines increasing with wavelength. Hα is shown as a solid line, [O III] λ5007 as a dashed line, and [O II] λ3727 as a dotted line. The airglow windows are shown as vertical shaded regions.

the sample would be directly star-formation-rate (SFR)-selected, except for the AGN contribution.

The search for emission-line galaxies using narrow-band filters is open to different lines at different redshifts. If the redshift of the source and the rest wavelength of the emission line act together to put the line inside the narrow-band filter, the galaxy will be selected as long as it is bright enough. The lines we expect to detect at low redshifts are Hα, [O III] λλ4959, 5007 and [O II] λ3727 (Tresse *et al.* 1999; Kennicutt 1992). At higher redshifts, Lyα can be detected with 8-m-class telescopes. In Figure 8.2 we can see how the redshift

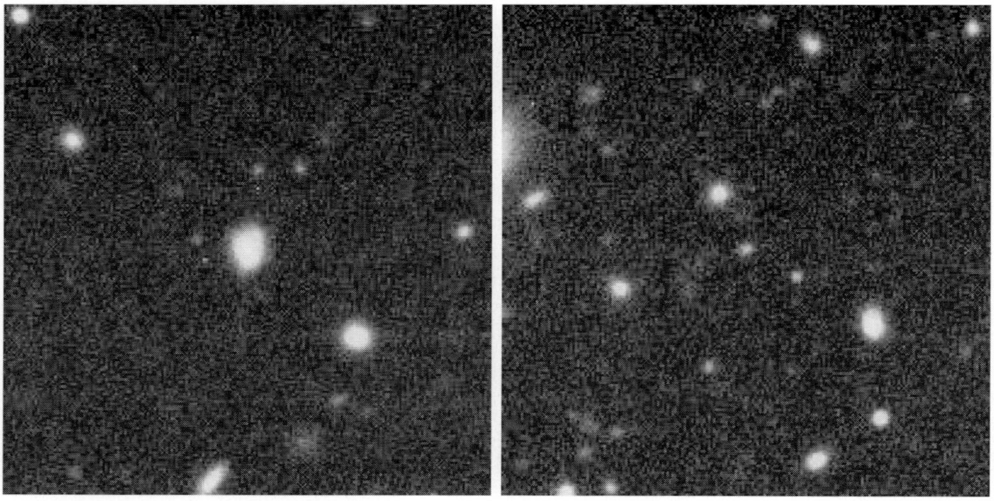

FIGURE 8.3. Two narrow-band objects from Pascual *et al.* (2007) selected on the 8200-Å airglow window. On the left is shown a $z = 0.24$ Hα emitter, on the right a $z = 0.6$ [O III] emitter. Images are composite U, g, r, 60 arcseconds wide, with north on top and east on the left.

of the source makes the different lines enter in the airglow windows. Also, some sample objects are shown in Figure 8.3.

Narrow-band imaging produces approximately volume-limited samples, since the narrow-observed bands correspond to small windows in redshift space. The objects are selected with a well-defined limit in equivalent width, and the line flux can be transformed into luminosity with some simple assumptions. Narrow-band imaging thus provides line luminosities for a volume-limited sample of emission-line galaxies. In the case of detecting lines used as star-formation tracers, the sample would be directly SFR-selected, except for the AGN contribution.

The problem with this approach is that stars contaminating the sample or contributions from different emission lines cannot be separated with narrow-band imaging alone. Additional assumptions (on the luminosity functions of ELGs, e.g. Jones & Bland-Hawthorn (2001)) or additional data (see, e.g., the color–color diagrams of Fujita *et al.* (2003)) are needed to complete the scenario.

The purpose of this chapter is to provide tools for handling narrow-band data. Although the notes are IRAF†-centric, they can easily be translated to other packages. Additionally, we assume that the images have been reduced up to the step of flat-fielding.‡ Section 8.2 deals with calibration and sky subtraction for narrow-band imaging of an extended source. Section 8.3 focuses on the problems that appear with narrow-band imaging in the airglow windows.

8.2. Extended-source imaging

In the case of extended-source images, such as, for example, nearby galaxies, several additional corrections are needed in order to wipe out the contributions from sources

† IRAF is distributed by the National Optical Astronomy Observatories, which is operated by the Association of Universities for Research in Astronomy, Inc. (AURA) under a cooperative agreement with the National Science Foundation.

‡ Documentation about generic image reduction and IRAF usage can be found, for example, at `http://iraf.net/docs`.

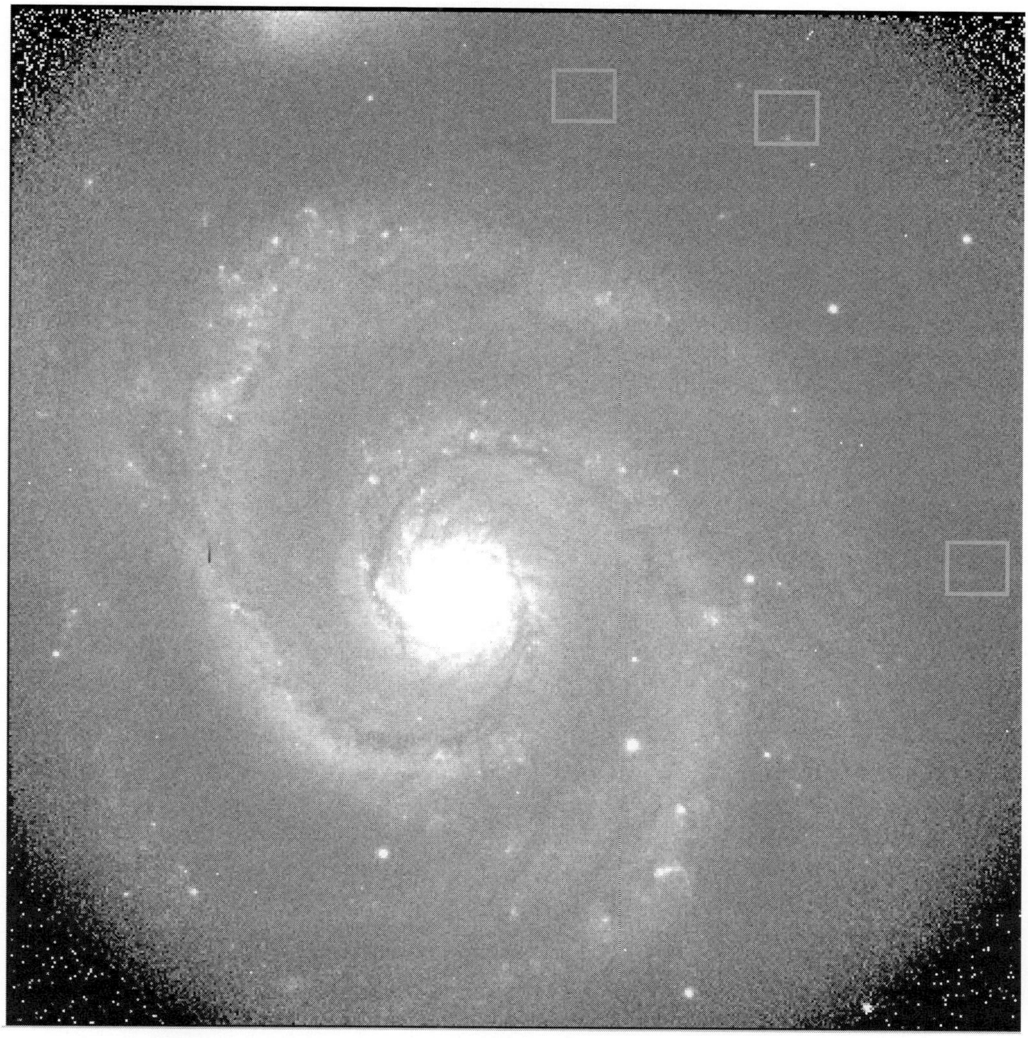

FIGURE 8.4. Optimal regions in a frame for carrying out the sky subtraction in order to avoid contamination from a galaxy (marked as rectangles).

different from the ones studied. The ADUs must also be converted into physical units. This is done through flux calibration.

8.2.1 *Sky subtraction*

The sky counts are not as important in narrow-band as in broad-band images. Nevertheless, for long-exposure images, such as galaxy images and weak-line images (O II λ3727, [S III] λ9069, etc.), it is necessary to subtract the sky contribution. The sky correction is done by collecting statistics in boxes that are not affected by the observed object (i.e. galaxy) near the corners of the image, as shown in Figure 8.4. The value for the sky is the mean of the mean values for each of the boxes. This is then subtracted from the image.

8.2.2 *Flux calibration*

In order to calibrate the images, fluxes from spectrophotometric stars are needed. We will use the stars proposed by Oke & Gunn (1983) and Oke (1990). The data can be

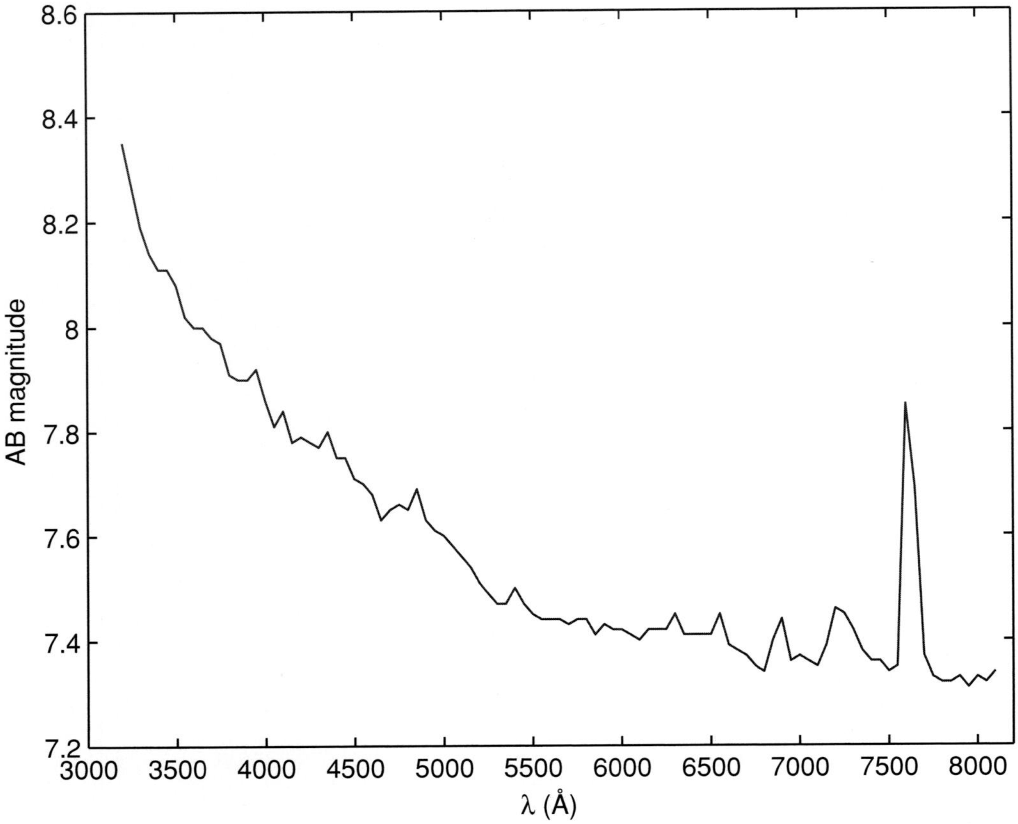

FIGURE 8.5. AB magnitude as a function of wavelength for the standard star HD192281.

obtained at `http://www.ing.iac.es/Astronomy/observing/manuals/html_manuals/tech_notes/tn065-100/workflux.html`. To obtain the magnitude, the data presented in the charts need only be interpolated.

The flux for the standard star is obtained through aperture photometry, using the task `phot` from IRAF (`noao.digi.dao.phot`). The spectral distribution of the AB magnitudes for the standard star HD192281 is shown in Figure 8.5.

8.2.3 *Equations for calibration*

Following Barth *et al.* (1994), the number of counts per second for the atmospheric extinction corrected line filter is

$$F_{\mathrm{g}}^{\mathrm{l}} = \kappa^{\mathrm{l}}(F^{\mathrm{c}} + T_{\mathrm{r}}^{\mathrm{l}} I^{\mathrm{l}}) \tag{8.1}$$

and that for the continuum filter is

$$F_{\mathrm{g}}^{\mathrm{c}} = \kappa^{\mathrm{c}} F^{\mathrm{c}} \frac{\Delta\nu_{\mathrm{c}}}{\Delta\nu_{\mathrm{l}}}, \tag{8.2}$$

where κ^{λ} is the telescope-plus-detector efficiency at λ, F^{c} is the continuum flux in the line filter, $T_{\mathrm{r}}^{\mathrm{l}}$ is the transmission of the line filter, I^{l} is the line flux, and $\Delta\nu_{\mathrm{c}}$ and $\Delta\nu_{\mathrm{l}}$ are the bandwidths of the continuum filter and the line filter, respectively.

FIGURE 8.6. An image of the spectrophotometric standard star BD1747. The star is marked with a circle.

The number of counts per second of the calibration star is

$$F_*^{\mathrm{l}} = \kappa^{\mathrm{l}} f_{\mathrm{l}}\, \Delta\nu_{\mathrm{l}} \tag{8.3}$$

and that for the continuum filter is

$$F_*^{\mathrm{c}} = \kappa^{\mathrm{c}} f_{\mathrm{c}}\, \Delta\nu_{\mathrm{c}}, \tag{8.4}$$

where f_λ is the monochromatic flux of the spectrophotometric standard star (Figure 8.6). Then

$$F_{\mathrm{g}}^{\mathrm{l}} = \frac{F_*^{\mathrm{l}}}{f_{\mathrm{l}}\, \Delta\nu_{\mathrm{l}}}(F^{\mathrm{c}} + T_{\mathrm{r}}^{\mathrm{l}} I^{\mathrm{l}}), \tag{8.5}$$

$$F_{\mathrm{g}}^{\mathrm{c}} = \frac{F_*^{\mathrm{c}} F^{\mathrm{c}}}{f_{\mathrm{c}}\, \Delta\nu_{\mathrm{l}}} \tag{8.6}$$

With this we are able to obtain I^{l}.

$$I^{\mathrm{l}} = \left(\frac{F_{\mathrm{g}}^{\mathrm{l}}}{F_{*}^{\mathrm{l}}} f_{\mathrm{l}} - \frac{F_{\mathrm{g}}^{\mathrm{c}}}{F_{*}^{\mathrm{c}}} f_{\mathrm{c}} \right) \frac{\Delta \nu_{\mathrm{l}}}{T_{\mathrm{r}}^{\mathrm{l}}} \tag{8.7}$$

Oke & Gunn (1983) give the following monochromatic magnitude for the calibration stars:

$$m_{\lambda}^{\mathrm{AB}} = -2.5 \log f_{\lambda} - 48.6. \tag{8.8}$$

With a few manipulations,

$$f_{\lambda} = 10^{-0.4(m_{\lambda}^{\mathrm{AB}}+48.6)}. \tag{8.9}$$

Moreover,

$$\frac{F_{\mathrm{g}}^{\lambda}}{F_{*}^{\lambda}} = 10^{0.4K_{\lambda}(x_{\mathrm{g}}^{\lambda}-x_{*}^{\lambda})} \frac{F_{\mathrm{G}}^{\lambda}}{F_{\mathrm{S}}^{\lambda}}, \tag{8.10}$$

where F_{G}^{λ} and F_{S}^{λ} are the counts per second at the object and at this calibration star, respectively, x_{g}^{λ} and x_{*}^{λ} are the air masses of the object and the calibration star, respectively, and K_{λ} is the atmospheric extinction for wavelength λ.

After making a few substitutions, we obtain

$$I^{\mathrm{l}} = c \frac{\Delta \lambda_{\mathrm{l}}}{\lambda_{\mathrm{l}}^{2} T_{\mathrm{r}}^{\mathrm{l}}} \left[10^{-0.4[m_{\mathrm{l}}^{\mathrm{AB}}+48.6-K_{\mathrm{l}}(x_{\mathrm{g}}^{\mathrm{l}}-x_{*}^{\mathrm{l}})]} \frac{F_{\mathrm{G}}^{\mathrm{l}}}{F_{\mathrm{S}}^{\mathrm{l}}} - \alpha 10^{-0.4[m_{\mathrm{c}}^{\mathrm{AB}}+48.6-K_{\mathrm{c}}(x_{\mathrm{g}}^{\mathrm{c}}-x_{*}^{\mathrm{c}})]} \frac{F_{\mathrm{G}}^{\mathrm{c}}}{F_{\mathrm{S}}^{\mathrm{c}}} \right], \tag{8.11}$$

where α is a factor that takes into account the differences between the continuum filter and the line filter. The method for the determination of this factor is explained in the following section.

8.2.4 *Continuum subtraction*

To obtain the flux of the line alone, the continuum contribution needs to be wiped out. This is done by using an image with a filter centered near the target line but not affected by it, so as to sample only the underlying continuum. One method for obtaining this is to multiply the continuum image by a factor near the unit (factor α in Equation (8.11)). This factor can be determined in two ways: by making the emission of the field stars null after subtraction, or by assuming that the inter-arm emission is null after the subtraction in the case of galaxies. The first method could lead to an overestimation or underestimation of the real value of the factor α due to differences between the spectral distribution of the field stars and that of the object. The second method would wipe out the diffuse emission from the object (as seen in Figure 8.7). So, depending on the aim of the research, one method would be more suitable than the other. Nevertheless, special care has to be taken with both methods in order to make the correction. The fluxes for several field stars (more than three) must be obtained, using aperture photometry, in the first method. In the second method, statistics must be collected for small boxes in the inter-arm zones for the line image and for the same inter-arm zones for the continuum image.

8.3. Imaging in the airglow-windows range

Fringing is one of the issues that appear when observing in the broad bands R, I, and Z, and in the narrow bands. In this section we explain a method for removing fringing using IRAF tasks. We also explain a simple method for selecting emission-line galaxies

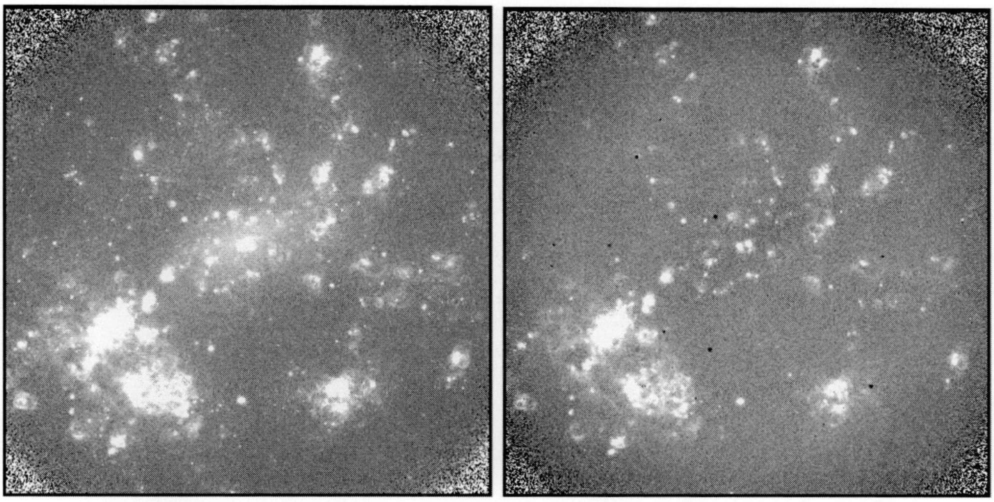

FIGURE 8.7. Hα images of NGC 4395 before (left panel) and after (right panel) continuum subtraction.

using the narrow-broad color excess. The selection process is studied in detail in Pascual *et al.* (2006).

8.3.1 *Fringing*

Fringing is produced by the multiple reflection and interference of light with the charge-coupled-device (CCD) substrate. When a monochromatic coherent light beam undergoes multiple reflection between two flat-parallel surfaces, the phase difference produces interference, either constructive or destructive, depending on the relationship between the wavelength and the distance between the two layers. If these are parallel, the resulting pattern is a series of bright and dark rings (Newton's rings). The effect arises when the wavelength is comparable to the thickness of the CCD. This implies wavelengths redward of ∼6000 Å and up to the limit of the CCDs, around ∼10 000 Å. In a real CCD, where the two layers are not parallel, the patterns are irregular.

When, instead of a monochromatic light, a wide range of wavelengths is used, constructive and destructive interferences tend to vanish locally. It is in narrow-band photometry, or when strong emission lines (nearly monochromatic) appear (such as those of the night-sky spectrum), that fringing is important. Note that a dome-flat image can have fringing if the illuminating lamp has emission lines or the filter is narrow. Figure 8.8 shows a sample I-band science image with a clearly visible fringe pattern. The effect is cumulative. This fringing must be removed by building a master fringe image. A method for doing this is explained below.

8.3.2 *Fringing removal using the package* `mscred`

This section describes how to remove fringing using tasks within the IRAF package `mscred` (Valdes 1998). The basic steps for removing fringing are creating a template of the fringe pattern, creating a mask that isolates the background in each exposure, creating a background map when there are background gradients across the exposure, determining the scale factor that best matches the fringe template to the exposure, and subtracting the scaled fringe template from the exposure.

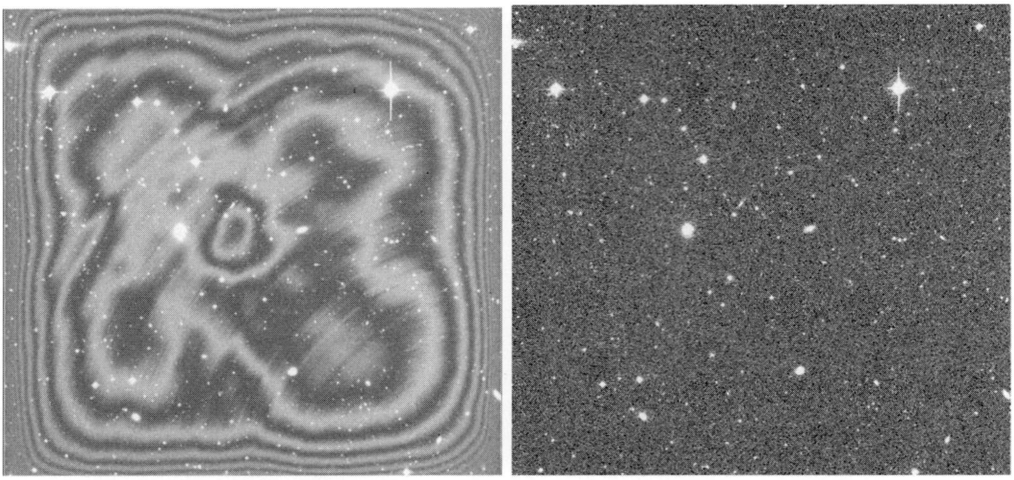

FIGURE 8.8. A science image containing the fringe pattern.

A fringe template is constructed from all the sky exposures which exhibit the same fringe pattern. The best result is obtained using dark-sky exposures where the fields have been dithered so that every pixel has several exposures that are uncontaminated by sources. The exposures are put together to make a fringe template using the `mscred.combine` task. The images are scaled to a common level and objects are excluded by rejection and masking techniques.

Object masks to be excluded during combining can be created with task `objmasks` from package `nproto`. Since making object masks (and background maps) for each exposure is needed for automatic computation and removal of the fringing, making the masks for creating the fringe template does not add unnecessary processing. In any case, the sky map is needed only if there is a significant sky gradient. The mean sky level is automatically accounted for during the fringe-removal step.

The following are typical parameters of the task `objmasks`:

```
cc> nproto
np> epar objmasks
images   =   @type_object        List of images or MEF files
objmasks=  obj_//@type_object    List of output object masks
(omtype =                numbers) Object mask type
(skys    = sky_//@type_object    ) List of input/output sky maps
...
(convolv=              block 3 3) Convolution kernel
(hsigma =                     3.) Sigma threshold above sky
(lsigma =                    10.) Sigma threshold below sky
```

The text file `type_object` is processed and new object masks and background maps are produced for each input image.

The template fringing is produced using the `mscred.combine` task. The critical parameter here is `masktyp`. The value of the parameter should be `!objmask`. This instructs `combine` to use the header keyword `OBJMASK` that `objmasks` inserts into the images.

The object masks and sky maps produced by `objmasks` are used in the task `rmfringe` to automatically determine the fringe scaling and subtract the pattern. The most important parameters of the task are

```
input    =  @filter_BB              List of input images
output   = f_//@filter_BB           List of output corrected images
fringe   = FringeBB                 Fringe or list of fringe patterns
masks    = !objmasks                List of object/bad data masks
...
(backgro= sky_//@filter_BB      ) Lisk of input image backgrounds
...
```

The task does not always remove the fringing successfully. An alternative task that can be used to remove the fringing with good results is `irmfringe`. The basis of the task is to find the scale factor interactively by visual inspection.

This task is prepared to work with multi-extension fits files; in order to use it with single-extension fits, this syntax is used:

```
input    =          r85132[0]  List of input mosaic exposures
output   =        r85132_f[0]  List of output mosaic exposures
template=         FringeBB[0]  Template mosaic exposure
```

8.4. Selection of candidates and line flux

Candidate line-emitting objects were selected using their excess narrow versus broad flux on a plot of $m_{\rm NB}$ versus $I - m_{\rm NB}$.

Catalogs of objects present in the images can be constructed with any object-detection software. Here we use `SExtractor` (Bertin & Arnouts 1996). The common approach to object detection is to use the *double-image mode*: the narrow-band frame is used as a reference image for detection and then the flux is measured both in the narrow- and in the broad-band image. Other methods can be used to construct a segmentation image where the object detection is performed. The broad- and narrow-band images can be added or combined (Szalay *et al.* 1999) to allow the recovery of objects present in only one of the images.

Narrow-band imaging not only allows object detection, but also makes it possible to estimate the flux and equivalent width of the emission line with simple assumptions.

The flux density in each filter can be expressed as the sum of the line flux and the continuum flux density (the line is covered by both filters):

$$f_\lambda^{\rm B} = f_\lambda^{\rm c} + \frac{f_{\rm L}}{\Delta_{\rm B}}, \qquad f_\lambda^{\rm N} = f_\lambda^{\rm c} + \frac{f_{\rm L}}{\Delta_{\rm N}} \tag{8.12}$$

with $f_\lambda^{\rm c}$ the continuum flux, $f_{\rm L}$ the line flux, $\Delta_{\rm B}$ and $\Delta_{\rm N}$ the broad- and narrow-band-filter effective widths and $f_\lambda^{\rm B}$ and $f_\lambda^{\rm N}$ the flux densities in the two filters. Then the line flux, continuum flux, and equivalent width can be expressed as follows:

$$f_{\rm L} = \Delta_{\rm N}\big(f_\lambda^{\rm N} - f_\lambda^{\rm B}\big)\frac{1}{1-\epsilon}, \tag{8.13}$$

$$f_\lambda^{\rm c} = f_\lambda^{\rm B}\frac{1-\epsilon f_\lambda^{\rm N}/f_\lambda^{\rm B}}{1-\epsilon}, \tag{8.14}$$

$$\mathrm{EW} = \frac{f_{\rm L}}{f_\lambda^{\rm c}} = \Delta_{\rm N}\left(\frac{f_\lambda^{\rm N} - f_\lambda^{\rm B}}{f_\lambda^{\rm B}}\right)\left(\frac{1}{1-\epsilon f_\lambda^{\rm N}/f_\lambda^{\rm B}}\right), \tag{8.15}$$

with $\epsilon = \Delta_{\rm N}/\Delta_{\rm B}$. These equations are also valid for the case of two contiguous narrow-band filters making $\epsilon = 0$ (Pascual *et al.* 2006).

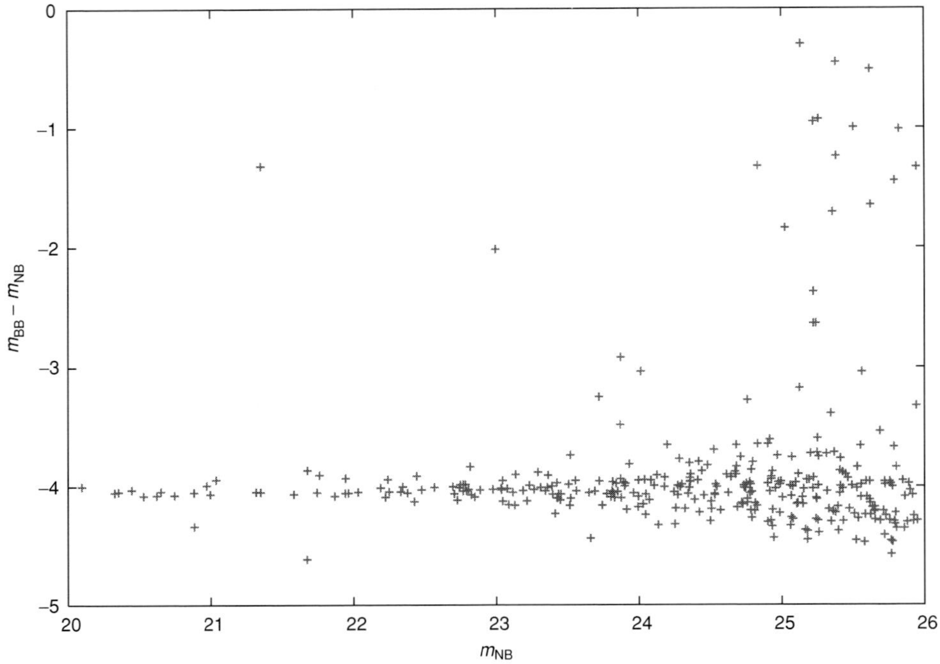

FIGURE 8.9. A magnitude–color selection diagram. Objects with excess emission in the narrow band appear in the upper part of the diagram.

The effective widths of the filters can be calculated to first order (assuming square filters) as the normalized integral of the transmittance of the filter profile (including the quantum efficiency of the detector). It agrees with the equivalent width of the band-transmittance profile W_0 of Fiorucci & Munari (2003).

Further refinement can be achieved by estimating the position of the emission line inside the filters. The computation of the effective width can be found in detail in Pascual *et al.* (2006).

Acknowledgments

S.P. would like to thank his collaborator Juán Carlos Muñoz-Mateos for providing the multicolor images of the galaxy UCM 2325+2318.

REFERENCES

Arnaboldi, M., Freeman, K. C., Okamura, S. *et al.* 2003, AJ, 125, 514

Barth, C. S., Cepa, J., Vilchez, J. M., Dottori, H. A. 1994, AJ, 108, 2069

Belley, J., Roy, J.-R. 1992, ApJS, 78, 61

Bertin, E., Arnouts, S. 1996, A&AS, 117, 393

Castro-Rodríguez, N., Aguerri, J. A. L., Arnaboldi, M. *et al.* 2003, A&A, 405, 803

Cedrés, B., Cepa, J. 2002, A&A, 391, 809

Fiorucci, M., Munari, U. 2003, A&A, 401, 781

Fujita, S. S., Ajiki, M., Shioya, Y. *et al.* 2003, ApJL, 586, L115

Jones, D. H., Bland-Hawthorn, J. 2001, ApJ, 550, 593

Kennicutt, R. C. 1992, ApJS, 79, 255

Okamura, S., Yasuda, N., Arnaboldi, M. *et al.* 2002, PASJ, 54, 883

Oke, J. B. 1990, AJ, 99, 1621

Oke, J. B., Gunn, J. E. 1983, ApJ, 266, 713

Pascual, S., Gallego, J., Zamorano, J. 2006, ArXiv astro-ph/0611121

Rauch, T. 1999, A&AS, 135, 487

Szalay, A. S., Connolly, A. J., Szokoly, G. P. 1999, AJ, 117, 68

Thompson, D., Djorgovski, S., Trauger, J. 1995, AJ, 110, 963

Tresse, L., Maddox, S., Loveday, J., Singleton, C. 1999, MNRAS, 310, 262

Valdes, F. G. 1998, in *Astronomical Data Analysis Software and Systems VII*, ed. R. Albrecht, R. N. Hook, & H. A. Bushouse (San Francisco, CA: Astronomical Society of the Pacific), pp. 53–57

van Zee, L., Salzer, J. J., Haynes, M. P., O'Donoghue, A. A., Balonek, T. J. 1998, AJ, 116, 2805

9. Long-slit spectroscopy

MIGUEL SÁNCHEZ-PORTAL AND ANA PÉREZ-GARCÍA

9.1. Long-slit spectroscopy

Spectroscopy is the main tool for studying the physics of astrophysical objects. Spectroscopic observations can be thought to be a method by which one samples the emitted spectral energy distribution (SED) from an astronomical source in wavelength bins of size $\Delta\lambda$. For example, narrow-band-filter photometry is a very-low-resolution spectroscopy technique. The required spectral resolution, i.e. the ability to separate the different spectral features (lines) in the source's SED, is driven by the scientific objectives. The basic parameter for characterizing a given spectrograph configuration is the resolving power. It is defined as $R = \lambda/\Delta\lambda$, where $\Delta\lambda$ is the difference in wavelength between two closely spaced spectral features, say two spectral lines of equal intensity, each with approximate wavelength λ.

There are various spectroscopic techniques, classified according to diverse criteria. We can distinguish three main groups.

- Considering the method of gathering the SED information, we have non-dispersive spectroscopy, such as that performed with X-ray charge-coupled devices (CCDs), whereby the detector is capable of providing information on the energy of the incoming photon; and dispersive spectroscopy, in which the instrument has a dispersive element such as a grating, prism, grism or echelle. In this class, we can include also Fabry–Pérot interferometry, tunable filters and Fourier-transform spectrographs.
- According to the "source selection" within the field of view (FoV) we have long-slit spectroscopy, slit-less spectroscopy, slit multi-object spectroscopy (MOS) and three-dimensional spectroscopy. Long-slit spectroscopy is oriented to a single source, point-like or extended, object. Slit-less spectroscopy consists in an objective prism or a simple grism in imaging cameras (for instance XMM/OM). A mask containing "slitlets" placed in the focal plane defines slit MOS (OSIRIS, GMOS, DEIMOS). Three-dimensional spectroscopy allows us to collect simultaneously the spectral and spatial information of an extended object. On the one hand we have the integral field units (IFUs) that use fiber bundles or image slicers, on the other the Fabry–Pérot spectrographs and tunable filters.
- According to the spectral resolution: if $R < 1000$, the spectrograph is said to be of low resolution. If R is higher than 1000 but lower than 5000, the spectrograph is of medium resolution. For $R > 10\,000$, it is said to be of high resolution.

9.2. Long-slit spectra

Figure 9.1 shows a typical spectrograph with the common components identified: an entrance slit, a disperser element (grisms, prisms), and a CCD detector. The entrance slit focuses the incoming light; it is used both to set the spectral resolution and to eliminate unnecessary background light. Spectrographs have also an internal lamp for the production of the flat-field and various-wavelength calibration emission-line sources. Both types of calibration lamps are included in the spectrograph in such a way that their light paths through the slit and onto the CCD detector match as closely as possible that

The Emission-Line Universe, ed. J. Cepa. Published by Cambridge University Press.

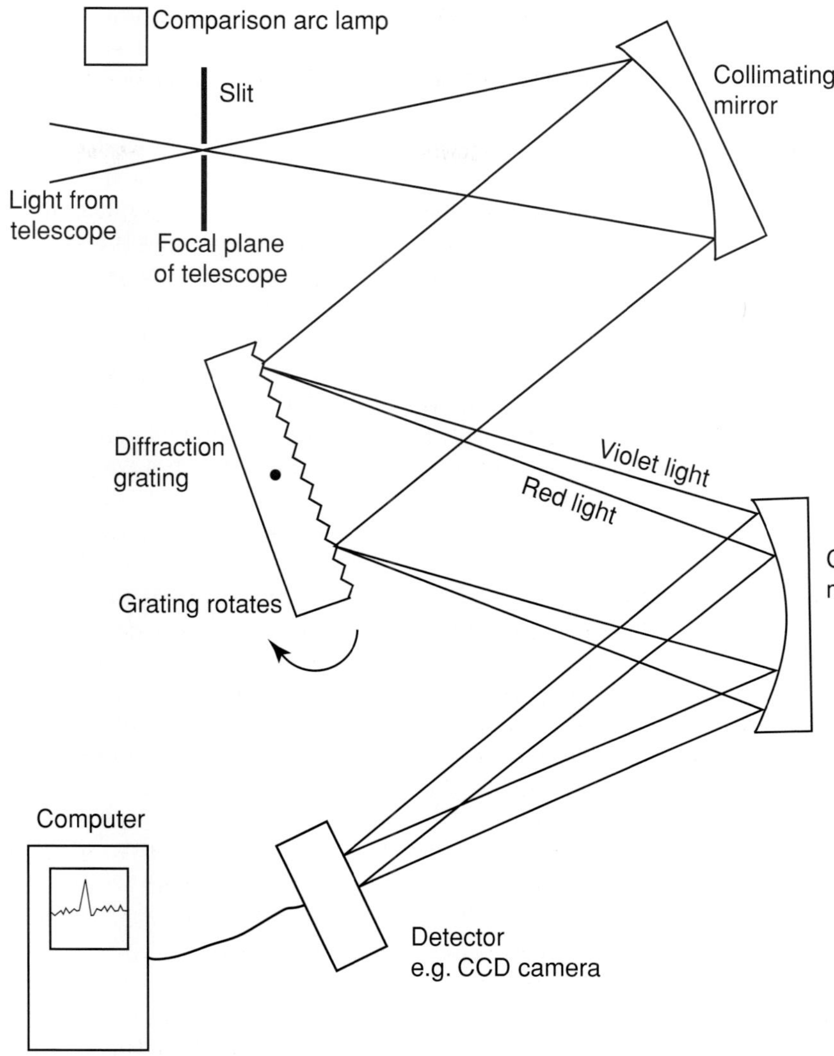

FIGURE 9.1. A schematic diagram of an astronomical spectrograph.

of the incoming telescope beam from an astronomical object. Some type of grating (or prism) is needed as a dispersive element.

Grating spectrographs usually cover 1000–2000 Å of the optical spectrum at a time with a typical resolution of 0.1–10 Å per pixel. To cover a larger spectral range, there are double spectrographs, consisting of two separate spectrographs, which share the incoming light that is divided by a dichroic prism into red and blue beams.

Two common measurements from an astronomical spectrum are those of the flux level as a function of wavelength and of the shape and strength of spectral lines. Absolute-flux observations should be made using a wide slit as the entrance aperture to the spectrograph, in order to assure that 100% of the source light is collected and to avoid problems related to telescope tracking, seeing changes, and differential refraction. Relative-flux measurements should be determinable to even better levels and the shape of a spectral line would be completely set by the instrumental profile of the spectrograph. Spectroscopy aiming to obtain relative fluxes generally makes use of a narrow slit in order to preserve the best possible spectral resolution. Calibration of the observed-object spectra is normally performed by obtaining spectra of flux-standard stars with the same instrumental setup, including the slit width. During the reduction process, these standard stars are used to assign relative fluxes to the object's spectrum counts as a function of wavelength.

Three factors unrelated to the CCD detector itself are important in the final spectrum result: tracking and guiding errors, seeing changes, and the spectrograph's slit angle. Observations of an astronomical object are obtained, in general, such that the slit matches the object-seeing disk. That is, the slit width is set wide enough to allow most of the point-spread function (PSF) (for a point source) to pass through but kept small enough to prevent as much of the sky background as possible from entering the spectrograph. Therefore, changes in the guiding or image seeing will cause more or less source light to enter through the slit. These effects cannot be corrected in our final image. Finally, when we observe an object, for a given distance to the zenith, a different image for each wavelength is formed in the telescope's focal plane. Also, for a selected wavelength, the location of the image in the focal plane depends on the distance from the zenith. These effects are caused by differential atmospheric diffraction (Filippenko 1982). All objects in the sky become slightly prismatic due to this differential atmospheric diffraction. If the spectrograph's slit is not placed parallel to the direction of atmospheric dispersion, light loss that is not chromatically uniform may occur in the slit. Atmospheric dispersion is caused by the variation of the angle of refraction of a light ray as a function of its wavelength, and the direction parallel to this dispersion is called the parallactic angle. Aligning the slit at the parallactic angle solves this problem. The parallactic angle can be determined as cos(object declination) $\times$ sin(parallactic angle) $=$ sign(hour angle) $\times$ cos(observer's latitude) $\times$ sin(object azimuth).

To get an idea of how important this can be, just take into account that even at a modest $\sec(z) = 1.5$, an image at 4000 Å is displaced towards the zenith by 1.1 arcsec relative to the image at 6000 Å. If we are trying to observe over this wavelength range using a 2-arcsec slit, we will suffer a large amount of light loss unless the slit is placed at the parallactic angle.

9.2.1 *Signal-to-noise calculations*

The equation for the signal-to-noise ratio (SNR) of a measurement made with a CCD is given by

$$\frac{S}{N} = \frac{N_*}{\sqrt{N_* + n_{\rm pix}(N_{\rm S} + N_{\rm D} + N_{\rm R}^2)}}, \tag{9.1}$$

where N_* is the total number of photons collected from the object of interest (signal). The noise term includes the square roots of N_*, plus $n_{\rm pix}$ (the number of pixels under consideration) times the contributions from $N_{\rm S}$ (the total number of photons per pixel from the sky), $N_{\rm D}$ (the total number of dark-current electrons per pixel, which in all CCDs is now negligible), and $N_{\rm R}^2$ (the total number of electrons per pixel resulting from the reading noise). For bright sources, $S/N \propto \sqrt{N}$, whereas for faint sources we must use the entire expression. For spectroscopic observations the largest noise contributors that will degrade the resulting signal-to-noise ratio are the background sky and how well the data have been flat-fielded.

In spectroscopy, it is possible to calculate the SNR for the continuum and for a given spectral line. In the continuum case, the number of pixels used in the SNR calculations can be determined by multiplying the continuum bandpass range over which the SNR

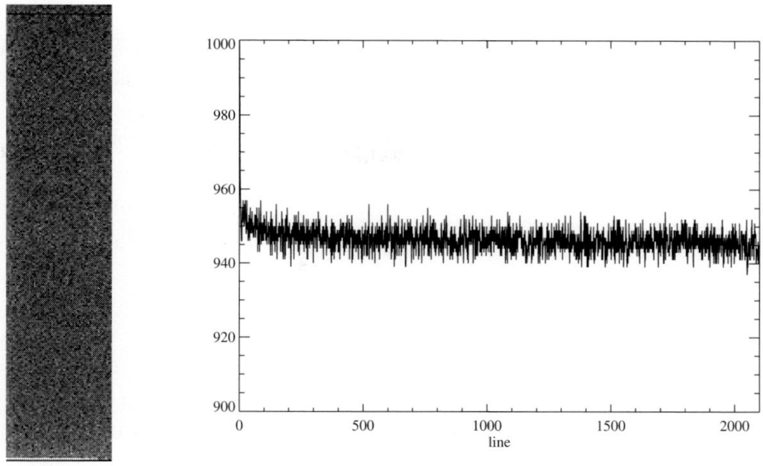

FIGURE 9.2. A bias frame (left) and plot of the raw spectrum of the same bias frame (right). Raw data are plotted as pixel-number (horizontal axis) versus ADU (vertical axis).

is desired by the finite width of the spectrum on the CCD. For example, a spectrograph might have an image scale of 0.7 Å per pixel and the imaged spectrum a width of five pixels on the array in the direction perpendicular to the dispersion. To calculate the SNR for the spectral continuum over a 200-Å bandpass, one would use $n_{\rm pix} = 1428$. In contrast, a narrow line with a full width at zero intensity of 40 Å would use $n_{\rm pix} = 286$. The SNR of the emission line would therefore be higher due to the smaller overall error contribution, and also because the emission-line flux is higher. Signal-to-noise calculations are useful in predicting observational values, such as the integration time.

9.3. Effects to correct in a long-slit spectrum

9.3.1 *Bias*

The bias level of a CCD camera consists of an added current intended to avoid negative values in the reading of the CCD. This zero level is independent of the integration duration of an exposure and should be theoretically constant for all pixels. In reality, there may some (very-low-level) variations in structure. To correct for this effect, every observing night a sufficiently large number of bias images, i.e. zero-length exposures with the shutter closed, is taken. Figure 9.2 shows an example of a bias frame.

Also, preceding read-out and after read-out of all pixels of the CCD, the read-out electronics will perform several additional cycles that do not correspond to physical pixels. These virtual pixels are referred to as overscan, and will be written as part of the CCD image. The overscan will be used to subtract the mean value of the zero level.

9.3.2 *Flat-field and illumination corrections*

Each pixel within the CCD array has its own unique light-sensitivity characteristics. Since these characteristics affect camera performance, they must be removed through calibration. The calibration image to correct this effect is the flat-field frame, that measures the response of each pixel in the CCD array to illumination and is used to correct for any variation in illumination over the field of the array (Figure 9.3). In spectroscopy, typically we use a quartz lamp or dome screen. This flat-field image requires a high SNR.

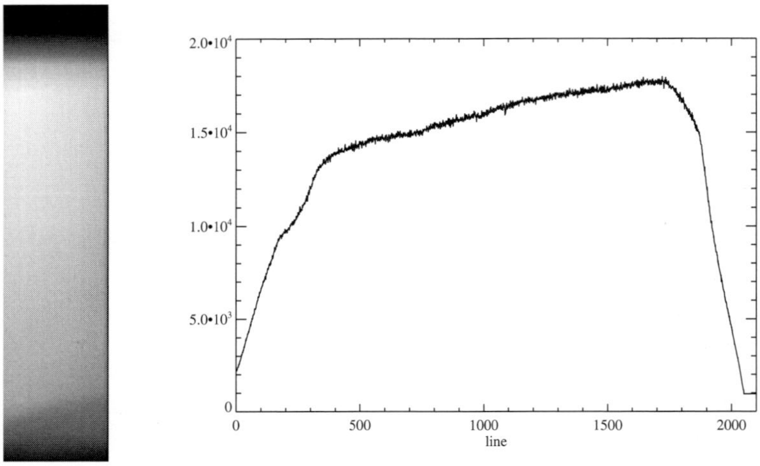

FIGURE 9.3. A quartz-lamp flat field (left) and plot of the raw spectrum of the same frame (right). Units are as in Figure 9.2.

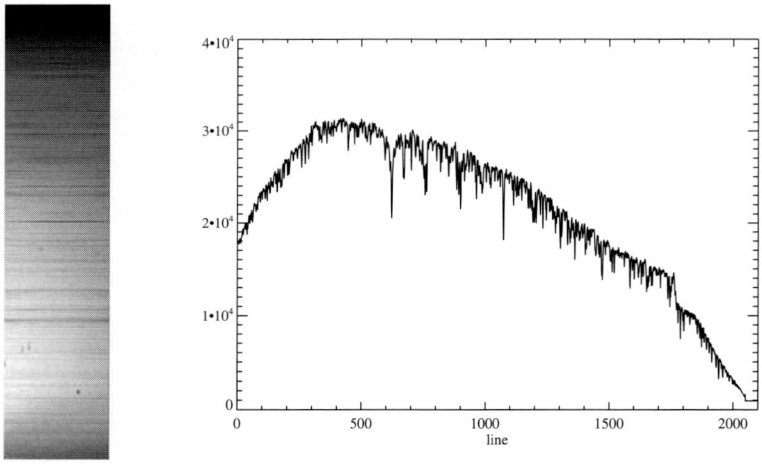

FIGURE 9.4. A sky flat field (left) and plot of the raw spectra of the same frame (right). In both, we can see absorption lines of the sky spectrum. Units are as in Figure 9.2.

Moreover, in long-slit spectroscopy variations in throughput appear along the slit. In order to correct for this effect, calibration images are obtained using twilight sky to obtain even illumination (Figure 9.4). At least a moderate SNR is required.

9.3.3 *Wavelength calibration*

At this point, the spectrum's dispersion axis will be measured in pixels that should be transformed into wavelength units. The procedure used to do this involves the observation of calibration arc spectra during the observing run. The idea is to match the pixel scale of the calibration arc lamps with their known wavelengths and then to apply this scaling procedure. In practice, we first identify features in a one-dimensional spectrum of the arc, repeating the process at every spatial position, and finally we fit a surface that transforms

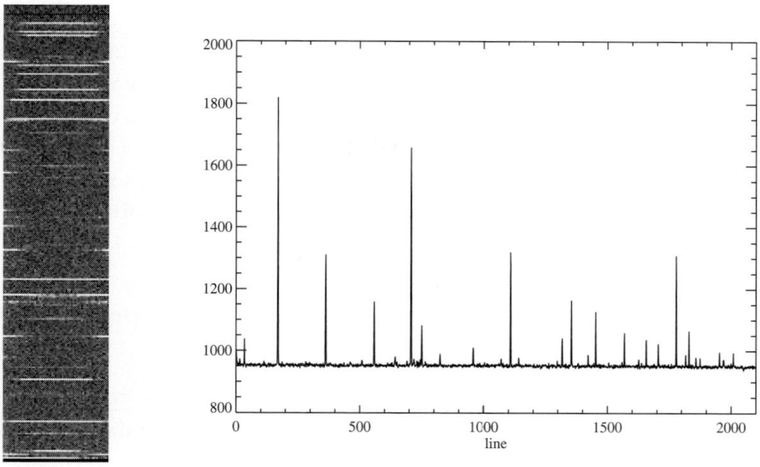

FIGURE 9.5. An arc lamp (left) and plot of the raw spectra of the same frame (right). In both images, we can see emission lines of the Cu–Ar–Ne lamp that will be used to transform pixels to wavelength. Units are as in Figure 9.2.

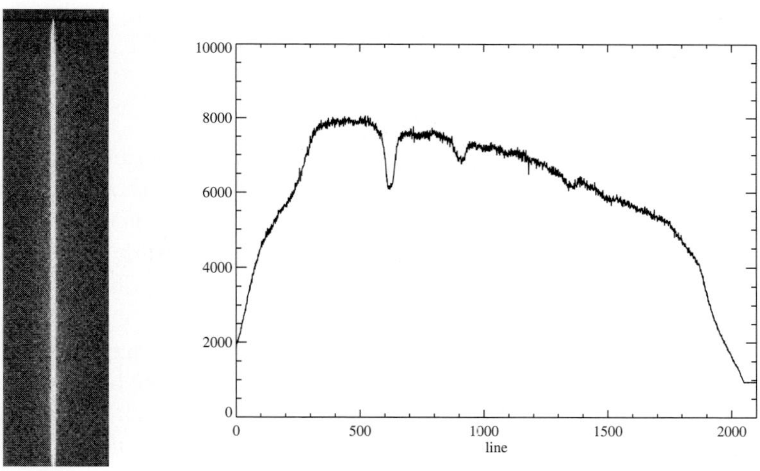

FIGURE 9.6. A standard flux star (left) and plot of the raw spectra of the same frame (right). Units are as in Figure 9.2.

pixels to wavelength coordinates in the arc. This transformation is also used to calibrate in wavelength our object data.

Selected lamps produce many lines along the spectrum; accurate line lists are available in manuals and software. The spectra of calibration sources such as Cu–Ar–Ne and Fe lamps contain numerous narrow emission lines of known wavelength (Figure 9.5). Arcs should be obtained as often as spectrograph stability demands.

9.3.4 *Flux calibration*

Flux standard images are taken only when flux-calibrated spectra are required. All standard star images have to be reduced following the steps described above. The goal is to convert ADUs or counts into flux units such as erg s^{-1} $Å^{-1}$. Most observatories and software-reduction packages have lists of suitable spectrophotometric-standard stars. Using the known fluxes of the standard stars, we convert pixel counts into relative or absolute fluxes. The difference between relative and absolute flux corresponds to the difference between a narrow and a large slit. The conversion of counts to flux is performed under the assumption that slit losses, color terms, transparency, and seeing for the standard-star observations and the object of interest were similar.

9.4. From raw data to calibrated data using IRAF procedures

9.4.1 *Bias correction*

The correction of bias level is similar to that applied in the image-reduction process. We select the bias images and combine all of them in order to minimize noise. To combine images we can use either the **imcombine** or the **zerocombine** task. Then, we analyze the bias image (**imstat, imexam**). If the image does not have structure, it is better to correct the zero level of the science and flat-field images using only the overscan region, since in that case we subtract a constant from the image and therefore we do not introduce noise into the images. The **ccdproc** task is used to correct for all effects present in a CCD exposure using various calibration images. To correct for the zero level, we can use both the bias image and the overscan. To determine the overscan and trim-strip sections, we can use the **implot** task with a flat-field image.

9.4.2 *Flat-field correction*

Spectroscopic flat-field images (to correct along the spatial direction) are obtained with white-lamp exposures. First, we check flat-field images, selecting those with the adequate level of counts. Then, we combine all lamp flat images meeting the count-level condition using **flatcombine**. If just one good lamp flat is found, we use it directly in the next step. We scale flat images with the mode value (it could also be done using the median value) before combining since they can have very different levels of counts.

To obtain the final flat-field image, we fit a smooth function along the spatial direction of our combined flat image; then we divide the combined flat image by the fitted function. This is accomplished with the **response** task from the *twodspec.longslit* package. It is usually necessary to use the *spline3* function and a high number of terms (>15). When we have obtained a normalized flat-field image, we can correct all science and sky flat images. Again, we use **ccdproc** using the output of **response** (the normalized flat image) as the correction flat-field image.

9.4.3 *Illumination correction*

We select all sky flat images and perform statistics to accept only those with the appropriate count level (i.e. rejecting those with levels of counts that are too low or too high). We then combine all **good** sky flat images to obtain an averaged sky flat image, with lower noise, using again the **flatcombine** task. The illumination calibration is obtained from this sky spectrum by fitting functions across the slit (the slit profiles) at a number of points along the dispersion, normalizing each fitted function to unity at the center of the slit, and interpolating the illumination between the dispersion points. The fitted data are formed by dividing the dispersion points into a set of bins and averaging the slit profiles within each bin. The **illumination** task creates the illumination-calibration image. The order of the fitted function should be low (generally two, three, or four). Finally, we correct all the science images by means of **ccdproc**, using the illumination-calibration image as the correction image.

9.4.4 *Wavelength calibration*

First, we select arc-lamp images and identify the lamp used (Cu–Ar, Cu–Ne, He–Ar, ...), grating, and central wavelength. The **identify** task (from *twodspec.longslit* packages) identifies features interactively in a one-dimensional spectrum. Before we execute the task, we look for the appropriate table of features in the *linelists*. In general, we fit a Legendre or Chebyshev function of order three or four. The output file is written in the database directory, named *idbfile*; it contains a list of identified features and parameters of the fitting function. Now it is necessary to identify feature lines across the whole image, using our identifications in the one-dimensional arc spectrum. To this end, we use a non-interactive task: **reidentify**. The output file overwrites *idbfile*, now including identified lines and a fit across the whole image.

The next step consists of fitting wavelength coordinates to pixel coordinates, that is, fitting a surface defining wavelength as a function of x and y coordinates or $\lambda = s(x, y)$. We use the interactive task **fitcoords**.

Using calibration files obtained from arc-lamp images, we can calibrate in wavelength our science images. The **transform** task transforms long-slit images into wavelength coordinates. Outputs of **transform** are wavelength-calibrated images.

9.4.5 *Sky subtraction*

To subtract the background of the object images, it is necessary to fit the sky at each wavelength value. We use another interactive task: **background** (from *twodspec.longslit* packages). We fit the background across each wavelength. In general, the order of the polynomial has to be low (one, two, or three); data points should be rejected as required in order to avoid taking into account the object flux in the fit, and the number of iterations should range between two and five.

9.4.6 *Flux calibration*

First, we need to collapse the two-dimensional spectra of the standard stars into a one-dimensional spectrum. To do this, we measure the spatial coverage of the star in the frame. We can use, for example, **implot** and select columns that contain the spectrum of the star. Now, we extract a one-dimensional spectrum by adding all the apertures including star flux, by means of the **apall** task. This task is located in the *apextract* package. Now, using the one-dimensional spectrum of the standard star, and the table listing its AB magnitudes, we will create a table with two entries: the real flux of the standard star (in erg cm^{-2} s^{-1} $Å^{-1}$) and the measured flux (in counts). We use the task **standard** located in the *onedspec* package. We need the image of the standard star, the name of the extinction file, the directory containing the standard flux tables, and the name of the star in the database. IRAF provides several tables of standard stars and the extinction curves for several observatories, placed in the *onedstds* directory.

From the output of **standard**, we determine the system's sensitivity as a function of wavelength and extinction functions using the **sensfunc** task. The last step is calibrating our spectra. The **calibrate** task is used for this. The input spectra are corrected for extinction and calibrated to a flux scale using sensitivity spectra produced by the **sensfunc** task. The output of **calibrate** will consist of our bidimensional spectrum, corrected for all effects and calibrated in wavelength and flux.

Now, we can proceed to analyze our spectra.

More information about how to reduce spectra with IRAF:

- **A User's Guide to Reducing Slit Spectra with IRAF**,† by P. Massey, F. Valdes, and J. Barnes (15 April 1992)
- **IRAF help page for the twodspec package**‡

REFERENCES

Filippenko, A. 1982 PASP, 94, 175

Howell, S. B. 2006 *Handbook of CCD Astronomy*, 2nd edn (Cambridge: Cambridge University Press)

† See URL http://star-www.rl.ac.uk/iraf/ftp/iraf/docs/spect.ps.Z.
‡ See URL http://www.stecf.org/scripts/irafhelp?twodspec.

10. Basic principles of tunable filters

HÉCTOR CASTAÑEDA AND ANGEL BONGIOVANNI

10.1. Introduction

Charles Fabry, who was born in 1867, specialized in optics and devised methods for the accurate measurement of interference effects. He worked with Alfred Pérot, during 1896–1906, on the design and uses of a device now known as the Fabry–Pérot interferometer, which was specifically designed for high-resolution spectroscopy, and is composed of two thinly silvered glass plates placed in parallel, producing interference due to multiple reflections.

In 1899 they described the *Fabry–Pérot interferometer* which enabled high-resolution observation of spectral features (Fabry & Pérot 1899). It was a significant improvement over the Michelson interferometer. The difference between the two lies in the fact that in the Fabry–Pérot design multiple rays of light reflected by the two plane surfaces are responsible for the creation of the observed interference patterns. The last sentence of the article reads *We must emphasize the simplicity of the apparatus used and the ease with which it can be mounted at the telescope. When the silvering has been carefully selected, the interference apparatus does not cause the loss of much light and permits the study of objects of very feeble brightness.*

10.2. Definition of a Fabry–Pérot interferometer

Basically, a Fabry–Pérot interferometer or *etalon* (from the French *étalon*, meaning "measuring gauge" or "standard") is typically made of a transparent plate with two reflecting surfaces, or two parallel highly reflecting mirrors (technically the former is an etalon and the latter is an interferometer, but the terminology is often used inconsistently). Its transmission spectrum as a function of wavelength exhibits peaks of large transmission corresponding to resonances of the etalon.

A Fabry–Pérot *interferometer* differs from a Fabry–Pérot *etalon* in that the distance between the plates can be tuned in order to change the wavelengths at which transmission peaks occur. Owing to the angle dependence of the transmission, the peaks can also be shifted by rotating the etalon with respect to the beam.

10.3. Optical principles

As stated before, a Fabry–Pérot filter consists of two plane-parallel transparent plates, which are coated with films of high reflectivity and low absorption. The coated surfaces are separated by a small distance to form a cavity that is resonant at specific wavelengths. Light entering the cavity undergoes multiple reflections, with the amplitude and phase of the resultant beams depending on the wavelength. At the resonance wavelengths, the resultant reflected beam interferes destructively with the light reflected from the first plate–cavity boundary and all the incident energy, in the absence of absorption, is transmitted. At other wavelengths, the Fabry–Pérot filter reflects almost all of the incident energy.

The Emission-Line Universe, ed. J. Cepa. Published by Cambridge University Press.

The varying transmission function of an etalon is caused by interference due to multiple reflections of light among the two reflecting surfaces. The phenomenon of constructive interference occurs if the transmitted beams are in phase, and this corresponds to a high-transmission peak of the etalon. If the transmitted beams are out of phase, destructive interference occurs and this corresponds to a transmission minimum. Whether the multiply reflected beams are in phase or not depends on the wavelength λ of the light, the angle at which the light travels through the etalon θ, the thickness of the etalon d and the refractive index of the material between the reflecting surfaces μ.

Notice that there can be a phase shift in the internal reflection that can be considered equivalent to a change in the cavity size, and can be modelled as an extra phase shift in the equation. The phase shift measures the degree of interference between the beams.

If both surfaces have a reflection coefficient R, the transmission function of the etalon is given by

$$\frac{I_{\text{transmitted}}}{I_{\text{incident}}} = \left(\frac{T}{1-R}\right)^2 \left[1 + \frac{4R}{(1-R)^2} \sin^2\left(\frac{2\pi\mu d\cos\theta}{\lambda}\right)\right]^{-1}, \tag{10.1}$$

where T is the transmission coefficient of each coating (the boundary of the plate cavity), R is the reflection coefficient, d is the plate separation, μ is the refractive index of the medium in the cavity (usually air, $\mu = 1$) and θ is the angle of incidence of incoming light.

The Fabry–Pérot filter transmits a narrow spectral band at a series of wavelengths given by

$$m\lambda = 2\mu d\cos\theta, \tag{10.2}$$

where m is an integer known as the order of interference.

Constructive interference occurs when the path-length difference is equal to half an odd multiple of the wavelength.

The maximum transmission ($I_{\max}$) occurs when the optical path-length difference ($2\mu d\cos\theta$) between each transmitted beam and the next is an integer multiple of the wavelength. In the absence of absorption, the reflectivity of the etalon R is the complement of the transmission, such that $T + R = 1$ (if there is absorption the maximum transmission of the system is less than 1).

The peak transmission of each passband is

$$I_{\max} = \left(\frac{T}{1-R}\right)^2 = \left(\frac{T}{T+A}\right)^2, \tag{10.3}$$

where A is the absorption and scattering coefficient of the coatings ($A = 1 - T - R$).

The optical transmission and reflection are determined completely by the length of the etalon, the index of refraction, the incidence angle of the light and the coatings of the surface.

The contrast between the maximum and minimum transmission intensities is

$$\frac{I_{\max}}{I_{\min}} = \left(\frac{T+R}{1-R}\right)^2. \tag{10.4}$$

10.3.1 *Bandwidth*

Each passband has a bandwidth ($\delta\lambda$), full width at half peak transmission, given by

$$\delta\lambda = \frac{\lambda(1-R)}{m\pi R^{1/2}}. \tag{10.5}$$

10.3.2 *Resolution*

The resolution or resolving power is one of the figures of merit of the etalon. The resolving power is the ratio of the central wavelength to the bandwidth, and this ratio depends on the order. The value is equal to the product of the order and the finesse:

$$\mathcal{R} \equiv \frac{\lambda}{\delta\lambda} = mF, \tag{10.6}$$

where m is the order and F the finesse of the etalon. If the finesse increases, the resolving power also increases. Thus, knowing the effective finesse, the resolution power can be obtained.

10.3.3 *The free spectral range*

The wavelength separation between adjacent transmission peaks is called the inter-order spacing or free spectral range (FSR) of the etalon, and is given by

$$\Delta\lambda = \frac{\lambda}{m}. \tag{10.7}$$

The FSR scales inversely with d, the distance between plates. For example, a typical high-resolution etalon has a distance between plates of the order of 100–500 μm (FSR about 15–50 Å). A tunable-filter etalon has distances between plates of 1–12 μm (FSR about 150–200 Å).

10.3.4 *Finesse*

The FSR is related to the full width at half maximum, $\delta\lambda$, of any one transmission band by a quantity known as the finesse:

$$F = \frac{\Delta\lambda}{\delta\lambda}. \tag{10.8}$$

The finesse is then the ratio of inter-order spacing and bandwidth. The finesse is the parameter that defines the performance of a real etalon.

This is commonly approximated for an ideal etalon by

$$F = \frac{\pi R^{1/2}}{1-R}. \tag{10.9}$$

Etalons with high finesse give sharper transmission peaks with lower minimum transmission coefficients.

The value of the finesse can be understood roughly as *the number of reflections an average photon makes before being transmitted through the system.*

There are three types of finesse values: reflection, defect and aperture finesse. The aperture finesse is negligible for most astronomical objects and observational modes. The final effective finesse is a combination of the three different types.

10.4. From theory to practical design

By 1970 the Astronomy Group of Imperial College London had begun the development of a piezo-tuned Fabry–Pérot system using capacitance sensors to servo stabilize the cavity. The first prototype servo-stabilized FP, known as CasFPer, was first tested in the 2.5-m Isaac Newton Telescope at Herstmonceux, Sussex. This first-generation device was piezo-tuned, mechanically aligned and capacitively stabilized (Atherton 1995).

Queensgate Instruments was born in 1979 in the basement of the Physics Department of Imperial College (the Astronomy Group of Imperial College moved in 1978 to a new building on Queens Gate). Most of the etalons operating in astronomical installations have been manufactured by this company (Queensgate is now called ICOS). It manufactures etalons of aperture up to 150 mm that are tunable over a few micrometres. The surface reflectivity varies from 92% with hard coatings to up to 96% with soft coatings, and the surface flatness quality reaches up to $\lambda/200$.

The Queensgate etalon served as the base for a series of Fabry–Pérot interferometers in operation in the last few decades.

- TAURUS: a Fabry–Pérot imaging device designed to obtain complete seeing-limited radial-velocity field maps of extended emission-line sources. A servo-controlled Fabry–Pérot interferometer was used with a focal reducer and a two-dimensional photon-counting (area-detector) system was used to obtain the velocity information. Three versions of the instrument were eventually built. TAURUS Mk1 (TAURUS-1) was originally available on the Isaac Newton Telescope. TAURUS Mk2 (TAURUS-2) was used with the William Herschel Telescope. Finally, a TAURUS instrument was in operation at the Anglo-Australian Telescope (Taylor & Atherton 1980).
- HIFI: Hawaii Imaging Fabry–Pérot Interferometer. This used a large-FSR etalon with high finesse and a charge-coupled device (CCD) at the image plane. The geometrical integrity of the CCD produced a clean Airy surface and bypassed the optoelectronic distortions of image intensifiers. The high quantum efficiency and linearity over a wide dynamic range in intensity proved to be essential in studies of extended narrow-line regions (Bland & Tully 1989).
- The Ohio State Imaging Fabry–Pérot Spectrometer was designed and built by the Ohio State University Astronomical Instrumentation Facility. It was designed for two-dimensional-imaging spectrophotometric and kinematic studies in the wavelength range of 3400–10 000 Å, and was used for both imaging Fabry–Pérot spectrophotometry and direct imaging in broad- and narrow-band filters (with the etalons removed from the beam). It used four Queensgate Instruments ET-50 etalons (Pogge *et al.* 1995).

10.4.1 *Gap-scanning etalons*

In order to manufacture a tunable Fabry–Pérot interferometer, which can change the central wavelength for a given order, there are three options:

- change the refractive index of the cavity,
- modify the angle θ and
- change the plate separation d.

In modern gap-scanning etalons, the parameter that is changed is the plate separation, that can be controlled to extremely high accuracy.

In recent years, these etalons have undergone considerable improvements. It is now possible to move the plates between any two discrete spacings at very high frequencies (200 Hz or better) with no hysteresis effects while maintaining $\lambda/200$ parallelism (measured at 633 nm). The etalon spacing is maintained by three piezoelectric transducers as discussed below.

10.4.2 *Piezoelectric transducers*

Piezoelectric materials undergo dimensional changes in an applied electric field. Conversely, they develop an electric field when strained mechanically. Under an applied electric field, a piezoelectric crystal deforms along all its axes. It expands in some directions and contracts in others. The dimensional change (expansion or contraction) of a piezoelectric material is a smooth function of the applied electric field. The material is stiff enough for the piezoelectric transducers (PZTs) to respond on submicrosecond timescales. The resolution is limited only by the precision with which the electric field can be controlled. For this reason, PZTs are commonly used for rapid switching and sensing, as indeed they are in the Queensgate etalons. However, all piezoelectric materials exhibit hysteresis, particularly in the relationship between the voltage applied and the amount of expansion. Thus, a servocontrol system is required to tune the spacing between two plates to high accuracy.

10.4.3 *Capacitance micrometry*

It has been shown that capacitance micrometry can be used to detect motions on scales as small as 10^{-15} m. Using this basic method, Queensgate Instruments developed a capacitance-bridge system to monitor the parallelism and spacing of a Fabry–Pérot etalon. Information from the capacitance bridge is used to drive PZTs in a closed-loop control system to maintain the parallelism and spacing. There are two X-channel and two Y-channel capacitors, and a fifth reference capacitor, which monitors the spacing with respect to a fixed reference capacitor in the circuit. The two etalon plates can be kept parallel to within an accuracy of $\lambda/200$ for many weeks at a time.

10.4.4 *Coatings and finesse*

The highly polished plates are coated for optimal performance over the wavelength of interest. Since it is difficult to create an efficient coating to operate over the complete range of visible spectra, the etalon coatings are optimized to selected wavelength ranges (i.e. "blue" or "red"). The reflectivity of the coating determines the shape and degree of order separation of the instrumental profile. This is fully specified by the coating's finesse, F, which has a quadratic dependence on the coating's reflectivity. At a finesse specification of $F = 40$ (which means that the separation between periodic profiles is 40 times the width of the instrumental profile) the profile is Lorentzian to a good approximation.

Notice that the analysis of Fabry–Pérot behaviour in Section 10.3 does not take into account the wavelength-dependent phase change inherent in reflections between the optical coatings on the inner plate surfaces. Such coatings reflect the design wavelength with zero phase change, but incur a lead and lag elsewhere.

10.5. The tunable filter

Whereas Fabry–Pérot interferometers are used for high-resolution spectral work, tunable filters are used for low-resolution imaging. In the case of a conventional Fabry–Pérot device, the etalon used large gap spacings (about 30–400 μm, giving resolution between 5000 and 17 000). The scan is done over a restricted range of spacings (equivalent to wavelengths).

A tunable filter is just a very-low-resolution Fabry–Pérot device, wherein the gap spacing is much smaller, for example, between 2 and 12 μm. The result is a low resolution ($R = 100$–1000). The scans are over a wider range of spacings, and the anti-reflection coatings are optimized for a broad range of wavelengths.

The tunable filters largely remove the need for buying arbitrary narrow and intermediate interference filters, since one can tune the bandpass and the centroid of the bandpass by selecting the plate spacing. Since tunable filters are periodic, the instrument requires a limited number of blocking filters.

FIGURE 10.1. Three etalons with different apertures and the CS-100 controller unit.

10.5.1 *History*

Some examples of etalons used in the tunable filter mode follow.

- The Goddard Fabry–Pérot Imager (GFPI). The system consisted of a piezoelectric scanning etalon and a servo-controller, both made by Queensgate Instruments, blocking filters and a Tektronic CCD detector. Besides being transportable, its most-notable characteristic was the relatively low spectral resolution (3–30 Å bandpass). It began operation in 1990, with a spectral coverage of 3800–10 500 Å, a nearly monochromatic field of view and a STIS SITe 2048 × 2048-pixel CCD. The current instrument has a choice of four Queensgate 50-mm-diameter piezoelectrically driven, capacitance-stabilized etalons, which can be tuned to any wavelength in the range 4000–10 000 Å, with resolution from 4 to 28 Å FWHM depending on etalon and wavelength (Gelderman *et al.* 1995).
- The Calar Alto Fabry–Pérot Spectrometer. A Fabry–Pérot etalon was used as a tunable narrow-filter ET-50 (Queensgate) plus CS100 electronic controller, both similar to the devices shown in Figure 10.1. The range is from 4200 to 8300 Å. The etalon plates are at 9.5 μm, order 30 at 6000 Å. Spacing can be varied for ±2.5 μm around the nominal value. The mean instrumental resolution is ∼400–600 km^{-1} FWHM (Meisenheimer & Hippelein 1992).
- The Taurus Tunable Filter (TTF) was developed in 1994–1995. The TTF is a pair of tunable narrow-band interference filters covering 3700–6500 Å (blue "arm") and 6500–9600 Å (red "arm"). It made it possible to have monochromatic imaging at the Cassegrain foci of the Anglo-Australian (3.9-m) and William Herschel (4.2-m) telescopes, with an adjustable passband of between 6 and 60 Å. Frequency switching with the TTF could be synchronized with movement of charge (charge shuffling) on the CCD. Unlike conventional Queensgate etalons, the TTF incorporated very large

FIGURE 10.2. The Fabry–Pérot red etalon of OSIRIS on the optical bench.

piezoelectric stacks (which determine the plate separation) and high-performance coatings over half the optical wavelength range. The plate separation could be varied between about 2 and 12 μm (Bland-Hawthorn & Jones 1989).

The Instituto de Astrofísica de Canarias has developed a new instrument that uses the technique of tunable filters: OSIRIS (Optical System for Imaging and low-intermediate Resolution Integrated Spectroscopy).

The management of the Gran Telescopio Canarias (GTC) decided on 11 March 1999 to sign a contract for the preliminary design of OSIRIS as chosen day-one GTC instrument for the optical wavelength range. On 29 July 1999, a contract was signed between the IAC and GRANTECAN for a preliminary design. After the preliminary design review, the contract for the development of the instrument was signed on 20 December 2000, for instrument delivery to the site to begin commissioning depending on the telescope schedule.

OSIRIS is an imaging system and a low-resolution long-slit and multi-object spectrograph for the GTC covering the wavelength range 0.365–1.0 μm with an unvignetted field of view of $8.53' \times 8.53'$ and $8.0' \times 5.2'$ in direct imaging and low-resolution spectroscopy, respectively. OSIRIS represents a new generation of instrumental observing techniques, that includes the concepts of tunable filters and charge shuffling on the CCD detectors (Cepa 1998).

The OSIRIS tunable filter (TF) is a pair of Fabry–Pérot etalons covering 365–670 nm (blue arm) and 620–1000 nm (red arm). Figure 10.2 shows the etalon of the red arm installed in OSIRIS. The OSIRIS TF offers monochromatic imaging with an adjustable bandwidth of between 0.6 and 6 nm. In addition, frequency switching with the TFs can be synchronized with movement of charge (charge shuffling) on the CCD, which has important applications to many astrophysical problems.

Future instruments in large telescopes that will make use of tunable filters are the following.

- The Maryland–Magellan Tunable Filter (MMTF). It will provide a high throughput over a broad range in wavelengths (5000–9200 Å), with a tunable bandpass of 10–100 Å over a 10 × 27-arcmin field. By frequency switching the etalon in synchronization with charge shuffling in the CCDs, the MMTF is expected to reach a sensitivity of 10^{-18} erg s^{-1} cm^{-2} at $\sim 3\sigma$ SNR in 1 h. It is to be installed on the IMACS instrument at the Magellan 6.5-m telescope.
- The Prime Focus Imaging Spectrograph of the SALT telescope project. The SALT project is a multinational collaboration to build a large telescope in South Africa similar to the Hobby–Eberly Telescope already in existence in West Texas. The project requires two 150-mm-aperture etalons and controllers.

10.6. Instrumental effects

10.6.1 *Monochromatic field*

The primary goal of a tunable filter is to provide a monochromatic field over as large a detector area as possible. However, the field of view is not strictly monochromatic. The effect is most acute at high orders of interference.

Wavelengths are longest at the centre and get bluer the further one moves off-axis. This change in wavelength, relative to the central wavelength when $\theta = 0$, can be written as

$$\frac{\delta\lambda}{\lambda} = \cos\left(\frac{\theta_{\rm sky} t_{\rm tel}}{f_{\rm coll}}\right) \approx -\frac{1}{2}\left(\frac{\theta_{\rm sky} f_{\rm tel}}{f_{\rm coll}}\right)^2, \qquad (10.10)$$

where $\theta_{\rm sky}$ is the angular distance on the sky away from the central axis of the etalon.

The monochromatic field will be the size of the Jacquinot spot, the central region of the ring pattern. In this region the wavelength changes by less than $\sqrt{2}$ times the etalon bandwidth, which verifies

$$\delta\lambda = \frac{\lambda}{Fm}, \qquad (10.11)$$

where F is the effective finesse of the etalon.

Then, the monochromatic field is a region subtending an angle $\phi_{\rm Jac}$ that can be written as

$$\phi_{\rm Jac}^2 = \frac{2\sqrt{2}}{Fm}. \qquad (10.12)$$

For a particular etalon, the size of the Jacquinot spot depends on the order m alone. The above equation shows how the spot covers increasingly larger areas on the detector as the filter is used at lower orders of interference. The absolute wavelength change across the detector remains the same, independently of order. However, its effect relative to the bandpass diminishes as m decreases.

10.6.2 *Ghosts*

A typical arrangement of a Fabry–Pérot instrument can have a significant number of flat surfaces, which can be a source for spurious reflections. For example, since the Airy function has a periodic behaviour, a narrow-band filter (known as an order sorter) must be used to eliminate unwanted interference orders. The narrow-band filter is placed in the converging beam before the collimator or after the camera lens, or in the collimated beam. The filter then generates ghost reflections in the instrument optics.

Another source of ghost reflections arises from the optical blanks which form the etalon, that can act as internally reflecting cavities. A possible solution is to have the outer surfaces wedge-shaped in order to deflect this spurious signal out of the beam.

In order to avoid ghosts the tunable filter could be allowed to tilt. Usually a tunable filter will be tilted only to angles of a few degrees.

10.7. Observing with tunable filters

10.7.1 *Order sorters*

A Fabry–Pérot filter (FPF) clearly gives a periodic series of narrow passbands. To use a FPF with a single passband, it is necessary to suppress the transmission from all the other bands that are potentially detectable. This is done by using conventional filters, called order sorters because they are used to select the required FPF order.

For example, with a finesse of about 40, and hence low resolution ($\lambda/\Delta\lambda = 300$), conventional broad-band UBVRI filters suffice. At high resolution ($\lambda/\Delta\lambda = 1000$), specifically designed interference filters should be used to subdivide the wavelength range of each arm.

Using an order-sorter filter and the TF etalon, single-order observations can be made at any wavelength, within the range of the filter, as long as the inter-order spacing is larger than the bandwidth of the filter. Also, observations can be made with smaller inter-order spacings if the order is close enough to the central wavelength of the filter that adjacent orders are outside the range of the filter.

The selection of order-sorter filters depends on the coating reflectivities and wavelength ranges of the TF arms. Additionally, there is also the consideration of sky emission lines. It could be desirable to choose the order-sorter filters such that they fall between the wavelengths of strong sky lines. Other points to consider are the range of tilting and the probable transmission profiles for each filter.

10.7.2 *Calibration*

The calibration of the TF is done using arc lamps. The TF is gap scanned, through at least one FSR, and the charge is shuffled between each exposure and the next. In this way, the spectrum of a lamp is obtained and the gap scanning is calibrated.

10.7.3 *Operation modes of tunable filters*

Examples of operation modes are the following.

- Tuning to a specific wavelength at a specific bandpass. This allows images of obscure spectral lines at arbitrary redshifts to be obtained. It is also possible to optimize the bandpass to accommodate the line dispersion and to suppress the sky background. The off-band frequency can be chosen so as to avoid night-sky lines and can be much wider so that only a fraction of the time is spent on the off-band image.
- Charge shuffling between off-band and on-band frequencies. If the CCD field is large enough, it is feasible to image the full field for two discrete frequencies. Or, for example, we can also choose a narrow bandpass for the on-band line and a much broader bandpass. Charge shuffling is a movement of charge along the CCD between exposures of the same frame, before the image is read out. An aperture mask ensures that only one section of the CCD frame is exposed at a time. For each exposure, the TF is systematically moved to different gap spacings in a process called frequency-switching. In this way, a region of sky can be captured at several different wavelengths

on one image. Alternatively, the TF can be kept at fixed frequency and charge shuffling performed to produce time-series exposures.

10.8. Imaging-data reduction with tunable filters

The tunable filters are versatile optical devices designed to provide, *inter alia*, monochromatic-field imaging over a large detector area.

The reduction process passes through different stages to convert raw data obtained using TF imaging into valuable scientific information. There is a handful of items of specialized and complete software to perform this kind of work, relying on two approaches: orthodox three-dimensional Fabry–Pérot spectroscopy (where the famous "data-cubes" are used for obtaining information about flux, velocity dispersion field, skewness, etc. for extended objects; it is a technique characterized by high resolution power and small tuning ranges) and properly named TF imaging, one kind of very-narrow-line, spectrally dynamic, photometry (or, in other words, (almost) full-field, low-resolution spectroscopy). For data reduction in Fabry–Pérot spectroscopy, originally the `TAUCAL` package (Lewis & Unger 1991) was conceived, as an add-on of `FIGARO`, for the TTF (Fabry–Pérot mode) data. The `IDL`-based `MATADOR` software is another possibility (Gavryusev & Muñoz-Tuñón 1996). More recently, the `ADHOC` software of Daigle *et al.* (2006) has been presented to the community. The routines associated with it are written in `C` and `IDL`, and the paper by Daigle *et al.* could be considered a useful information source about the instrumental and systematic difficulties related to this technique. Another practical article about Fabry–Pérot data reduction was published by Gordon *et al.* (2000). Some problems broached in this paper are common to the TF imaging scenario.

A complete software package for TF imaging-data reduction, described in Jones *et al.* (2002) is `TFred`. We strongly recommend reading this work before undertaking reduction work with TF images. This software is essentially based on `IRAF` procedures and `Fortran77`-specific programs. It was originally created for reduction of data from the TTF for imaging of point-like sources. This modality is suitable when emission-line surveying with large angular and wavelength coverage is an observational priority. The software is optimized for an efficient selection of emission-line objects. A full description of this observing system can be found in the TTF manual (which is available on-line). There is also a website associated with `TFred` and TTF facts (reduction tips and script sources): `http://www.aao.gov.au/local/www/jbh/ttf/`.

A basic reduction scheme for obtaining a concise idea about the processing stages and the procedures involved is shown below. Additionally, a detailed set of instructions for a hands-on reduction of a working example can be obtained by contacting the authors.

10.9. Definitions

A `scan` is a series of images taken at each step of a sequence in the separation of etalon plates. It is not so in the case presented here, but a scan should be recorded at least three times for each observing run and sky region. By `stack` we mean one or more scans, with the same spectral range, in any processing stage. At a given (fixed) order of interference, there is a bijection between the gap between etalon plates and the image's effective wavelength (with the limitations pointed out below). This relationship is established prior to the scientific observations.

At the telescope the gap is measured as Z-step values (in arbitrary microprocessor units, or mpu). A *phase effect* occurs in all filters of this type and it can be represented by a spatial gradient of the wavelength. This implies the presence of an *optical axis* (not

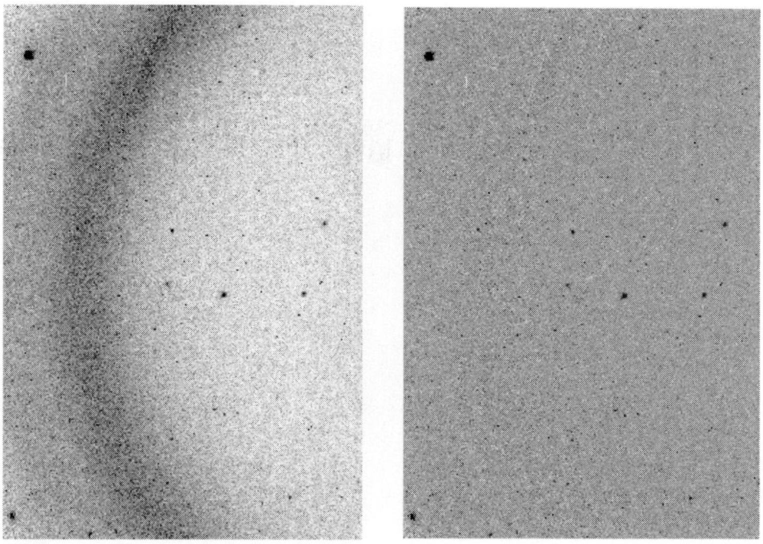

FIGURE 10.3. Two sections of the same image before (left) and after (right) subtraction of the OH sky rings.

necessarily coincident with the field centre), which is used to measure the phase effect: off-axis rays pass through the etalon at a slightly different angle from that for on-axis rays, resulting in a small wavelength shift to the blue with the optical axis moving further away on the detector surface.

The *sky rings* are broad diffuse circles produced by OH night-sky glow. Depending on the spectral region, the band will be more or less evident. Night-sky rings appear as circular bands around the optical axis as a consequence of the phase effect. In Figure 10.3, a typical fingerprint of these rings on a deep image of the HDF-North field is represented.

Ghost images are spurious objects that appear due to multiple reflections of bright objects in or near the observed field. A couple of representative examples were presented above. One of them is shown in Figure 10.4. It is a ghost produced by a bright galaxy (M82) imaged using the Taurus 2 camera (on the William Herschel telescope). For their effective subtraction – leaving intact the real objects – the scan sampling must be done by imposing a dithering of the telescope. The ghosts appear as opposites of the real objects with respect to the optical axis.

10.10. Procedure

The analysis of multi-object TF data has three stages:

- preparation of images;
- object detection and selection of real/spurious objects; and
- flux calibration.

The procedures of each stage are summarized in the following. Initially the raw-scan frames must be prepared for analysis. This includes the removal of the bias level and pixel-to-pixel variations, and the fitting and subtraction of night-sky rings. Images are then aligned with respect to a common reference frame and trimmed. A duplicate set of frames is created and degraded to a common worst seeing. Then frames are co-added into combined scans, of which there are different versions arising from the use of

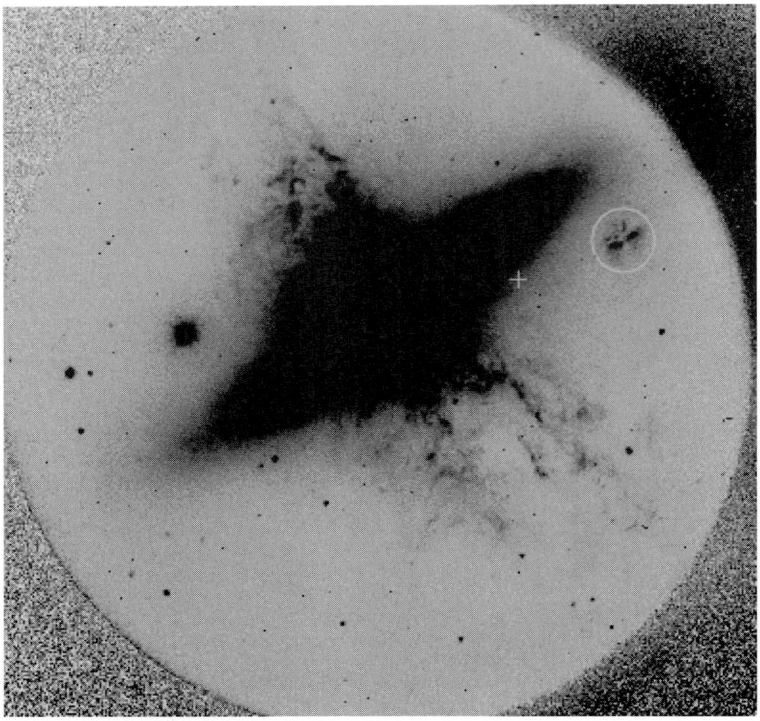

FIGURE 10.4. A typical ghost image (circled) of the galaxy M82. The cross marks the projection of the optical axis. In this case it is easy to identify the spurious object. Otherwise, it is essential to have at one's disposal at least two images with a noticeable offset with regard to the optical axis for an unquestionable identification of the ghosts.

median filtering or straight summation. Versions are also generated from the smoothed (seeing-degraded) and unsmoothed frames. For each of the final combined scans noise-edge masks are computed and applied to the image corners. This last step is necessary in order to stop the object-detection software from setting a detection threshold that is biased by the statistics of pixels outside the image area. The scheme that follows is an adapted version of the on-line instructions written by Heath Jones (ANU/AAO), which are available at the TTF URL: `http://www.aao.gov.au/local/www/jbh/ttf/`. The specific procedures are pointed out in terms of the software tasks. For the sake of completeness, some important comments were added.

10.10.1 *Preliminary steps*

- Wavelength calibrate Z-step versus wavelength. Ideally, this has already been done at the telescope, before the scan was taken (`tsample`).
- Create bias and flat-field images (`imsum`, `tff`). This step can also be performed using standard `IRAF` routines.
- Bias-subtract, flat-field, flip and subtract OH sky rings and other telluric features from images (`tpipe`).
- Align images and trim (`toffsets`, `imalign`). Usually, translation of the images with regard to a reference image suffices.
- Find the centre and radius of the best circle to match the circular field-aperture stop, if present (`tcircle`).

- Create a zero corner mask (if required) using the dimensions found by `tcircle` (`tmask`).
- Measure the changes in seeing/PSF through the stack (`tfwhm`).
- Convolute the full stack to the worst seeing (`tgauss`).
- Combine images to produce two scans: one cleaned (for good object detection) and the other straight-summed (for good object photometry) (`tsingle`).
- Create a deep image from all individual narrow-band frames, which is useful for identifying object-free sky regions (`tdeep`).
- Obtain locations of many sky regions, free of objects (`tregions`).
- Create a zero edge mask (`tmask`), if it is required.
- Measure the sky noise (`tnoise`), create noise-edge masks and apply these (with the zero edge mask) to each respective stack (`tmulti`). Ideally, the aperture edge should be rendered undetectable.

10.10.2 *Detection and selection of candidates*

By this stage there are three stacks: (i) a summed (uncleaned) scan with the original variable seeing; (ii) a summed (uncleaned) scan with frames smoothed to a common "worst" seeing; and (iii) a cleaned scan with frames smoothed to this common seeing. From here, the steps of object detection, photometry and selection are straightforward and efficient. Most of the subsequent analysis works with object catalogues rather than images, so processing is fast. In summary, do the following.

- Detect objects and carry out photometry in each of four pre-selected apertures (`tsex`). Objects are detected using the uncleaned, unsmoothed images. Photometry, however, is done on the matching uncleaned but smoothed frames. As stated in its name, the task implies the use of `SExtractor` (Bertin & Arnouts 1996) software.
- Create a file of the central wavelengths that correspond to each image (`twavelengths`).
- Sort individual object photometry into TF spectra and `SExtractor` additional parameters (`tespect`).
- Compile a cosmic-ray/ghost pixel-location file for all images in the stack (`tpull`).
- Photometrically register frames and extract double-detection emission-line candidates (with and without continuum). Write these to separate candidate catalogues (`tscale`).
- Check raw candidate catalogues for cases of double-counting by the object-detection software. More often than not, double-counting is not found, so the task is run more as a check (`tdouble`).
- Inspect double-detection candidates to ensure that no cosmic rays or ghosts have evaded detection. Remove them from the candidate catalogues if any are found (`tcull`).
- Extract single-detection candidates from catalogues (`tesone`).
- Check candidate aperture corrections and replace object fluxes with those from the next-largest aperture in cases where selected apertures are too small (`tmodap`).

10.10.3 *Flux calibration*

With selection of emission-line candidates finalized, all that remains is for the measured fluxes to be calibrated in terms of physical units. Initially, standard-star scans (taken on the same night as the science frames) must be reduced in order to obtain flux-calibration constants for that night. These calibrations can then be applied to the object catalogues in a single step. In summary, do the following.

- Bias-subtract and flat-field standard-star frames, using appropriately trimmed versions of the corresponding bias and flat-field frames obtained (`tstar`).
- Carry out photometry of standard stars and correction for air mass and the effective passband width. Also correct wavelengths for phase effects associated with the position of the standard star in the beam (`tflux`).
- Flux calibrate emission-line candidates to physical flux units (using standard-star calibration constants). Correct for air mass and Galactic extinction (`tcalibrate`).
- Calibrate the fluxes of single-detection candidates (`tonecal`).
- Finally, create strip-mosaic and chart images of final candidates (`tmosaic`).

This is an exhaustive description of the process employed to reduce TF images with point-like candidates for emission-line objects. Nevertheless, some procedures can be used for the reduction of images from very-extended sources obtained using TFs. A new version of the `TFred` package, especially designed to reduce the data for the OTELO project (`http://www.iac.es/proyect/otelo`), is at present in preparation.

REFERENCES

Atherton, P. D. 1995, in *Tridimensional Optical Spectroscopic Methods in Astrophysics*, ed. S. G. Comte & M. Marcelin (San Francisco, CA: Astronomical Society of the Pacific), p. 50

Bertin, E., Arnouts, S. 1996, A&ASS, 117, 393

Bland, J., Tully, R. B. 1989, AJ, 98, 723

Bland-Hawthorn, J., Jones, D. H. 1989, PASA, 15, 44

Cepa, J. 1998, ApSS, 263, 369

Daigle, O. *et al.* 2006, MNRAS, 368, 1016

Fabry, C., Pérot, A. 1899, Ann. Chim. Phys., 7, 115

Gavryusev, V., Muñoz-Tuñón, A. 1996, in *Astronomical Data Analysis Software and System V.* ed. G. H. Jacoby & J. Barnes (San Francisco, CA: Astronomical Society of the Pacific), p. 76

Gelderman, R., Woodgate, B. E., Brown, L. W. 1995, in *Tridimensional Optical Spectroscopic Methods in Astrophysics*, ed. S. G. Comte & M. Marcelin (San Francisco, CA: Astronomical Society of the Pacific), p. 89

Gordon, S. *et al.* 2000, MNRAS, 315, 248

Jones, D. H., Shopbell, P. L., Bland-Hawthorn, J. 2002, MNRAS, 329, 759

Lewis, J. R., Unger, S. W. 1991, *TAURUS Data and How to Reduce It* (Cambridge: Royal Greenwich Observatory)

Meisenheimer, K., Hippelein, H. 1992, AA, 264, 455

Pogge, R. W. *et al.* 1995, PASP, 107, 1226

Taylor, K., Atherton, P. D. 1980, MNRAS, 191, 675